**Grade 3**

# Addison-Wesley Mathematics

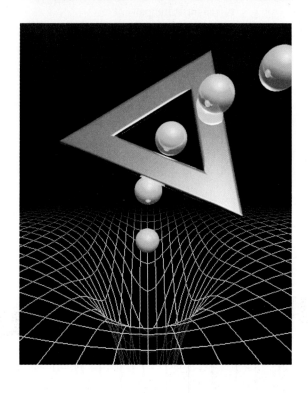

Robert E. Eicholz    Phares G. O'Daffer    Randall I. Charles
Sharon L. Young    Carne S. Barnett    Charles R. Fleenor

Stanley R. Clemens    Carol A. Thornton
Andy Reeves    Joan E. Westley

## ◢◣ Addison-Wesley Publishing Company

*Menlo Park, California* ■ *Reading, Massachusetts* ■ *New York*
*Don Mills, Ontario* ■ *Wokingham, England* ■ *Amsterdam* ■ *Bonn*
*Sydney* ■ *Singapore* ■ *Tokyo* ■ *Madrid* ■ *San Juan*

## PROGRAM ADVISORS

**John A. Dossey**
Professor of Mathematics
Illinois State University
Normal, Illinois

**David C. Brummett**
Educational Consultant
Palo Alto, California

**Irene Medina**
Mathematics Coordinator
Tom Browne Middle School
Corpus Christi, Texas

**Freddie Renfro**
K-12 Mathematics
   Coordinator
La Porte Independent
   School District
La Porte, Texas

**William J. Driscoll**
Chairman
Department of
   Mathematical Sciences
Central Connecticut State
   University
Burlington, Connecticut

**Rosalie C. Whitlock**
Educational Consultant
Stanford, California

**Bonnie Armbruster**
Associate Professor
Center for the Study of
   Reading
University of Illinois
Champaign, Illinois

**Betty C. Lee**
Assistant Principal
Ferry Elementary School
Detroit, Michigan

## CONTRIBUTING WRITERS

*Betsy Franco*
*Marilyn Jacobson*
*Marny Sorgen*
*Judith K. Wells*

*Mary Heinrich*
*Ann Muench*
*Connie Thorpe*

*Penny Holland*
*Gini Shimabukuro*
*Sandra Ward*

## EXECUTIVE EDITOR

*Diane H. Fernández*

*Cover Photo credit: Orion Press/West Light*

TI-12 Math Explorer™ is a trademark of Texas Instruments.

ISBN 0-201-27300-4

5 6 7 8 9 10 11 12 - VH - 95 94 93 92

# Contents

3

**TIME AND MONEY**

## 4

### ADDITION

## 5

### SUBTRACTION

6

## DATA, GRAPHS, AND PROBABILITY

## 10

### GEOMETRY

## 13

### MULTIPLICATION

## 14

### METRIC MEASUREMENT

## 15

### FRACTIONS AND DECIMALS

## 16

### DIVISION

### RESOURCE BANK AND APPENDIX

Dear Student:

Get ready for an adventure in mathematics. You will learn many new and exciting things. You will have fun!

This year, you will be using numbers in the thousands! You will also discover how to do many new things with numbers, like make estimates, round off numbers, and collect and organize data. You will explore Roman numerals, decimals, and learn more about multiplication and division. You will even be introduced to some new ways to solve problems.

You will see how mathematics can be used everyday. We will talk about money, hobbies, clothes, sports and other things you enjoy. You will work in groups or with a partner to solve problems. That makes mathematics fun for everyone!

Good luck this year. We know you will enjoy all the new ideas you will learn.

From your friends at Addison-Wesley.

# 1

1 Beethoven began to write music when he was a child. What did Beethoven do when he was 4 years old? What happened when he was 12 years old?

# ADDITION AND SUBTRACTION CONCEPTS AND FACTS

**2** Beethoven began to lose his hearing when he was a young man, but he still wrote music. How old was he when he started to become deaf?

**3** The last symphony that Beethoven wrote took him 20 years to plan and write. How many symphonies did Beethoven write in all?

**4** **Use Critical Thinking** Make up the first 2 lines of a third verse to the song, "Over in the Meadow." Follow the pattern in the first two verses.

3

# Problem Solving
## Understanding Addition and Subtraction

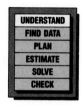

UNDERSTAND
FIND DATA
PLAN
ESTIMATE
SOLVE
CHECK

**LEARN ABOUT IT**

These actions help you understand addition and subtraction.

The actions below also help you decide what operation to use. Show the actions with counters. Complete a number sentence about the action.

**Put together**

$$2 + 3 = 5$$
addend  addend  sum

**Take away**

$$5 - 2 = 3$$
sum  addend  addend

**Problem**

| **Action** | **Operation** |

6 frogs are on a lily pad. 2 of them jump in the water.

**Take away**

**Subtract**
$$6 - 2 = |||||$$

5 green frogs and 3 brown frogs are matching up for a game. How many more green than brown frogs?

**Compare**

**Subtract**
$$5 - 3 = |||||$$

There are rocks for 7 frogs to sit on. 4 frogs are sitting on rocks. How many more frogs can sit?

**Find the missing part**

**Subtract**
$$7 - 4 = |||||$$

**TRY IT OUT**

Use the counters to find the answer. Choose the correct number sentence.

1. 6 frogs sat on a mudbank. 4 were big frogs. The rest were small. How many were small?

   $$6 + 4 = 10 \qquad 6 - 4 = 2$$

2. 5 frogs swam in the pond. 2 more frogs joined them. How many frogs were swimming?

   $$5 + 2 = 7 \qquad 5 - 2 = 3$$

4

Show each action with your counters. Tell whether you add or subtract. Copy and complete the number sentence.

**1.**

5 ladybugs    3 more ladybugs
sat on a leaf.    joined them.

How many ladybugs in all?

5 ⫼ 3 = ⦀

**2.**

3 turtles rest    4 turtles swim
on the shore.    in the water.

How many more turtles are in the water?

4 ⫼ 3 = ⦀

**3.**

6 snails crawled on a rock. 2 crawled off. How many snails were left on the rock?

6 ⫼ 2 = ⦀

**4.**

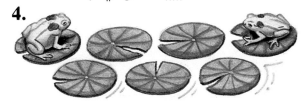

There are lily pads for 7 frogs. 2 frogs hopped on lily pads. How many more frogs can come?

7 ⫼ 2 = ⦀

**5.**

4 birds sit in a tree. 5 more birds fly to the tree. How many birds are in the tree?

4 ⫼ 5 = ⦀

**6.**

8 dragonflies fly to 5 flowers. Each fly wants to land on a flower. How many more flies than flowers?

8 ⫼ 5 = ⦀

▶ **COMMUNICATION**

Write a story for each sentence.

**7.** 4 + 3 = 7    **8.** 9 − 3 = 6

*More Practice, page 500, set A*

# Mental Math
## Counting On and Counting Back

**LEARN ABOUT IT**

You can use **counting** to add or subtract small numbers.

**EXPLORE** **Play a Game with Number Cards**
Rico and Lindsay are playing "Get Off the Bridge." Make 8 game cards like theirs and take turns playing the game. Start with your markers on the bridge at either 6 or 7.

Without looking, choose a card. Move your marker to follow the directions. First player off the bridge wins.

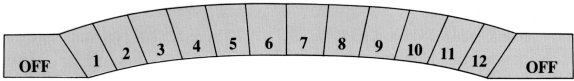

**TALK ABOUT IT**

1. Suppose Rico is on 7 and draws the card Add 3 .
   Where will he be then? If the card is Subtract 0 ?

2. How can you use counting to make the moves?

When you use counting, start with the larger number.

| 9, 10, 11, **12** | 7, 8, **9** | 11, 10, 9, **8** | 10, 9, **8** |
|:---:|:---:|:---:|:---:|
| $9 + 3 = 12$ | $2 + 7 = 9$ | $11 - 3 = 8$ | $10 - 2 = 8$ |

**TRY IT OUT**

Add or subtract.

**1.** $2 + 8$   **2.** $11 - 2$   **3.** $9 + 2$   **4.** $8 - 3$   **5.** $0 + 7$   **6.** $9 - 3$

6

Add or subtract.

1. $9 - 2$     2. $1 + 8$     3. $3 + 8$     4. $12 - 3$     5. $3 + 7$

6. $11 - 2$     7. $7 + 2$     8. $7 - 0$     9. $3 + 9$     10. $10 - 3$

11. $9 + 2$     12. $9 - 1$     13. $2 + 8$     14. $5 + 0$     15. $8 - 2$

16.
$$\begin{array}{r} 10 \\ -\ 2 \\ \hline \end{array}$$
17.
$$\begin{array}{r} 1 \\ +9 \\ \hline \end{array}$$
18.
$$\begin{array}{r} 9 \\ +3 \\ \hline \end{array}$$
19.
$$\begin{array}{r} 6 \\ -2 \\ \hline \end{array}$$
20.
$$\begin{array}{r} 0 \\ +5 \\ \hline \end{array}$$
21.
$$\begin{array}{r} 8 \\ -3 \\ \hline \end{array}$$

22.
$$\begin{array}{r} 6 \\ +3 \\ \hline \end{array}$$
23.
$$\begin{array}{r} 11 \\ -\ 3 \\ \hline \end{array}$$
24.
$$\begin{array}{r} 2 \\ +8 \\ \hline \end{array}$$
25.
$$\begin{array}{r} 7 \\ -3 \\ \hline \end{array}$$
26.
$$\begin{array}{r} 8 \\ -0 \\ \hline \end{array}$$
27.
$$\begin{array}{r} 2 \\ +9 \\ \hline \end{array}$$

28. Find the sum of 7 and 3.       29. Subtract 2 from 9.

30. Subtract 0 from 5.       31. Find the sum of 3 and 4.

**MATH REASONING** Is the ⫼ covering a $+$ or a $-$?

32. $9$ ⫼ $2 = 11$      33. $9$ ⫼ $3 = 6$      34. $11$ ⫼ $2 = 9$      35. $38$ ⫼ $3 = 41$

**PROBLEM SOLVING**

36. Rico was at 9 on the bridge. In his next two turns he got ⟨Add 2⟩ and ⟨Subtract 3⟩. Where was he then?

37. **Problems with More Than One Answer** Lindsay is on the bridge at 11. How can she win on her next turn?

**MENTAL MATH**

Use counting to find each sum or difference.

38. $25 - 3$      39. $37 + 2$      40. $49 + 2$      41. $50 - 2$      42. $41 - 3$

43. $3 + 58$      44. $81 - 2$      45. $2 + 28$      46. $90 - 3$      47. $68 + 3$

*More Practice, page 500, set B*

# Finding Sums
## Using Strategies

**Doubles**

### EXPLORE  Study the Pictures

The pictures are models of some special sums called doubles. Can you use the pictures to give each sum?

$4 + 4$    $5 + 5$    $6 + 6$

### TALK ABOUT IT

1. Why do you think these sums are called doubles?

2. How do the pictures help you remember the doubles?

$7 + 7$     $8 + 8$     $9 + 9$

Here is a special strategy you can use if you remember the doubles sums.

### Doubles Plus 1

( 1 more than 5 + 5 )   ( 1 more than 7 + 7 )   ( 1 more than 8 + 8 )

$6 + 5 = 11$       $7 + 8 = 15$       $9 + 8 = 17$

### TRY IT OUT

Add.

| | | | | | |
|---|---|---|---|---|---|
| **1.** $7 + 7$ | **2.** $6 + 5$ | **3.** $7 + 8$ | **4.** $6 + 6$ | **5.** $5 + 4$ | **6.** $9 + 3$ |
| **7.** $4 + 4$ | **8.** $7 + 6$ | **9.** $8 + 8$ | **10.** $2 + 7$ | **11.** $5 + 5$ | **12.** $9 + 9$ |

| | | | | | |
|---|---|---|---|---|---|
| **13.** $\begin{array}{r} 6 \\ +5 \\ \hline \end{array}$ | **14.** $\begin{array}{r} 8 \\ +3 \\ \hline \end{array}$ | **15.** $\begin{array}{r} 2 \\ +9 \\ \hline \end{array}$ | **16.** $\begin{array}{r} 7 \\ +3 \\ \hline \end{array}$ | **17.** $\begin{array}{r} 3 \\ +3 \\ \hline \end{array}$ | **18.** $\begin{array}{r} 6 \\ +6 \\ \hline \end{array}$ |

Add.

**1.** $7 + 8$     **2.** $8 + 3$     **3.** $5 + 4$     **4.** $2 + 9$     **5.** $7 + 7$

**6.** $7 + 3$     **7.** $9 + 9$     **8.** $2 + 2$     **9.** $6 + 7$     **10.** $3 + 6$

**11.** $6 + 6$     **12.** $3 + 3$     **13.** $9 + 8$     **14.** $8 + 8$     **15.** $5 + 6$

| | | | | | | |
|---|---|---|---|---|---|---|
| **16.** $\begin{array}{r}4\\+5\\\hline\end{array}$ | **17.** $\begin{array}{r}7\\+7\\\hline\end{array}$ | **18.** $\begin{array}{r}3\\+9\\\hline\end{array}$ | **19.** $\begin{array}{r}8\\+7\\\hline\end{array}$ | **20.** $\begin{array}{r}2\\+9\\\hline\end{array}$ | **21.** $\begin{array}{r}3\\+8\\\hline\end{array}$ | **22.** $\begin{array}{r}5\\+3\\\hline\end{array}$ |
| **23.** $\begin{array}{r}2\\+7\\\hline\end{array}$ | **24.** $\begin{array}{r}3\\+3\\\hline\end{array}$ | **25.** $\begin{array}{r}8\\+9\\\hline\end{array}$ | **26.** $\begin{array}{r}9\\+9\\\hline\end{array}$ | **27.** $\begin{array}{r}7\\+3\\\hline\end{array}$ | **28.** $\begin{array}{r}5\\+5\\\hline\end{array}$ | **29.** $\begin{array}{r}9\\+3\\\hline\end{array}$ |
| **30.** $\begin{array}{r}6\\+6\\\hline\end{array}$ | **31.** $\begin{array}{r}6\\+5\\\hline\end{array}$ | **32.** $\begin{array}{r}3\\+4\\\hline\end{array}$ | **33.** $\begin{array}{r}1\\+9\\\hline\end{array}$ | **34.** $\begin{array}{r}7\\+6\\\hline\end{array}$ | **35.** $\begin{array}{r}8\\+8\\\hline\end{array}$ | **36.** $\begin{array}{r}4\\+4\\\hline\end{array}$ |

**APPLY**

**MATH REASONING**

**37.** Since $28 + 28 = 56$, we know that $29 + 28 = $ |||||.

**38.** Since $297 + 297 = 594$, we know that $298 + 297 = $ |||||.

**PROBLEM SOLVING**

**39.** Miko had 6 red apples. Heather had 7 yellow apples. How many apples did they have all together?

**40. Data Hunt** List 4 objects in your home or classroom that show doubles.

► **USING CRITICAL THINKING**

Use your answers to Exercises 1 through 36 to write *odd* or *even* to finish the sentence.

**41.** The sum of an odd and an even number is an ||||| number.

**42.** The sum of two even numbers is an ||||| number.

**43.** The sum of two odd numbers is an ||||| number.

# Finding Sums
## Another Strategy

Make 10    Add Extra

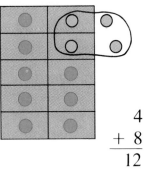

8
+ 5
13

10 and 3 extra

### EXPLORE Try the Strategy

To find certain sums that are 10 or greater, you can use a strategy called **Make 10, Add Extra.** Looking at a 10-frame helps with this strategy. You can draw a 10-frame or imagine one.

### TALK ABOUT IT

**1.** How many of the 5 were used to fill up the frame? How many extras were there?

**2.** Tell or show how you would use a 10-frame to find $7 + 4$.

When you use the 10-frame to add two numbers, put the larger number in the frame first.

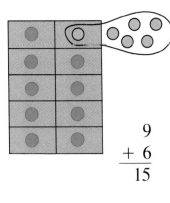

9
+ 6
15

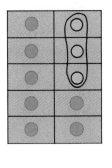

7
+ 3
10

4
+ 8
12

**TRY IT OUT**

Add.

**1.** $9 + 5$    **2.** $5 + 7$    **3.** $4 + 9$    **4.** $4 + 7$    **5.** $2 + 8$

**6.**   6        **7.**   9    **8.**   3    **9.**   8    **10.**   6    **11.**   7    **12.**   8
    + 8           + 7           + 7           + 5           + 9           + 5           + 7

Add.

1. $8 + 4$    2. $8 + 8$    3. $7 + 5$    4. $4 + 7$    5. $9 + 8$

6. $5 + 8$    7. $6 + 4$    8. $9 + 5$    9. $8 + 6$    10. $5 + 7$

11. $3 + 9$    12. $5 + 6$    13. $5 + 5$    14. $7 + 7$    15. $7 + 4$

16. $\begin{array}{r} 6 \\ +8 \\ \hline \end{array}$  17. $\begin{array}{r} 7 \\ +5 \\ \hline \end{array}$  18. $\begin{array}{r} 4 \\ +4 \\ \hline \end{array}$  19. $\begin{array}{r} 4 \\ +6 \\ \hline \end{array}$  20. $\begin{array}{r} 8 \\ +7 \\ \hline \end{array}$  21. $\begin{array}{r} 5 \\ +8 \\ \hline \end{array}$  22. $\begin{array}{r} 4 \\ +8 \\ \hline \end{array}$

23. $\begin{array}{r} 6 \\ +6 \\ \hline \end{array}$  24. $\begin{array}{r} 7 \\ +4 \\ \hline \end{array}$  25. $\begin{array}{r} 7 \\ +9 \\ \hline \end{array}$  26. $\begin{array}{r} 3 \\ +8 \\ \hline \end{array}$  27. $\begin{array}{r} 3 \\ +4 \\ \hline \end{array}$  28. $\begin{array}{r} 9 \\ +9 \\ \hline \end{array}$  29. $\begin{array}{r} 2 \\ +3 \\ \hline \end{array}$

30. Find the sum of 4 and 8.    31. Find the sum of 6 and 7.

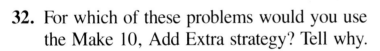

**MATH REASONING**

32. For which of these problems would you use
    the Make 10, Add Extra strategy? Tell why.

    A. $6 + 8$    B. $4 + 3$    C. $30 + 21$    D. $9 + 3$

**PROBLEM SOLVING**

33. Adam caught 9 catfish and 5
    trout. How many fish did he
    catch all together?

34. Mrs. Wong caught 11 fish. She
    threw back 3 fish that were too
    small to keep. How many fish
    did she keep?

Add or subtract.

35. $3 + 4$    36. $7 - 3$    37. $7 - 4$    38. $4 + 3$    39. $7 - 0$

40. $\begin{array}{r} 4 \\ +4 \\ \hline \end{array}$  41. $\begin{array}{r} 5 \\ +4 \\ \hline \end{array}$  42. $\begin{array}{r} 8 \\ -4 \\ \hline \end{array}$  43. $\begin{array}{r} 9 \\ -8 \\ \hline \end{array}$  44. $\begin{array}{r} 6 \\ -3 \\ \hline \end{array}$  45. $\begin{array}{r} 3 \\ -1 \\ \hline \end{array}$  46. $\begin{array}{r} 6 \\ +4 \\ \hline \end{array}$

*More Practice, page 500, set D*

# Using Critical Thinking

Jon and Steven were playing a game called Nim.

## GAME RULES

1. Start with 11 counters.
2. At your turn you must pick up 1, 2, or 3 counters.
3. To win, make your partner have to pick up the last counter.

There were 6 counters left and it was Jon's turn. "Wow!" said Jon, "I've got you now!"

## TALK ABOUT IT

1. How many counters can Jon choose to pick up?
2. What must Jon do to win?
3. How many counters should Jon pick up? Explain.

## TRY IT OUT

Play Nim with a friend. Try to use what you have learned.

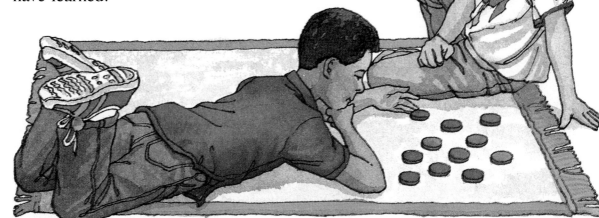

12

# MIDCHAPTER REVIEW/QUIZ

Add or subtract.

1. $5 + 2$     2. $8 - 3$     3. $7 - 1$     4. $3 + 0$     5. $9 + 3$

6. $\begin{array}{r} 5 \\ + 5 \\ \hline \end{array}$     7. $\begin{array}{r} 5 \\ + 6 \\ \hline \end{array}$     8. $\begin{array}{r} 8 \\ + 8 \\ \hline \end{array}$     9. $\begin{array}{r} 9 \\ + 8 \\ \hline \end{array}$     10. $\begin{array}{r} 11 \\ - 3 \\ \hline \end{array}$

11. $9 - 3$     12. $6 + 1$     13. $6 + 6$     14. $8 - 1$     15. $0 + 4$

16. $\begin{array}{r} 8 \\ + 6 \\ \hline \end{array}$     17. $\begin{array}{r} 9 \\ + 7 \\ \hline \end{array}$     18. $\begin{array}{r} 6 \\ + 7 \\ \hline \end{array}$     19. $\begin{array}{r} 2 \\ - 1 \\ \hline \end{array}$     20. $\begin{array}{r} 7 \\ + 5 \\ \hline \end{array}$

21. $\begin{array}{r} 9 \\ + 9 \\ \hline \end{array}$     22. $\begin{array}{r} 9 \\ + 0 \\ \hline \end{array}$     23. $\begin{array}{r} 5 \\ + 9 \\ \hline \end{array}$     24. $\begin{array}{r} 13 \\ - 3 \\ \hline \end{array}$     25. $\begin{array}{r} 4 \\ - 3 \\ \hline \end{array}$

26. $9 + 4$     27. $6 + 8$     28. $7 + 9$     29. $9 + 5$     30. $4 + 8$

31. $\begin{array}{r} 0 \\ + 6 \\ \hline \end{array}$     32. $\begin{array}{r} 7 \\ + 7 \\ \hline \end{array}$     33. $\begin{array}{r} 9 \\ - 5 \\ \hline \end{array}$     34. $\begin{array}{r} 10 \\ - 3 \\ \hline \end{array}$     35. $\begin{array}{r} 4 \\ + 4 \\ \hline \end{array}$

## PROBLEM SOLVING

36. A pet shop has 3 puppies and 7 kittens for sale. How many more kittens are there than puppies?

37. On the playground, 6 children are on the swings and 5 are on the slide. How many children are playing?

38. Giorgio's team has 9 points. They score 6 more points. How many points do they have now?

39. In a family with 7 children, 2 of the children are boys. How many of the children are girls?

40. Nick's team is 3 points ahead of Ana's team. Ana's team has a score of 5. What is Nick's team score?

41. The pet shop had 6 rabbits. The store owner sold 4 rabbits. How many are left?

# Finding Differences
## Using Strategies

**EXPLORE** **Solve to Understand**

Can you find strategies to help you solve each of these problems?

Linda knows 9 verses of the song "Skip to My Lou." She sang 7 verses. How many verses did she have left to sing?

Gary knows 12 verses of "The Wheels on the Bus." He taught his brother 7 verses. How many verses does he have left to teach?

**TALK ABOUT IT**

1. How can you find the difference $9 - 7$ if you don't already know it?

2. Can you explain a way to figure out $12 - 7$?

When numbers are close together, you can **count up** to find the difference.

$$\begin{array}{r} 8 \\ -6 \\ \hline 2 \end{array} \quad (6, 7, 8)$$

$$\begin{array}{r} 11 \\ -8 \\ \hline 3 \end{array} \quad (8, 9, 10, 11)$$

You can **subtract from 10 and add the extra** to find the difference. Think of a 10-frame. Subtract inside first.

$$\begin{array}{r} 12 \\ -7 \\ \hline 5 \end{array}$$

3 and 2 are 5.

$$\begin{array}{r} 15 \\ -8 \\ \hline 7 \end{array}$$

2 and 5 are 7.

**TRY IT OUT**

Subtract. Use the strategies if you need help.

**1.** $10 - 7$　　**2.** $14 - 6$　　**3.** $8 - 6$　　**4.** $12 - 9$　　**5.** $13 - 8$　　**6.** $17 - 9$

14

Subtract.

**1.** $14 - 8$     **2.** $11 - 9$     **3.** $15 - 9$     **4.** $14 - 12$     **5.** $13 - 6$

**6.** $10 - 9$     **7.** $17 - 9$     **8.** $12 - 9$     **9.** $7 - 4$     **10.** $14 - 7$

**11.** $12 - 7$     **12.** $9 - 6$     **13.** $13 - 7$     **14.** $16 - 7$     **15.** $10 - 7$

**16.** $\begin{array}{r} 13 \\ -\ 8 \\ \hline \end{array}$  **17.** $\begin{array}{r} 16 \\ -\ 8 \\ \hline \end{array}$  **18.** $\begin{array}{r} 8 \\ -5 \\ \hline \end{array}$  **19.** $\begin{array}{r} 11 \\ -\ 7 \\ \hline \end{array}$  **20.** $\begin{array}{r} 12 \\ -\ 5 \\ \hline \end{array}$  **21.** $\begin{array}{r} 15 \\ -\ 7 \\ \hline \end{array}$

**22.** $\begin{array}{r} 11 \\ -\ 8 \\ \hline \end{array}$  **23.** $\begin{array}{r} 14 \\ -\ 6 \\ \hline \end{array}$  **24.** $\begin{array}{r} 17 \\ -\ 8 \\ \hline \end{array}$  **25.** $\begin{array}{r} 13 \\ -\ 9 \\ \hline \end{array}$  **26.** $\begin{array}{r} 9 \\ -7 \\ \hline \end{array}$  **27.** $\begin{array}{r} 11 \\ -\ 6 \\ \hline \end{array}$

**APPLY**

**MATH REASONING** What number was subtracted? Think about counting.

**28.** $12 - ? = 9$     **29.** $27 - ? = 25$     **30.** $40 - ? = 37$     **31.** $31 - ? = 29$

**PROBLEM SOLVING**

**32.** Connie's class is going to learn 12 verses of ''Froggie Went A-Courting.'' So far, they have learned 8 verses. How many more do they need to learn?

**33. Data Bank** The composer Beethoven was 4 years old when he began piano lessons. How many years later did he have some music published? See page 468.

▶ **ESTIMATION**

**34.** Suppose you can grow 1 bushel of soy beans in the small field. About how many can you grow in the large field?

You can check your estimate by cutting pieces of paper about the size of the small field. About how many are needed to cover the large field?

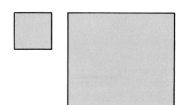

# Finding Differences
## Using Addition

**Whole**

$\boxed{13}$

**Part** $\boxed{8}$   **Part** $\boxed{5}$

You can **use related addition facts** to find differences.

**EXPLORE  Use Snap Cubes**

Work in groups. Snap together 13 cubes to show the whole. Break it into parts to show the check fact $5 + 8$. This also shows you the difference for $13 - 5$.

$$\begin{array}{r} 13 \\ -5 \\ \hline 8 \end{array} \quad \text{because} \quad \begin{array}{r} 5 \\ +8 \\ \hline 13 \end{array}$$

**check fact**

Use snap cubes to show the whole and parts for $14 - 8$. Find the check fact first. Then give the difference.

**TALK ABOUT IT**

$$\begin{array}{r} 14 \\ -8 \\ \hline \end{array} \qquad \begin{array}{r} 8 \\ + \\ \hline 14 \end{array}$$

**check fact**

1. What are the parts and what is the whole for $14 - 8$? How does addition help you find the difference?

2. Show or draw an example of your own. What are the whole and parts? What is the check fact? What is the difference?

Think about the check fact to help you find a difference.

$$\begin{array}{r} 13 \\ -5 \\ \hline 8 \end{array}$$

Think: $\begin{array}{r} 5 \\ +\mathbf{8} \\ \hline 13 \end{array}$

Think: $6 + \mathbf{9} = 15$

$15 - 6 = 9$

Think: $7 + \mathbf{4} = 11$

$11 - 7 = 4$

Subtract. Use check facts to find the differences.

**1.** $12 - 5 = $ ‖‖‖   **2.** $12 - 9 = $ ‖‖‖   **3.** $17 - 8 = $ ‖‖‖

Subtract.

1. $11 - 7$     2. $14 - 7$     3. $10 - 7$     4. $13 - 8$     5. $15 - 7$

6. $16 - 9$     7. $10 - 6$     8. $15 - 9$     9. $12 - 5$     10. $8 - 5$

11. $\begin{array}{r} 9 \\ -0 \\ \hline \end{array}$   12. $\begin{array}{r} 15 \\ -7 \\ \hline \end{array}$   13. $\begin{array}{r} 12 \\ -6 \\ \hline \end{array}$   14. $\begin{array}{r} 10 \\ -5 \\ \hline \end{array}$   15. $\begin{array}{r} 12 \\ -8 \\ \hline \end{array}$   16. $\begin{array}{r} 13 \\ -7 \\ \hline \end{array}$

17. $\begin{array}{r} 11 \\ -7 \\ \hline \end{array}$   18. $\begin{array}{r} 4 \\ -4 \\ \hline \end{array}$   19. $\begin{array}{r} 10 \\ -2 \\ \hline \end{array}$   20. $\begin{array}{r} 8 \\ -4 \\ \hline \end{array}$   21. $\begin{array}{r} 13 \\ -9 \\ \hline \end{array}$   22. $\begin{array}{r} 11 \\ -6 \\ \hline \end{array}$

## APPLY

**MATH REASONING** Find the difference without subtracting.

23. Since $12 + 17 = 29$, we know that $29 - 17 = $ ‖‖‖.

24. Since $38 + 29 = 67$, we know that $67 - 29 = $ ‖‖‖.

**PROBLEM SOLVING**

Think about two number cubes, each marked with the numbers 4, 5, 6, 7, 8, and 9.

25. Brenda rolled the two number cubes. Her sum was 17. What numbers did she roll?

26. **Problems with More Than One Answer** Name the different ways Brenda could get a sum of 13.

▶ **COMMUNICATION**

Use addition or subtraction facts to write an example for each sentence.
Example: A number plus itself is a double. $4 + 4 = 8$

27. The sum of a number and 0 is that number.

28. The difference for a number and 0 is that number.

29. A number minus itself is 0.

*More Practice, page 500, set F*

# Fact Families

## LEARN ABOUT IT

Addition and subtraction are related. Look at the **fact family** for a whole number and its parts.

### EXPLORE Use Number Cubes

Roll two number cubes labeled with the numbers 4, 5, 6, 7, 8, and 9. Use the numbers as addends. How many different addition and subtraction number sentences can you write using these two addends and their sum?

### TALK ABOUT IT

**1.** Why can you write 2 addition sentences if the addends are different?

**2.** Is there a related subtraction sentence for each addition sentence?

Using 2 addends and their sum, you can write a fact family.

Addend         Addend         Sum

 + = 16

| Fact Family |
|---|
| $9 + 7 = 16$ |
| $7 + 9 = 16$ |
| $16 - 7 = 9$ |
| $16 - 9 = 7$ |

## TRY IT OUT

Write a fact family for each pair of addends.

**1.** Addends: 6 and 7      **2.** Addends: 9 and 8      **3.** Addends: 8 and 8

18

Write a fact family for each pair of addends.

**1.** Addends: 5 and 6     **2.** Addends: 4 and 9     **3.** Addends: 7 and 5

Subtract. Think about related facts.

**4.** $10 - 7$     **5.** $15 - 9$     **6.** $13 - 8$     **7.** $12 - 4$     **8.** $17 - 9$

**9.** $16 - 8$     **10.** $14 - 5$     **11.** $12 - 7$     **12.** $11 - 4$     **13.** $15 - 6$

**14.** $\begin{array}{r} 11 \\ -\ 9 \\ \hline \end{array}$   **15.** $\begin{array}{r} 15 \\ -\ 8 \\ \hline \end{array}$   **16.** $\begin{array}{r} 13 \\ -\ 6 \\ \hline \end{array}$   **17.** $\begin{array}{r} 12 \\ -\ 5 \\ \hline \end{array}$   **18.** $\begin{array}{r} 9 \\ -6 \\ \hline \end{array}$   **19.** $\begin{array}{r} 11 \\ -\ 7 \\ \hline \end{array}$

## APPLY

**MATH REASONING** Do not subtract. Just tell which difference is greatest.

**20.** $\begin{array}{r} 12 \\ -4 \\ \hline \end{array}$ $\begin{array}{r} 12 \\ -\ 8 \\ \hline \end{array}$ $\begin{array}{r} 12 \\ -\ 7 \\ \hline \end{array}$    **21.** $\begin{array}{r} 15 \\ -\ 9 \\ \hline \end{array}$ $\begin{array}{r} 15 \\ -\ 6 \\ \hline \end{array}$ $\begin{array}{r} 15 \\ -\ 8 \\ \hline \end{array}$

**PROBLEM SOLVING**

**22.** Bev rolled two number cubes. One number was 4 and her sum was 11. What was the other addend?

**23. Data Bank** Six years after he began piano lessons, Beethoven began to play the organ. How old was he then? See page 468.

## ⊞ MIXED REVIEW

Find the sum or difference.

**24.** $8 - 7$     **25.** $6 - 0$     **26.** $5 - 5$     **27.** $3 + 9$     **28.** $8 + 4$

**29.** $3 + 7$     **30.** $9 - 5$     **31.** $3 - 1$     **32.** $5 + 5$     **33.** $8 + 8$

**34.** $6 + 7$     **35.** $7 - 6$     **36.** $7 + 4$     **37.** $9 - 1$     **38.** $7 - 0$

**39.** $\begin{array}{r} 1 \\ +9 \\ \hline \end{array}$   **40.** $\begin{array}{r} 6 \\ +4 \\ \hline \end{array}$   **41.** $\begin{array}{r} 9 \\ -7 \\ \hline \end{array}$   **42.** $\begin{array}{r} 2 \\ +8 \\ \hline \end{array}$   **43.** $\begin{array}{r} 10 \\ -\ 8 \\ \hline \end{array}$

# Sums with Three Addends

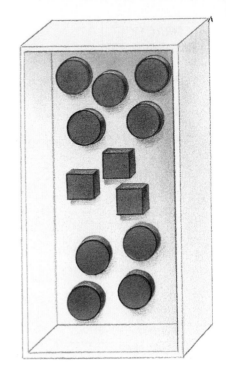

### EXPLORE  Use Objects

To find the sum of three addends, you add them two at a time.

There are three sets of colored shapes in the box, 4 blue circles, 5 red circles, and 3 red squares. Are there different ways to find the total number of pieces in the box?

### TALK ABOUT IT

1. How many circle pieces are there? How many squares? How many pieces in all?

2. How many red pieces are there? How many blue? How many pieces in all?

3. When you add three numbers, does it matter which two you add first?

When you add 3 or more numbers, you can add any 2 numbers first.

$3 + 4 = 7$ and 2 more make 9

$3 + 4 + 2 = 9$

$4 + 2 = 6$
$3 + 6 = 9$

$3 + 4 + 2 = 9$

Add down.

$$\begin{array}{r} 2 \\ 3 \\ +8 \\ \hline 13 \end{array}$$

5 and 8 more

Add up.

$$\begin{array}{r} 2 \\ 3 \\ +8 \\ \hline 13 \end{array}$$

11 and 2 more

Look for sums of 10.

$$\begin{array}{r} 2 \\ 3 \\ +8 \\ \hline 13 \end{array}$$

10 and 3 more

## TRY IT OUT

1.
$6 + 2 + 4$

2. $3 + 5 + 2$

3. $\begin{array}{r} 8 \\ 1 \\ +1 \\ \hline \end{array}$

4. $\begin{array}{r} 3 \\ 7 \\ +2 \\ \hline \end{array}$

5. $\begin{array}{r} 5 \\ 4 \\ +5 \\ \hline \end{array}$

Add.

| | | | | | |
|---|---|---|---|---|---|
| **1.** $\begin{array}{r} 5 \\ 2 \\ +5 \\ \hline \end{array}$ | **2.** $\begin{array}{r} 4 \\ 5 \\ +4 \\ \hline \end{array}$ | **3.** $\begin{array}{r} 6 \\ 3 \\ +7 \\ \hline \end{array}$ | **4.** $\begin{array}{r} 6 \\ 3 \\ +5 \\ \hline \end{array}$ | **5.** $\begin{array}{r} 6 \\ 4 \\ +8 \\ \hline \end{array}$ | **6.** $\begin{array}{r} 4 \\ 5 \\ +3 \\ \hline \end{array}$ |
| **7.** $\begin{array}{r} 5 \\ 2 \\ +4 \\ \hline \end{array}$ | **8.** $\begin{array}{r} 8 \\ 2 \\ +7 \\ \hline \end{array}$ | **9.** $\begin{array}{r} 3 \\ 6 \\ +2 \\ \hline \end{array}$ | **10.** $\begin{array}{r} 1 \\ 7 \\ +9 \\ \hline \end{array}$ | **11.** $\begin{array}{r} 7 \\ 2 \\ +3 \\ \hline \end{array}$ | **12.** $\begin{array}{r} 8 \\ 2 \\ +4 \\ \hline \end{array}$ |

**13.** $2 + 3 + 1$     **14.** $4 + 7 + 6$     **15.** $2 + 8 + 5$     **16.** $5 + 8 + 5$

**17.** $6 + 7 + 3$     **18.** $5 + 1 + 6$     **19.** $2 + 3 + 2$     **20.** $4 + 2 + 3$

Use just one number in each problem. Fill in the
|||| to make a true sentence. Example: $2 + 2 + 2 = 6$

**21.** $\text{||||} + \text{||||} + \text{||||} = 9$     **22.** $\text{||||} + \text{||||} + \text{||||} = 15$     **23.** $\text{||||} + \text{||||} + \text{||||} = 12$

**MATH REASONING**

**24.** Show three different ways to make 10 with
three numbers. Example: $8 + 1 + 1 = 10$

**PROBLEM SOLVING**

**25.** Tanya missed school 3 days in
January, 4 days in February and
2 days in March. How many
days did she miss in these
months?

**26. Missing Data** What do you
need to know to answer the
question? Matthew missed
school 7 days in the first 3
months of the year. How many
days did he miss in March?

**ALGEBRA**

**27.** Suppose $\square + \square = 10$. Then what is $\square + \square + 4$?

**28.** Suppose $\triangle + \triangle = 8$. Then what is $3 + \triangle + \triangle$?

*More Practice, page 501, set A*

# Problem Solving
## Introduction

| UNDERSTAND |
| FIND DATA |
| PLAN |
| ESTIMATE |
| SOLVE |
| CHECK |

**LEARN ABOUT IT**

Problems that can be solved by adding or subtracting are called one-step problems. They can be solved using the strategy **Choose the Operation.**

> 17 students tried out for the school play, *Annie*. 9 students did not get singing roles. The rest did. How many did get singing roles?

The checklist can help you solve problems.

**Understand the situation.**

> I want to find how many people got singing roles in the play.

**Find data needed.**

> 17 students tried out. 9 students did not get singing roles.

**Plan the solution.**

> I know the whole and one part, so I will subtract to find the missing part.

**Estimate the answer.**

> $17 - 9$ is about $20 - 10$, or 10.

**Solve the problem.**

> $17 - 9 = 8$

**Check the answer.**

> 8 is close to the estimate of 10, so 8 is a reasonable answer.

**TRY IT OUT**

Pick the number sentence you could use to solve the problem. Do not solve.

1. In the school auditorium, there are 650 seats downstairs. There are 45 seats in the balcony. How many seats are there in all?

   **A.** $650 + 45$     **B.** $650 - 45$

2. 27 of the characters in the play are children and 8 are adults. How many more characters in the play are children than adults?

   **A.** $27 - 8$     **B.** $27 + 8$

22

Pick the number sentence you could use to solve
the problem. Do not solve.

**1.** The first act of *Annie* was 45
minutes long. The second act was
30 minutes long. How long was
the play?

    **A.** $45 - 30$    **B.** $45 + 30$

**2.** The music club sold 125 tickets
for the first show. The club sold
140 tickets for the second show.
How many more tickets were
sold for the second show?

    **A.** $140 - 125$    **B.** $140 + 125$

**3.** How many more times was *Annie*
performed on Broadway than *The
Wiz?*

    **A.** $2377 + 1672$
    **B.** $2377 - 1672$

| Number of Performances on Broadway | |
|---|---|
| Annie | 2,377 |
| Oklahoma! | 2,212 |
| The Wiz | 1,672 |
| The Sound of Music | 1,443 |

Solve.

**4.** Luisa is 48 inches tall. In the part
of Miss Hannigan, she needs to
wear shoes with 2-inch heels.
How tall is she with the heels on?

**5.** Bob had 12 lines in the play. In
rehearsal, he made a mistake on
3 of his lines. How many lines
did he say correctly?

**6.** There are 9 songs in the first
act and 8 songs in the second
act. How many songs are in
the play?

**7.** There were 3 students working
the lights and 12 changing
the scenery. How many more
students were working on scenery
than on lights?

▶ **WRITE YOUR OWN PROBLEM**

Write problems that can be answered
using these number sentences.

  **8.** $6 + 9 = 15$

  **9.** $15 - 7 = 8$

**10.** $64 - 25 = 39$

*More Practice, page 512, set B*

# Group Decision Making

UNDERSTAND
FIND DATA
PLAN
ESTIMATE
SOLVE
CHECK

**Group Skills:**

Listen to Others
Encourage and Respect Others
Explain and Summarize
Check for Understanding
Disagree in an Agreeable Way

Often in this book you will be asked to put your heads together to come up with ideas to solve problems. This chart shows some skills that will help you work as a team.

Why do you think it is important for members of your group to listen to each other? Work with your group to think of 3 reasons. Remember to listen to others when you do the activity with the cards.

## Cooperative Activity

For this problem, your group will need five cards with the numbers 1 to 5 written on them.

Use the clues to put the cards in a new order. Look for more than one way to solve each problem. Write down the ways that you find.

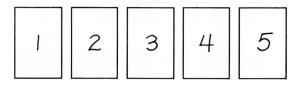

### Problem 1

Clue 1: The number on the second card is 2 more than the number on the third card.

Clue 2: The last number is the least.

Clue 3: The sum of the second and fourth numbers is 7.

### Problem 2

Clue 1: The first number is 2 more than the second number.

Clue 2: The last number is 2 more than the fourth number.

Clue 3: The sum of the three numbers in the middle is 6.

1. How many ways did you find to solve each problem? Show the ways you found.

2. Do you think there are any more ways to solve these problems? Explain.

3. Was it hard to do this activity together? Tell why or why not.

4. Did people listen to each other's ideas? Ask each member of your group how he or she feels. Talk about how you could improve your group skills.

# WRAP UP

## Make a Good Match

Match the operation to the action.

**Take Away**

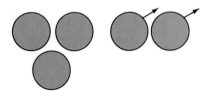

1. take away     **a.** addition
2. put together     **b.** subtraction
3. compare
4. find the missing part

Match the example to a strategy that you could use to find the answer.

**Put Together**

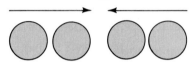

5. $8 + 7$     **a.** counting on
6. $5 + 3$     **b.** make 10, add extra
7. $9 - 3$     **c.** doubles plus one
8. $12 - 4$     **d.** counting back
9. $6 + 8$     **e.** subtract from 10, add extra

**Compare**

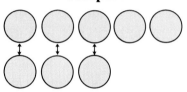

## Sometimes, Always, Never

Decide which word should go in the blank, <u>sometimes</u>, <u>always</u>, or <u>never</u>. Explain your choices.

**Find the Missing Part**

10. A sum _____ is made up of two addends.

11. A fact family _____ is made up of four number sentences.

12. You _____ estimate to get an exact answer.

## Project

How many of each kind of pet were at the Pine Tree Pet Hotel on Wednesday? How many pets all together were at the hotel on Wednesday? HINT: First, figure out how many of each kind of pet were there on Tuesday.

| Day | Cats | Dogs | Birds |
|---|---|---|---|
| Monday | 8 | 4 | 2 |
| Tuesday | 4 more | 5 more | 2 more |
| Wednesday | 6 fewer | 2 fewer | 8 more |

26

# CHAPTER REVIEW/TEST

## Part 1    Understanding

Write a number sentence for each picture.

**1.**

**2.**

**3.** What number was subtracted?
$25 - |||||| = 23$

**4.** What number was added?
$39 + |||||| = 40$

**5.** What is left if you remove 3 snap cubes? Give the check fact.

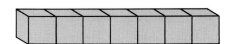

**6.** Which fact does not belong in this fact family? Why?

    **A.** $1 + 3 = 4$    **B.** $4 - 3 = 1$

    **C.** $3 + 4 = 7$    **D.** $4 - 1 = 3$

## Part 2    Skills

Add or subtract.

**7.**    $\begin{array}{r} 7 \\ +\,2 \\ \hline \end{array}$    **8.**    $\begin{array}{r} 8 \\ -\,1 \\ \hline \end{array}$    **9.**    $\begin{array}{r} 9 \\ +\,7 \\ \hline \end{array}$    **10.**    $\begin{array}{r} 5 \\ +\,6 \\ \hline \end{array}$    **11.**    $\begin{array}{r} 15 \\ -\,2 \\ \hline \end{array}$    **12.**    $\begin{array}{r} 4 \\ +\,5 \\ \hline \end{array}$

**13.** $14 - 6$    **14.** $8 + 5$    **15.** $4 + 0 + 1$    **16.** $7 + 7 + 2$

## Part 3    Applications

**17.** Ralph picked 7 apples and 1 pear. How many pieces of fruit did he have in all?

**18.** Linda ate 9 grapes and 4 cherries. How many pieces of fruit did she eat?

**19.** Jim's orchard can hold 12 pear trees. He has 7 trees now. How many more trees can he plant?

**20.** **Challenge** Fred has as many apple trees as pear trees. He has 8 cherry trees and 6 pear trees. How many trees has he in all?

27

# ENRICHMENT
## A Mental Math Game of 15

Play the 15 Game with a partner.

Use a grid like this for each game.

Here are the rules.

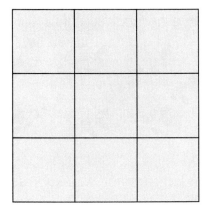

1. The first player uses the odd digits: 1, 3, 5, 7, and 9.

   The other player uses the even digits: 2, 4, 6, and 8.

2. Players take turns writing a digit in a square. Players may use a digit only once.

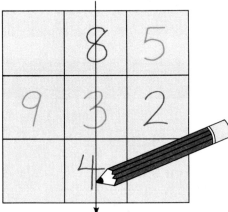

3. To win, complete a row (→), a column (↓), or a diagonal (↙ ↘) so that the sum of the three numbers is 15.

The winner of this game is the player using even digits.

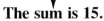

**The sum is 15.**

# CUMULATIVE REVIEW

Add or subtract.

1.  4
   $+2$

   **A.** 4     **B.** 2
   **C.** 6     **D.** 8

2.  7
   $-3$

   **A.** 4     **B.** 3
   **C.** 10    **D.** 5

3. $5 + 3$

   **A.** 7     **B.** 8
   **C.** 2     **D.** 9

4. $8 - 2$

   **A.** 5     **B.** 10
   **C.** 6     **D.** 4

5. $4 + 6$

   **A.** 4     **B.** 2
   **C.** 9     **D.** 10

6. Which belongs in the fact family for addends 2 and 4?

   **A.** $4 - 2 = 2$
   **B.** $6 + 4 = 10$
   **C.** $6 - 2 = 4$
   **D.** $6 + 2 = 8$

7. $17 - 6$

   **A.** 10    **B.** 9
   **C.** 11    **D.** 23

8. $10 - 3$

   **A.** 13    **B.** 7
   **C.** 8     **D.** 6

9. $2 + 7 + 1$

   **A.** 10    **B.** 11
   **C.** 4     **D.** 9

10. In the school concert the first grade sang 3 songs, the second grade sang 4 songs, and the third grade sang 6 songs. How many songs in all did these grades sing?

   **A.** 7     **B.** 10
   **C.** 13    **D.** 9

11. Ian sold 13 tickets to the school concert. He sold 5 adult tickets. The rest that he sold were student tickets. How many student tickets did Ian sell?

   **A.** 13    **B.** 5
   **C.** 18    **D.** 8

# 2
# PLACE VALUE

**M**ATH AND SCIENCE

DATA BANK

Use the Science Data Bank on page 471 to answer the questions.

1 How many pounds would the largest beet and pear weigh together?

2 How many pounds would the largest pear, apple, and lemon weigh together?

3 How much more did the largest carrot weigh than the largest onion?

4 **Use Critical Thinking** Which of the largest fruits and vegetables weighed more than 100 pounds?

# Reading and Writing 2- and 3-Digit Numbers

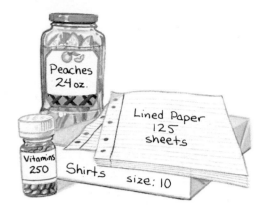

**EXPLORE** Use a Place Value Model

Put 9 hundreds, 9 tens, and 9 ones blocks in a pile. Pick ten pieces so that you have some of each kind. On a calculator, display the number they make. Repeat for five numbers.

**2 hundreds  3 tens 5 ones**

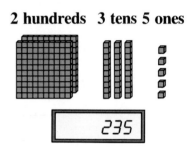

**TALK ABOUT IT**

1. Did all of your numbers have 3 digits? Why?

2. What could you say about your number if you picked up only hundreds and ones? only tens and ones? only hundreds and tens?

**Show:**

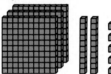

**Say:**   "three hundred twenty-five"   "eighty-two"   "two hundred eight"
**Write:**         325                          82              208

You can find the value of each digit by thinking about its place in the number.

| Hundreds | Tens | Ones |
|----------|------|------|
| 4        | 3    | 8    |

438 has a value of 4 hundreds, 3 tens and 8 ones.

$$400 \quad + \quad 30 \quad + \quad 8$$

Use place value models. Show, display, and write each number.

1. 3 hundreds, 6 tens, and 4 ones
2. 9 tens and 9 ones
3. 4 hundreds and 2 tens
4. 1 hundred and 7 ones

Tell how many hundreds, tens, and ones. Then
write the number.

1.    2.    3.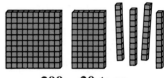

Draw pictures of place value models for each.
Then write the number.

**4.** three hundred
seventy-three

**5.** six hundred
ninety

**6.** two hundred
seven

Write two numbers from the displays in which
the 8 has the given value.

| 283 | 508 | 830 |
|-----|-----|-----|
| 805 | 681 | 498 |

**7.** 80          **8.** 800          **9.** 8

**MATH REASONING** Suppose you want to show each
number using only tens and ones models. Tell how
many tens and ones you need. Use the fewest ones
possible.

**200 = 20 tens**

**10.** 300 = ||||| tens

**11.** 307 = ||||| tens ||||| ones

**12.** 348 = ||||| tens ||||| ones

**13.** 709 = ||||| tens ||||| ones

**PROBLEM SOLVING**

**14.** Todd said, "I'm thinking of a
number that has the same
number of ones, tens, and
hundreds. The sum of the digits
is 18." What is Todd's number?

**15.** Tami said, "In my number, the
hundreds are 1 more than the
tens. The tens are 1 more than
the ones. The sum of the digits
is 6." What is Tami's number?

▶ **ALGEBRA**

Find the missing number that makes each sentence true.

**16.** $400 + ||||| + 8 = 458$     **17.** $396 = ||||| + 90 + 6$     **18.** $200 + 30 + ||||| = 237$

*More Practice, page 512, set C*

# Counting and Order

Numbers have an order. When we count we use this order.

**EXPLORE** **Discover a Relationship**

Work with a partner. Outline a 10 by 10 square on graph paper. Number the hundred squares from 301 to 400 as in the figure. Have a classmate cover any five squares. Try to tell your partner what numbers are covered. Try it again with different numbers until both you and your partner can tell the numbers quickly.

| 301 | 302 | 303 | 304 | 305 | 306 | 307 | 308 | 309 | 310 |
|-----|-----|-----|-----|-----|-----|-----|-----|-----|-----|
| 311 | 312 |     |     |     |     |     |     |     |     |
| 321 |     |     |     |     |     |     |     |     |     |
|     |     |     |     |     |     |     |     |     |     |
|     |     |     |     |     |     |     |     |     |     |
|     |     |     |     |     |     |     |     |     |     |
|     |     |     |     |     |     |     |     |     |     |
|     |     |     |     |     |     |     |     |     |     |
|     |     |     |     |     |     |     |     |     |     |
|     |     |     |     |     |     |     |     |     | 400 |

**TALK ABOUT IT**

1. How can looking at the number just before a hidden number help you find the number?

2. Can you find a hidden number by looking at the number just after the number? How?

3. Terrell said, "Counting from 301 to 400 is almost like counting from 1 to 100. Do you agree or disagree with Terrell? Why?

You can use counting to find numbers just before or just after given numbers.

## Examples

330 comes after 329.     699 comes before 700.

Write the number after.     **1.** 29     **2.** 629     **3.** 400     **4.** 399

Write the number before.     **5.** 40     **6.** 340     **7.** 399     **8.** 400

Write the number that comes after.

**1.** 90     **2.** 99     **3.** 128     **4.** 109     **5.** 289     **6.** 799

Write the number that comes before.

**7.** 71     **8.** 50     **9.** 100     **10.** 535     **11.** 460     **12.** 700

Give the next four numbers.

**13.**

592   593   594   595   596   597   598   ___   ___   ___   ___

Write the number of the next page.

**14.**     **15.**     **16.**

**MATH REASONING** Write the number that is between.

**17.** 80 and 82      **18.** 419 and 421      **19.** 899 and 901

**PROBLEM SOLVING**

**20. Missing Data** Put in numbers to make the total true. Sally bought 12 apples. ||||| were red apples and ||||| were yellow apples.

Find the sum or difference.

**21.** $10 - 2$    **22.** $6 + 5$    **23.** $6 + 6$    **24.** $6 + 7$    **25.** $8 + 6$

**26.** $8 + 7$    **27.** $9 - 8$    **28.** $9 - 1$    **29.** $4 + 5 + 5$    **30.** $3 + 4 + 9$

**31.** $\begin{array}{r} 8 \\ +8 \\ \hline \end{array}$    **32.** $\begin{array}{r} 7 \\ +7 \\ \hline \end{array}$    **33.** $\begin{array}{r} 8 \\ +9 \\ \hline \end{array}$    **34.** $\begin{array}{r} 10 \\ -3 \\ \hline \end{array}$    **35.** $\begin{array}{r} 9 \\ -3 \\ \hline \end{array}$

# Skip Counting Patterns

**LEARN ABOUT IT**

**EXPLORE** Use a Calculator

Make a hundreds chart. Try these skip counting activities.

Skip-count by 2s starting at 0.

[ON/AC] 0 [+] 2 [=] [=] [=]  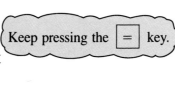 Keep pressing the [=] key.
Color yellow each number that shows on the display.

Skip-count by 2s starting at 1.

[ON/AC] 1 [+] 2 [=] [=] [=]  Keep pressing the [=] key.
Color each number green.

**TALK ABOUT IT**

1. Did you color all of the numbers on the chart?
2. Did any number get colored twice?
3. What are the ones digits in the yellow squares? What are the ones digits in the green squares?

Whole numbers can be separated into two sets.

Even numbers have a 0, 2, 4, 6, or 8 in the ones place.

Odd numbers have a 1, 3, 5, 7, or 9 in the ones place.

**0, 2, 4, 6, 8, 10, 12, ...**     **1, 3, 5, 7, 9, 11, 13, ...**

**TRY IT OUT**

Use a calculator to skip count.

1. Skip-count by 5s starting at 0.
   [ON/AC] 0 [+] 5 [=] [=] [=]
   Record each number to 100.
   What pattern do you notice?

2. Skip-count by 10s starting at 74.
   [ON/AC] 74 [+] 10 [=] [=] [=]
   Record 74 and the next nine numbers you get. Which digit changed each time? Why?

Write the names of the students who use these patterns. There can be more than one student for each pattern.

| Our Number Patterns | |
| --- | --- |
| Allan | 75, 80, 85, 90, ____ |
| Brenda | 75, 85, 95, 105, ____ |
| Carter | 150, 155, 160, 165, ____ |
| Delma | 73, 75, 77, 79, ____ |
| Emmet | 73, 173, 273, 373, ____ |
| Frieda | 150, 152, 154, 156, ____ |
| Gina | 6, 106, 206, 306, ____ |
| Hank | 56, 66, 76, 86, ____ |

**1.** Counting by 2s

**2.** Counting by 5s

**3.** Counting by 10s

**4.** Counting by 100s

Give the next four numbers for each pattern.

**5.** Frieda's pattern     **6.** Brenda's pattern     **7.** Gina's pattern

Write the names of the people who use these patterns.

**8.** all odd numbers     **9.** all even numbers     **10.** both even and odd numbers

**MATH REASONING**  Answer odd or even.

**11.** The sum of any two even numbers is ___?___

**12.** The sum of any two odd numbers is ___?___

**13.** The sum of an odd and an even number is ___?___

**PROBLEM SOLVING**

**14. Data Hunt** Find the number of dimes in a roll. How many cents is the roll of dimes worth?

**15.** Margaret has 5 pennies, 8 dimes, and 14 nickels. How many fewer dimes than nickels does she have?

▶ **MENTAL MATH**

Skip counting can help you find sums using mental math. Think about counting on by tens.

Count on by tens to find these sums.

**16.** 37 + 20     **17.** 56 + 30     **18.** 25 + 10     **19.** 20 + 63     **20.** 18 + 40

# Problem Solving
## Draw a Picture

UNDERSTAND
FIND DATA
PLAN
ESTIMATE
SOLVE
CHECK

### LEARN ABOUT IT

Some problems are easier to solve if you draw a
picture. You can use the data in the problem to
help you draw the picture.

> Four students were in the Spoon and Raw Egg
> Race. Amy finished ahead of Derek. Derek was
> between Emilou and Amy. Bill finished ahead
> of Amy. In what order did the students finish?

First I'll draw a line to show the race.

I'll show Amy ahead of Derek.

Then I'll show Derek between Emilou and Amy.

I'll put Bill ahead of Amy.

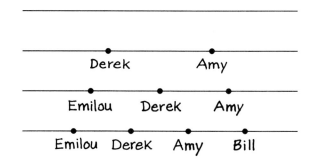

Bill was first. Then came Amy, Derek, and Emilou.

### TRY IT OUT

At the basketball contest, students lined up from
tallest to shortest. Kenji is taller than Michiko.
Jeff is shorter than Michiko. Barb is taller than
Kenji. Who is the tallest and who is the shortest?

1. Since Kenji is taller than Michiko,
   could Michiko be the tallest?

2. Who is shorter than Michiko?

3. Copy and finish drawing the
   picture to solve the problem.

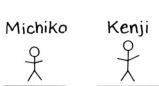

38

Draw a picture to help you solve each problem.

1. In the obstacle race, the tunnel was after the wading pool. The rope swing was between the tunnel and the tires. The tires were after the rope swing. What was the order of the obstacles?

2. Four teams ran the Potato Sack Relay Race. The Reds beat the Blues. The Yellows came in after the Greens. The Blues beat the Yellows and Greens. Tell the order in which the teams finished.

Choose a strategy from the strategies list or other strategies you know to solve these problems.

3. For the Tricycle Relay there were two teams. There were 8 students on each team. How many students were in the race?

4. There were 5 towns on the way to the surfing contest. Birch was the nearest to the ocean. Oak was after Pine. Maple was between Ash and Oak. Ash was between Maple and Birch. Give the order of the towns.

5. Ely was the handball champ. He beat 6 players the first day and 7 players the next day. How many players did he beat in all?

6. Seventeen students entered the Walk-on-Your-Hands Race. Only 8 made it to the finish line. How many did not finish?

| Some Strategies |
|---|
| Act It Out |
| Use Objects |
| Choose an Operation |
| Draw a Picture |

7. In the Tug-of-War, the Reds pulled 3 Greens across the line. The Greens pulled 1 Green back to their side. The Reds pulled 7 more Greens across the line to end the game. How many Greens were there?

Reds     Greens

8. How many winners will there be in all in the baseball hitting contest?

| Baseball Hitting Contest | |
|---|---|
| 2 First prizes | Park tickets |
| 3 Second prizes | Free pizza |
| 4 Third prizes | Big bear |
| 4 Fourth prizes | T-shirt |

*More Practice, page 512, set D*

**39**

# Comparing and Ordering Numbers

**EXPLORE** Use a Place Value Model

At the school carnival, students tried to guess the number of beans in a jar without going over. The actual number was 378. Was Gillian's guess too high?

Guesses for
Number of Beans

Casey        301
Christina 394
Gillian    382
Joshua    328
Loren      368

**TALK ABOUT IT**

1. How can you use place value blocks to prove that Gillian went over?
2. Can you use counting to prove Gillian's guess was high?
3. Who was closest without going over?

You can **compare** numbers by looking at the digits that are in the same places in the numbers.

| | |
|---|---|
| Start at the left. Find the first place where the digits are different. | 382     378 <br> The tens place is different. |
| Compare these digits. | 8 is greater than 7. <br> 7 is less than 8. |
| The numbers compare the way the digits compare. | 382 is greater than 378. <br> 382 > 378 |
| The arrow > or < points to the smaller number. | 378 is less than 382. <br> 378 < 382 |

You can **order** a set of numbers by comparing them two at a time.

**TRY IT OUT**

Write > or < for each ⫿⫿⫿.

1. 426 ⫿⫿⫿ 451          2. 374 ⫿⫿⫿ 378          3. 302 ⫿⫿⫿ 296

Write > or < for each ||||.

**1.** 36 |||| 38

**2.** 47 |||| 37

**3.** 59 |||| 62

**4.** 234 |||| 236

**5.** 754 |||| 750

**6.** 623 |||| 643

**7.** 874 |||| 774

**8.** 396 |||| 412

**9.** 601 |||| 599

Which number cards would you put in each box?

**10.** greater than 749

**11.** less than 285

**12.** between 350 and 700

**13.** List these numbers from least to greatest. 816, 798, 820

**14.** List these numbers from greatest to least. 92, 65, 69, 80

**MATH REASONING** Write a number that you would put in each box.

**15.** a little more than 400

**16.** a lot more than 400

**17.** a little less than 400

**18.** between 375 and 380

**PROBLEM SOLVING**

**19. Unfinished Problem** Joshua put 8 bags of beans in a jar. Gillian put in 3 bags. Write and answer an addition or subtraction question about this data.

**20.** Roll a number cube 3 times. Write the greatest number you can with your digits. Write the least number. Write a number between the other numbers.

▶ **USING CRITICAL THINKING**

**21.** Use the clues to find the number. It is greater than 344. It is less than 352. It is an even number. You say it when you count by fives.

*More Practice, page 501, set D*

# Ordinal Numbers

EXPLORE  **Use Math Language**
Six students are waiting in a line to buy tickets. How can you explain where different students are standing in the line?

**TALK ABOUT IT**

1. How many students are in front of Don? behind Don?

2. Who is first in line? last in line?

3. Who will get a ticket right before Cal? just after Cal?

Ann
Bev
Cal
Don
Eve
Flo

We use **ordinal numbers** to tell the position of something in an ordered set. Here are the first nine ordinal numbers from left to right.

| first | second | third | fourth | fifth | sixth | seventh | eighth | ninth |
|-------|--------|-------|--------|-------|-------|---------|--------|-------|
| 1st   | 2nd    | 3rd   | 4th    | 5th   | 6th   | 7th     | 8th    | 9th   |

**TRY IT OUT**

1. What position in line is Don?

2. Who is in 6th position in line?

3. If Gus gets in line behind Flo, what position will he be in?

4. If someone is 9th in line, how many people are waiting in front of that person?

| Other ordinal numbers | |
|-------|-------------|
| 10th  | tenth       |
| 11th  | eleventh    |
| 12th  | twelfth     |
| 13th  | thirteenth  |
| 14th  | fourteenth  |
| 15th  | fifteenth   |
| 16th  | sixteenth   |
| . . . |             |
| 20th  | twentieth   |
| 21st  | twenty-first |

**42**

*More Practice, page 512, set E*

# MIDCHAPTER REVIEW/QUIZ

Draw a picture for each. Then write the number.

**1.** two hundred seventy-five

**2.** forty-six

**3.** three hundred eight

**4.** 3 tens, 5 hundreds, and 0 ones

Write the number that comes before.

**5.** 286        **6.** 99        **7.** 600        **8.** 731

Write the next four numbers.

**9.** 0, 10, 20, 30, ___, ___, ___, ___

**10.** 324, 325, 326, 327, ___, ___, ___, ___

**11.** 32, 34, 36, 38, ___, ___, ___, ___

**12.** 65, 70, 75, 80, ___, ___, ___, ___

**13.** 696, 697, 698, ___, ___, ___, ___

**14.** 55, 155, 255, 355, ___, ___, ___, ___

Write > or < for each ⦚⦚.

**15.** 23 ⦚⦚ 18        **16.** 157 ⦚⦚ 75        **17.** 236 ⦚⦚ 263

**18.** 101 ⦚⦚ 99        **19.** 789 ⦚⦚ 790        **20.** 816 ⦚⦚ 698

**21.** List the numbers from greatest to least. 499, 556, 402, 560

## PROBLEM SOLVING

**22.** Nina, Beth, and Mary guessed how many beans were in a jar. Nina's guess was higher than Mary's guess and lower than Beth's guess. What number did each girl guess?

Guesses
572
605
578

**23.** Gabe lives next door to Eddy. Annie lives next door to Lisa and next door to Gabe. How far apart do Eddy and Lisa live?

**24.** Leilani used place value blocks to show 3-digit numbers that had 2 more tens than hundreds. Write some numbers that she could show. What is the biggest number she could show?

43

# Rounding to the Nearest Ten

Lettuce 62 days

Peppers 72 days

Carrots 78 days

Tomatoes 74 days

## LEARN ABOUT IT

### EXPLORE  Use a Number Line

Darrell counted the number of days from planting seeds to harvesting vegetables. Which of the numbers of days are closer to 70 than to 60 or 80?

```
+--•--•--•--•--•--•--•--•--•--•--•--•--•--•--•--•--•--•--•--•-->
   50          55          60          65          70
```

## TALK ABOUT IT

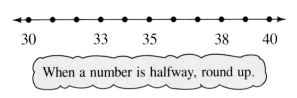

Pumpkins 75 days

Beans 64 days

Watermelon 79 days

Corn 68 days

1. What numbers did you select? Explain why.

2. Is 75 closer to 70 or 80 or is it halfway between?

We can **round** numbers to tell **about how many**. To round a number to the nearest ten, find the closest number ending in zero.

Round **up** to the next ten if the one's digit is **5 or more**.

Round to the **nearest ten**.

$$33 \rightarrow 30$$
$$38 \rightarrow 40$$
$$35 \rightarrow 40$$

Round **down** to the nearest ten if the one's digit is **less than 5**.

```
+--•--•--•--•--•--•--•--•--•--•-->
  30    33   35      38   40
```

When a number is halfway, round up.

## TRY IT OUT

Round to the nearest ten. Use the number line.

```
+--•--•--•--•--•--•--•--•--•--•--•-->
  0   10   20   30   40   50   60   70   80   90   100
```

**1.** 25          **2.** 83          **3.** 94          **4.** 49          **5.** 57          **6.** 19

Round to the nearest ten. Which red number is the better choice?

**1.** 33 → **30** or **40**    **2.** 57 → **50** or **60**    **3.** 85 → **80** or **90**

Round to the nearest ten.

**4.** 25 → ___    **5.** 14 → ___    **6.** 73 → ___    **7.** 55 → ___

**8.** 49 → ___    **9.** 21 → ___    **10.** 66 → ___    **11.** 82 → ___

Use the numbers on this chalkboard.

**12.** Which number rounds up to 50?

**13.** Which two numbers round to 70?

**14.** Which three numbers round to 60?

**MATH REASONING**

**15.** List the five numbers less than 30 that round up to 30.

**16.** List the four numbers greater than 30 that round down to 30.

**PROBLEM SOLVING**

**17.** Dave said, "I have saved about $20." List all the dollar amounts that Dave may have saved.

**18. Science Data Bank** Which world's record vegetable weight is closest to 30 pounds? See page 471.

DATA BANK

**MIXED REVIEW**

Add or subtract.

**19.** $5 + 5 + 4$    **20.** $17 - 7$    **21.** $17 - 8$    **22.** $6 + 8 + 2$    **23.** $16 - 7$

**24.** $8 + 7 + 3$    **25.** $7 + 9 + 1$    **26.** $14 - 9$    **27.** $12 - 5$    **28.** $4 + 4 + 1$

**29.** $11 - 5$    **30.** $18 - 9$    **31.** $13 - 7$    **32.** $15 - 8$    **33.** $1 + 9$

# More About Rounding

You can use what you know about rounding to
the nearest ten to help you round to the nearest
hundred or nearest dollar.

**EXPLORE  Think About the Situation**
Tickets to the circus cost $4.95 each. There were
812 tickets sold. Look at the newspaper headline.

**TALK ABOUT IT**

1. Do you think the newspaper headline is
   reporting fairly? Why?

2. Why do you think they rounded the numbers?

3. What do you think the headline would be if
   689 people paid $4.25 each for tickets?

Here is how to round to the nearest hundred or to
the nearest dollar.

| Round up to the next hundred if the tens digit is 5 or more. | Keep the hundreds digit the same if the tens digit is less than 5. | Round money as if the dollar and cent notation were not there. |
| --- | --- | --- |

| Nearest Hundred | Nearest Dollar |
| --- | --- |
| 335 → 300 | $3.35 → $3.00 |
| 350 → 400 | $3.50 → $4.00 |
| 382 → 400 | $3.82 → $4.00 |

**TRY IT OUT**

Round to the nearest hundred or nearest dollar.

**1.** 256    **2.** $3.69    **3.** 482    **4.** $2.79    **5.** 629    **6.** $2.20

Round to the nearest hundred or nearest dollar.
Which red number is the better choice?

1. 749 → **700** or **800**           2. 350 → **300** or **400**

3. 514 → **500** or **600**           4. $4.98 → **$4.00** or **$5.00**

Round to the nearest hundred or nearest dollar.

5. 470    6. 504    7. 185    8. 333    9. 650    10. 981

11. 249    12. 782    13. 516    14. 839    15. 136    16. 163

17. $8.52    18. $0.75    19. $2.49    20. $3.56    21. $1.25    22. $7.89

## APPLY

**MATH REASONING** Make up a number to finish the
sentence.

23. Tickets were about $8.00. The
exact price was ‖‖‖.

24. About 300 people attended. The
exact number was ‖‖‖.

## PROBLEM SOLVING

25. The Sports Club hoped to sell
12 hundred tickets to the game.
So far they have sold about 8
hundred. About how many more
tickets do they need to sell?

26. In the ticket line, Rio was ahead
of Tyler. Bea was behind Tyler
and ahead of Sue. Pam was last.
What is the students' order in
the line?

▶ **ESTIMATION**

Choose the estimate that seems most reasonable.

27. About how many
people went to the
school play?

3 people
30 people
300 people

28. About how long
did the basketball
game last?

1 minute
10 minutes
100 minutes

29. About how much
were the game
tickets?

2 cents
20 cents
200 cents

# Problem Solving
## Understanding the Question

| UNDERSTAND |
| FIND DATA |
| PLAN |
| ESTIMATE |
| SOLVE |
| CHECK |

To solve a problem, you must understand what the question is asking. Sometimes it helps to put the question in your own words or to ask it in a different way.

> Kara bought 5 glow-in-the-dark stickers at the drugstore. She bought 7 more at the hobby shop. How many stickers did Kara buy in all?

First I'll read the question.

How many stickers did Kara buy in all?

Then I'll ask the question in a different way.

What was the total number of stickers Kara bought?

Read the problem. Then choose the question that asks the same thing.

1. Megan's new skateboard had 4 stickers on it. Judy's old skateboard had 11 stickers. How many more stickers are on Judy's skateboard?

   A. How many fewer stickers are on Megan's skateboard?

   B. How many stickers are on both skateboards?

Write a question that you can answer using addition or subtraction.

2. Roberto had 3 baseball banners. His brother Juan had 8 banners.

3. Ginny has 5 puffy animal stickers and 8 Snoopy stickers.

48

1. Yori got some stickers on the clothes he bought. He collected 6 on his birthday clothes and 5 on his new school clothes. How many stickers did he collect in all?

2. Teresa pasted 13 googly-eyed stickers on one page of her sticker book. Then she pasted on 3 more. How many stickers are on the page now?

3. Julio put 4 football banners in a row. The Bears came after the Oilers. The 49ers came between the Bears and the Oilers. The Dolphins came before the Oilers. Tell in what order Julio put the banners.

4. Joanne made constellations on her ceiling using glow-in-the-dark star stickers. How many more stars did she use on the Great Dog than on the Little Bear?

5. Ada had 12 stickers. She tore 3 by mistake and threw them away. How many stickers did she have left?

6. For her birthday party, Pat bought 6 stickers for party favors and 12 stickers for game prizes. How many more stickers did she buy for game prizes?

7. Joanne put the Charioteer on her ceiling and Queen Cassiopeia on her wall. How many stars did she use in all?

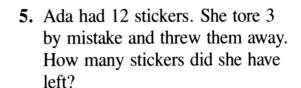

| Constellation | Number of Stars |
|---|---|
| Canis Major, the Great Dog | 13 |
| Leo, the Lion | 15 |
| Auriga, the Charioteer | 8 |
| Cassiopeia, the Queen | 5 |
| Ursa Minor, the Little Bear | 7 |
| Aquila, the Eagle | 11 |

8. **Understanding the Operations** Tell what operation you would use. Use objects to solve the problem.

   Jane had 5 stickers of football team helmets. She needed 9 more to make a set. How many stickers are there in a set?

*More Practice, page 513, set A*

**49**

# Reading and Writing 4-Digit Numbers

**LEARN ABOUT IT**

**EXPLORE** **Use a Place Value Model**

Work in groups. Use hundreds and tens to show 1,000 three different ways.

**10 hundreds is the same as 1 thousand.**

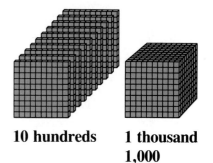

**10 hundreds**      **1 thousand 1,000**

**TALK ABOUT IT**

1. Could you show 1,000 using tens only? Tell why.

2. Could you show 1,000 using ones only? Tell why.

You can show, say, and write 4-digit numbers.

**Show:**

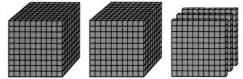            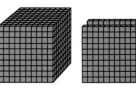

**Say:**   "two thousand, three hundred eighteen"      "one thousand, two hundred four"

**Write:**          2,318                    1,204

You can find the value of each digit by thinking about its place in the number.

3,572 has a value of 3 thousands, 5 hundreds, 7 tens, and 2 ones.

| Thousands | Hundreds | Tens | Ones |
|-----------|----------|------|------|
| 3 ,       | 5        | 7    | 2    |

$3,000$ + $500$ + $70$ + $2$

**TRY IT OUT**

Use place value models. Show, say, and write each number.

1. 2 thousands, 4 hundreds, 5 tens, and 9 ones

2. 2 hundreds and 7 ones

3. 1 thousand, 8 tens, and 4 ones

4. 1 thousand and 3 tens

Tell how many thousands, hundreds, tens, and
ones. Then write the number.

**1.**

**2.**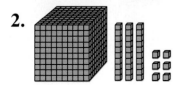

Write two numbers from the box in
which the digit 3 has the given value.

| 3,500 | 2,038 |
| 2,734 | 2,374 |
| 1,389 | 3,418 |

**3.** 30      **4.** 300      **5.** 3,000

Write the numbers from the box that match.

**6.** greater than 3,478    **7.** less than 2,724    **8.** between 2,724 and 3,478

**APPLY**

### MATH REASONING

Give the number that is 1,000 more.    **9.** 2,034    **10.** 1,840   **11.** 798

Give the number that is 1,000 less.    **12.** 4,500    **13.** 5,708   **14.** 3,762

### PROBLEM SOLVING

**15.** Think of a digital clock as showing 3- and
4-digit numbers. What 4-digit clock number
has the greatest digit sum? the least?

▶ **CALCULATOR**

Change the digit by adding or subtracting just
once. The other digits must stay the same. Tell
what you did for each exercise.

**16.** Change the 3 in 4,387 to 0.      **17.** Change the 2 in 2,875 to 0.

# Understanding Thousands

**LEARN ABOUT IT**

**EXPLORE** **Think About the Situation**

Work in groups. Did you ever count the kernels on an ear of corn? Your group can show 1,000 kernels. Draw rows of Os for kernels.

Make 10 rows of Os on a piece of paper. Draw 10 Os in each row. Put your paper in a stack with 9 other papers so that a stack has 10 papers.

**TALK ABOUT IT**

1. How many Os do you think are in a stack of 10 papers?

2. Imagine having 10,000 Os. How many of these stacks would you need?

A **period** is a group of three digits. Use a comma between the thousands period and the ones period.

| Thousands Period | | | Ones Period | | |
|---|---|---|---|---|---|
| Hundred Thousands 7 | Ten Thousands 2 | Thousands 4 , | Hundreds 8 | Tens 7 | Ones 3 |

700,000 + 20,000 + 4,000 + 800 + 70 + 3

Read 724,873 as
**"seven hundred twenty-four thousand,** eight hundred seventy-three."

**TRY IT OUT**

Read these numbers aloud.

**1. 6,**000      **2. 30,**000      **3. 58,**400      **4.** 119,300      **5.** 346,491

52

Write the digit in the given place for 604,891.

**1.** thousands      **2.** hundreds      **3.** tens

**4.** ones      **5.** hundred thousands      **6.** ten thousands

Write two numbers from the basket
in which the digit 4 has the given value.

**7.** 400      **8.** 40,000

**9.** 4,000      **10.** 400,000

Write the number. Use a comma to separate the
thousands period and the ones period.

**11.** thirty thousand, eight hundred
seven

**12.** five hundred twelve thousand,
six hundred twenty-six

**13.** two hundred eight thousand,
ninety

**14.** two hundred eight thousand,
nine

**MATH REASONING**

**15.** Write a sentence to tell what **place value**
means.

**PROBLEM SOLVING**

**16.** Use these digits to write four
different 6-digit numbers greater
than 500,000.

6   3   4   2   0   7

**17. Science Data Bank** List the
seeds in order from the fewest
seeds in a pound to the most
seeds in a pound. See page 471.

▶ **USING CRITICAL THINKING**

**18.** Trace this line. Then make a mark to show
where you think the number 10,000 would go.

0   1,000

# Data Collection and Analysis
## Group Decision Making

UNDERSTAND
FIND DATA
PLAN
ESTIMATE
SOLVE
CHECK

**Doing a Survey**

**Group Skill:**
Encourage and Respect Others

Is your favorite cartoon character a beagle, a fat cat, or a freckle-faced little girl? What comic strip characters do people like best?

You can ask people questions to find out. This is called making a survey.

### Data Collection

1. Have your group choose three comic strip characters from the newspaper. Write the names of your three characters on a piece of paper.

2. Ask 15 people to choose their favorite character from your list. Record each person's answer on a list like this.

| Name | Favorite Cartoon Character |
|------|----------------------------|
| Kate | _____ |
| Vinh | _____ |
| _____ | _____ |
| _____ | _____ |

## Organizing Data

**3.** Fold a large piece of paper into three equal parts.

**4.** Cut pictures of your cartoon characters out of a newspaper. Paste each character in one section of your folded paper.

**5.** Cut out a square for each person in your survey. Paste a square on your folded paper to show which cartoon character each person chose.

## Presenting Your Analysis

**6.** Which comic strip character did the people answering your questions like the best? Why do you think so?

55

# WRAP UP

## What's the Number?

Write how you would say each number. Give the
value of each digit in the number.

**1.** 50      **2.** 82      **3.** 167      **4.** 205      **5.** 790
**6.** 1,012    **7.** 3,406    **8.** 4,020    **9.** 132,280    **10.** 200,047

## Sometimes, Always, Never

Decide which word should go in the blank,
<u>sometimes</u>, <u>always</u>, or <u>never</u>. Explain your choices.

**11.** A digit in the thousands period _____ has a value
greater than a digit in the ones period.
**12.** You _____ can tell if a number is odd or even by
looking at the digit in the ones place.
**13.** An odd number _____ has a 5 in the ones place.
**14.** You _____ round numbers to the nearest ten.

## Project

Make up a twenty-sentence story using ordinal
numbers from **first** to **twentieth.** To begin,
choose an animal or object that you like. Imagine
that there are 20 of them. Make up a sentence for
each that tells what it did or said. You may want
to put the actions or sayings in alphabetical order,
too. Here is how you might start.

The **first** boat sailed to Africa. The **second** boat sailed to Bermuda....

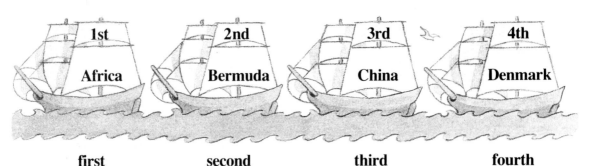

first        second        third        fourth

# CHAPTER REVIEW/TEST

## Part 1    Understanding

1. What do you write in the tens place for five hundred nine?

2. What are the positions of the persons before and after the ninth person in line?

3. Name the skip-counting pattern for both groups of numbers. 7, 9, 11, 13, 15 and 8, 10, 12, 14, 16

In which two numbers in the box does the digit 7 have the given value?

| | |
|---|---|
| 307,862 | 677,831 |
| 735,400 | 226,739 |
| 6,752,103 | 173,027 |

4. 7,000    5. 70,000    6. 700,000

## Part 2    Skills

7. List from greatest to least. 798, 897, 97, 879

8. List the next three numbers after 99.

Write the missing numbers.

9. 230, 235, 240, |||||, 250

10. 438, |||||, |||||, |||||, 442

Write *true* or *false*.

11. 46 < 55    12. 719 < 687    13. 84 rounds to 90.    14. 247 rounds to 200.

## Part 3    Applications

15. Sylvia collected $6.53 for the band trip. About how many dollars did she collect?

16. Ron buys 9 tapes. He gives 5 to Li. Write a question that you can answer using subtraction.

17. How many days are between the 9th day and the 15th day of the month?

18. **Challenge**  Sara was first in a race. Rob finished ahead of Abby but behind Sara. Joel beat Rob. Who was in third place?

# ENRICHMENT
## Roman Numerals

Many years ago, the Romans used letters to write their numbers.

Three of the letters were I, V, and X. Look at the chart. It shows you the standard numerals 1 to 20 and the Roman numerals for each.

Write the Roman numerals for each number.

**1.** 1      **2.** 5      **3.** 10

**4.** 8      **5.** 11      **6.** 14

Write the standard numerals for each Roman numeral.

**7.** III      **8.** IV      **9.** IX

**10.** X      **11.** XV      **12.** XIX

The numbers 21 and 22 are written as XXI and XXII.

**13.** Write the Roman numerals from 23 to 31.

Use Roman numerals.

**14.** What is your age?

**15.** Give the number of the month in which you were born.

**16.** On what day of the month were you born?

**17.** Design a clock using Roman numerals.

| Number | Roman Numeral | What It Means |
|--------|---------------|---------------|
| 1 | I | 1 |
| 2 | II | 1 + 1 |
| 3 | III | 1 + 1 + 1 |
| 4 | IV | 5 − 1 |
| 5 | V | 5 |
| 6 | VI | 5 + 1 |
| 7 | VII | 5 + 2 |
| 8 | VIII | 5 + 3 |
| 9 | IX | 10 − 1 |
| 10 | X | 10 |
| 11 | XI | 10 + 1 |
| 12 | XII | 10 + 2 |
| 13 | XIII | 10 + 3 |
| 14 | XIV | 10 + 4 |
| 15 | XV | 10 + 5 |
| 16 | XVI | 10 + 6 |
| 17 | XVII | 10 + 7 |
| 18 | XVIII | 10 + 8 |
| 19 | XIX | 10 + 9 |
| 20 | XX | 10 + 10 |

# Cumulative Review

Add or subtract.

**1.** $9 + 1$

    **A.** 9     **B.** 11

    **C.** 8     **D.** 10

**2.** $8 - 5$

    **A.** 2     **B.** 3

    **C.** 12     **D.** 13

**3.** $4 + 5$

    **A.** 8     **B.** 9

    **C.** 10     **D.** 11

**4.** $\begin{array}{r} 7 \\ -\ 3 \\ \hline \end{array}$

    **A.** 10     **B.** 7

    **C.** 4     **D.** 9

**5.** $6 + 9$

    **A.** 12     **B.** 17

    **C.** 14     **D.** 15

**6.** $14 - 8$

    **A.** 6     **B.** 8

    **C.** 7     **D.** 9

**7.** Which sum or difference could help you solve $4 + 5$?

    **A.** $4 + 4 + 1$     **B.** $5 - 4$

    **C.** $9 - 5 - 1$     **D.** $9 + 4$

**8.** $3 + 8 + 4$

    **A.** 7     **B.** 12

    **C.** 11     **D.** 15

**9.** $\begin{array}{r} 17 \\ -\ 9 \\ \hline \end{array}$

    **A.** 8     **B.** 9

    **C.** 7     **D.** 10

**10.** Which number comes next? 343, 344, 345, _____?

    **A.** 340     **B.** 342

    **C.** 336     **D.** 346

**11.** Choose the correct number sentence. 7 bears sat on the ice. 5 were small bears. The rest were big. How many were big?

    **A.** $7 + 5 = 12$     **C.** $12 - 7 = 5$

    **B.** $5 + 7 = 12$     **D.** $7 - 5 = 2$

**12.** Warren's family owns 4 cats. They got Licks before they got Tabby. They got Mookie before Licks. They got Swipes after Tabby. Which cat did the family get first?

    **A.** Tabby     **B.** Licks

    **C.** Mookie     **D.** Swipes

# 3

## TIME AND MONEY

**M**ATH AND HEALTH AND FITNESS

### DATA BANK

Use the Health and Fitness Data Bank on page 481 to answer the questions.

1 Look at the picture graph. Who hit the greatest number of home runs? How many home runs did that player hit?

**2** Look at the picture graph again. Who hit the least number of home runs? How many home runs did that player hit?

**3** How many more home runs did Babe Ruth hit in 1927 than Willie McCovey hit in 1969?

**4** **Use Critical Thinking** You are pitching. The batter facing you has been hitting a curve ball 4 out of every 5 times. Should you throw her a curve ball?

# Telling Time

## EXPLORE  Use Skip Counting

The **hour** hand on a clock is the short hand. It takes 1 hour, or 60 minutes, to move from one number to the next.

The **minute** hand is the long hand. It takes 5 minutes to move from one number to the next.

5 minutes

minute hand

hour hand

1 hour

## TALK ABOUT IT

1. What number does the minute hand point to when it is 20 minutes after an hour?

2. How many minutes does it take the minute hand to move from the 7 to the 10?

3. How many minutes does it take the minute hand to go around one time? Skip count by 5s.

You read nine-fifteen, 15 minutes past 9, or quarter past nine.

You read nine-thirty, 30 minutes past 9, or half past nine.

You read nine forty-five, 45 minutes past 9, or quarter to ten.

## TRY IT OUT

Write each time two different ways.

1.

2.

3.

4.

Write each time as you would see it on a digital clock.

**1.**    **2.**    **3.**    **4.**

**5.**    **6.**    **7.**    **8.**

**APPLY**

**MATH REASONING** Decide if the estimated time is reasonable.

**9.** It takes about 5 minutes to make your bed.

**10.** It takes about 30 minutes to make a peanut butter sandwich.

**11.** It takes about 50 minutes to take a bath and get dressed.

**PROBLEM SOLVING**

**12.** Does this clock show breakfast time?

**13. Data Hunt** Find out what time two of your classmates eat dinner. Draw two clocks to show the times. Skip count by 5s to find the difference.

▶ **MENTAL MATH**

Skip count by 5s to find how many minutes are shaded.

**14.** Nick left school at 3:00. What time did Nick get home?

**15.** Ruby began eating lunch at 12:00. What time did she finish lunch?

# Telling Time to the Minute

**EXPLORE** **Study the Clocks**

Scott's watch shows when he and Karin sat down to see the movie.

The minute hand moves from one mark to the next in 1 minute.

 **1 minute later** →  **1 minute later** →

Movie
Starts
4:20

**TALK ABOUT IT**

1. How many minutes after the hour did Scott and Karin sit down?

2. Did they get there before or after the movie started?

Write **a.m.** for times between 12:00 midnight and 12:00 noon.

Write **p.m.** for times between 12:00 noon and 12:00 midnight.

| Breakfast | Lunch | Supper |
|---|---|---|
|  |  |  |
| 7:22 a.m. | 12:11 p.m. | 6:31 p.m. |

Give each time. Write a.m. or p.m.

**1.** Art class     **2.** School ends     **3.** Recess     **4.** Science class

Write each time with a.m. or p.m.

**1.** Soccer game

**2.** School begins

**3.** Wake up

**4.** Math group

**5.** Library time

**6.** Recess

**7.** Study group

**8.** Piano lesson

**APPLY**

**MATH REASONING** Which is the best estimate of the time?

**9.** `7:54`
    **A.** about 7:30
    **B.** about 8:00

**10.** `10:23`
    **A.** about 10:00
    **B.** about 10:30

**PROBLEM SOLVING**

**11.** Jan said, "When it is 8:36, the minute hand is between the 6 and the 7." Do you you agree?

**12.** Is it light or dark outside at 1:00 a.m.? Is it light or dark outside at 4:00 p.m.?

▶ **USING CRITICAL THINKING**

**13.** The class picture was taken at 2:45. The clocks show when four students arrived. Who was early? Who was on time? Who was late?

Dana

Ling

Ryan

Mary

# Mental Math
## Time

**EXPLORE  Solve to Understand**

The soccer game started at 5:00. Jeff arrived 1 hour and 35 minutes later. What time did Jeff arrive?

A    B    C

**TALK ABOUT IT**

1. How much time has passed from clock A to clock B? Explain.

2. How much time has passed from clock B to clock C? How does the dotted hand help you?

You can use mental math to find the time something ends.

| 1 hour | 30 minutes | 5 minutes |
|---|---|---|
|  |  |  |
| Imagine the hour hand moving from one number to the next. | Imagine the minute hand moving halfway around the clock. | Imagine the minute hand moving from one number to the next. |

The game started at 5:00. It ended 1 hour and 45 minutes later. What time did it end?

66

## PRACTICE

Soccer practice started at 3:00. The list shows how long each child played.

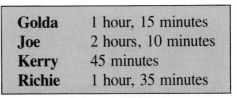

| Golda | 1 hour, 15 minutes |
| Joe | 2 hours, 10 minutes |
| Kerry | 45 minutes |
| Richie | 1 hour, 35 minutes |

Tell what time each child stopped playing.

**1.** Kerry      **2.** Richie

**3.** Golda      **4.** Joe

## APPLY

**MATH REASONING** Did they take pictures for more or less than 1 hour?

**5.** Jane took pictures of the soccer team from 2:00 to 2:48.

**6.** Garnett took pictures of the team from 3:00 to 4:10.

**PROBLEM SOLVING**

**7.** Abdul's family left the soccer field at 6:00. It took them 24 minutes to drive home. What time did they arrive home?

**8. Health and Fitness Data Bank** How many more home runs did McCovey hit than Evans? See page 481.

## MIXED REVIEW

**9.** Skip-count by fives from 60 to 100.

**10.** Skip-count by twos from 19 to 31.

**11.** Skip-count by tens from 157 to 207.

Use the numbers from the box.

**12.** In which numbers does the digit 4 have a value of 400?

**13.** Which numbers are greater than 2,450?

**14.** Which numbers are less than 450?

**15.** Which numbers are between 500 and 1,500?

| 347 | 2,431 |
| 4,127 | 824 |
| 1,814 | 1,476 |
| 2,645 | 408 |

*More Practice, page 514, set A*

# Reading a Calendar

**January**

| Sunday | Monday | Tuesday | Wednesday | Thursday | Friday | Saturday |
|--------|--------|---------|-----------|----------|--------|----------|
|        |        |         | 1         | 2        | 3      | 4        |
| 5      | 6      | 7       | 8         | 9        | 10     | 11       |
| 12     | 13     | 14      | 15        | 16       | 17     | 18       |
| 19     | 20     | 21      | 22        | 23       | 24     | 25       |
| 26     | 27     | 28      | 29        | 30       | 31     |          |

**EXPLORE  Study the Calendar**

January 16 is on Thursday. It is the third Thursday in January. The second Tuesday is the 14th day of January.

**TALK ABOUT IT**

1. What is the date of the fourth Friday in January on this calendar?

2. Use an ordinal number to tell which Monday is the 20th of January.

A calendar shows the 12 months of the year in order. Count to find the number for the month.

You can use three numbers to write dates. For July 23, 1984, write 7/23/84.

January
February
March
April
May
June
July
August
September
October
November
December

the 7th month     the 23rd day     1984

**7  /  23  / 84**

**PRACTICE**

Use the January calendar. Give the date for these days.

**1.** the fourth Friday     **2.** the second Tuesday     **3.** the fifth Thursday

Give the day of the week for these dates.

**4.** January 31          **5.** January 12          **6.** January 9

Give the name of the month.

**7.** the 3rd month  **8.** the 11th month  **9.** the 6th month  **10.** the 8th month

**11.** Sharon wrote her birthday as 1/28/83. What is the month? the day? the year?

**12.** Write your own birthday using three numbers.

*More Practice, page 501, set G*

# MIDCHAPTER REVIEW/QUIZ

Use the clock faces to answer the questions.

   **A**       **B**       **C**       **D**       **E**       **F**

1. What time is shown on each clock?

2. Which two clocks show about 4 o'clock?

3. Is clock **D** closer to quarter past 3 or half past 3?

4. If clock **E** shows the time right now, what time will it be in 20 minutes? What time was it 20 minutes ago?

Use the calendar to answer the questions.

5. The first day of one of the months is a Tuesday. What is the third day of that month?

6. What are the names of the months between the third month and the eighth month?

## PROBLEM SOLVING

7. Liz was born at 11:23 a.m. on 4/26/84. Was she born before or after noon? What month and year was she born?

8. Marisa will begin her evening paper route in 1 hour. If clock **F** (above) shows the time right now, what time will she begin? (Is that a.m. or p.m.?)

9. Joan practiced piano from 3:45 until 4:30. Was that more than 1 hour or less than 1 hour?

10. Soccer practice lasted 2 hours and 20 minutes. It ended at 6:30. What time did it begin?

# Problem Solving
## Extra Data

| UNDERSTAND |
| FIND DATA |
| PLAN |
| ESTIMATE |
| SOLVE |
| CHECK |

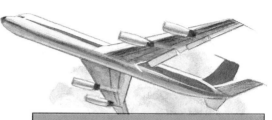

### LEARN ABOUT IT

Some problems have more data than
you need to solve them. You must
sort through the data carefully.

| City | Gate | Departure |
|------|------|-----------|
| Cleveland | 4 | 7:30 p.m. |
| Dallas | 7 | 9:10 p.m. |
| Miami | 10 | 8:15 p.m. |

Suki's family got to the airport at 7:00 p.m. to
catch their plane to Miami. The plane was
leaving at 8:15 p.m. While Suki and her family
were waiting, they spent $4 on magazines.
How long did they have to wait before their
plane left?

I'll read the question again.

How long did they have to wait?

I'll hunt for the data I need.

They got to the airport at 7:00 p.m.
The plane leaves at 8:15 p.m.

Some data is extra.

They spent $4.

I'll solve the problem using the data
I need.

From 7:00 p.m. to 8:15 p.m. is 1
hour and 15 minutes.

Suki's family waited for 1 hour and 15 minutes.

### PRACTICE

Solve. Tell the extra data in each problem.

**1.** All 6 people in Diana's family
went on a trip to Dallas. It took 2
hours to drive to the airport and 5
hours to fly to Dallas. How many
hours did the trip take?

**2.** Jefferson saw the movie *Phantom
Busters* on an airplane to St.
Louis. The movie started at
10:00 a.m. and lasted 1 hour and
35 minutes. The movie cost $4.
What time did the movie end?

Solve. Use skills and strategies you know.

**1.** The airport clock looked like this:

Jeremiah's watch said this:

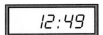

Find the difference in minutes.

**2.** On the trip to Hawaii, Brenda had 2 bags. Her parents had 3 bags. Her grandmother had 3 bags. How many bags did they have in all?

**3.** Samuel brought $13 to the airport. He saw a mug for $6 and a poster for $4. If he bought the mug, how much money would he have left?

**4.** The Lockheed L1011 is full. There are 386 people in their seats. Use the table to find how many people still have to sit down.

**5.** In the Boeing 747, Tara's family was sitting in a row. Mom was to the right of Dad. Dad was between Tara and Mom. Doug was to the left of Tara. Tell the order in which they were sitting.

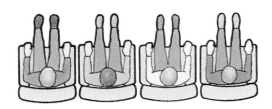

**6.** On a holiday weekend, Nina's family got in line to buy tickets at 9:00. They waited 45 minutes because there were 22 people ahead of them. What time did they buy their tickets?

| How Many People in the Plane? | |
| --- | --- |
| Airplane | Number of People |
| Boeing 707 | 219 |
| Boeing 727 | 189 |
| Boeing 747 | 550 |
| Concorde | 100 |
| Lockheed L1011 | 400 |

**7. Missing Data** This problem has missing data. Tell what data is needed to solve the problem.

There are 125 empty seats in the airplane flying to Rapid City. How many seats are in the airplane all together?

# Making a Purchase

**EXPLORE** **Think About the Situation**

Ron counted his money to see if he had enough to buy a World Series pennant.

 $1.00  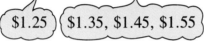 $1.25  $1.35, $1.45, $1.55   $1.60, $1.65   $1.66

**TALK ABOUT IT**

1. Did Ron have enough money to buy the pennant? How do you know?

2. How did Ron use skip counting to count money?

When counting money, start with the bills and coins with the greatest value.

| half dollar | quarter | dime | nickel | penny |
|---|---|---|---|---|
|  |  |  |  |  |
| 50 cents | 25 cents | 10 cents | 5 cents | 1 cent |
| 50¢ | 25¢ | 10¢ | 5¢ | 1¢ |

**TRY IT OUT**

1. How much money does Gillian have? Is it enough to buy the pennant?

2. What bills and coins could she give the clerk?

Give the value of each set of coins. Choose an item that you could buy with each amount.

1.

2.

Write these amounts in words.

3. $1.35        4. $0.87        5. $1.07        6. $0.03

**APPLY**

**MATH REASONING** Use play money to show two ways to make each amount.

7. 37¢        8. $1.14        9. $0.82

**PROBLEM SOLVING**

10. Roger had 3 dimes, 2 quarters, 1 nickel, and 4 pennies. Does he have enough money to buy popcorn for 80¢?

11. **Fitness Data Bank** John was at the World Series. He had 3 quarters. Did he have enough money to buy a bag of peanuts? See page 481.

**MIXED REVIEW**

Round to the nearest ten.

12. 49        13. 31        14. 54        15. 26        16. 45        17. 50

Round to the nearest hundred or dollar.

18. $2.49        19. 831        20. 654        21. $8.26        22. 345        23. $7.50

Write a fact family for each pair of addends.

24. 2 and 5        25. 9 and 7        26. 8 and 4        27. 3 and 3

*More Practice, page 514, set C*

# Estimating Amounts of Money

**EXPLORE** **Think About the Situation**

Suppose you wanted to buy a birthday card that cost $1.00. You looked at the change in your coin purse.

I need 4 quarters to make $1.00.
I have 3 quarters and some other coins.

## TALK ABOUT IT

**1.** How can you decide quickly, without counting, if you have enough? Explain.

**2.** Is it always necessary to count money exactly?

Estimating amounts of money is easier if you remember these relationships.

I need 4 quarters to make $1.00.
I have that much and more.

Do you have more or less than $1.00?

**1.**    **2.**

Is each amount more than or less than $1.50?

**1.**

**2.**

Estimate if you have enough to buy each card.

**3.**

**4.**

**APPLY**

**MATH REASONING**

**5.** Jenny was counting pennies from her piggy bank. What could she do to make it easy to count the pennies by skip counting?

**PROBLEM SOLVING**

**6.** Yuan had 2 quarters, 2 dimes, and 2 pennies. Did he have more or less than $0.59?

**7. Talking About Your Solution**
Willie had a bag of quarters, dimes, and nickels that he earned from a popcorn stand. What is a good way to pile the money to count it? Explain.

▶ **USING CRITICAL THINKING**

This piece of rope costs $1.00. About how much do you think the other pieces cost?

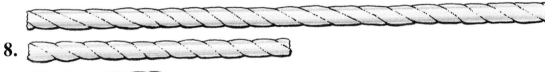

**8.**

**9.**

**10.**

# Counting Change

### EXPLORE  Solve to Understand

Ernie gave the clerk $1.00 to pay for a flower for his mother. The clerk counted out the **change.** How much change did Ernie get from $1.00?

**TALK ABOUT IT**

1. What is the first amount the clerk said?
2. What coins did the clerk give Ernie as she said 84¢ and 85¢?
3. Why did the clerk stop calling out amounts when she got to $1.00?

**Examples**   Ernie gave 75¢.          Ernie gave $1.25.

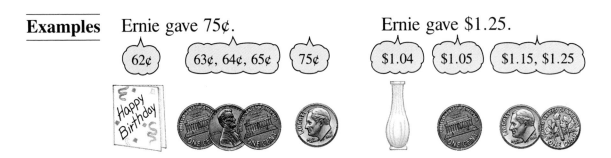

Use play money to count out change. Work with a partner.

1. Give the clerk $1.00 for a 92¢ flower. The change is 3 pennies and 1 nickel.

2. Give the clerk $1.00 for a 63¢ card. The change is 2 pennies, 1 dime, and 1 quarter.

76

Count the change. Write the numbers the clerk
would say.

**1.** You gave the clerk 3 quarters.

62¢ ___ ___ ___ ___

**2.** You gave the clerk $1.00.

83¢ ___ ___ ___ ___

**MATH REASONING** Tell what coins are missing.

**3.** You gave the clerk 75¢.

64¢   65¢   75¢

 ?

**4.** You gave the clerk $1.00.

78¢   79¢   80¢   90¢   $1.00

?   ?   ?

**PROBLEM SOLVING**

**5.** Lizzy bought some seeds for
$0.95. She gave the clerk $1.00.
How much change did she get
back?

**6.** Bert bought a garden plant for
44¢. He gave the clerk 2
quarters. Write what the clerk
said when he gave back the
change.

▶ **ALGEBRA**

**7.** Marcy bought one of these items. She paid for
it with a $1 bill and got this much change
back. What did she buy?

Greenbeans   Tomatoes   76¢   Corn   59¢   66¢   Peas   86¢

*More Practice, page 511, set E*

# Problem Solving
## Make an Organized List

| UNDERSTAND |
| FIND DATA |
| PLAN |
| ESTIMATE |
| SOLVE |
| CHECK |

### LEARN ABOUT IT

To solve some problems, you may need to make a list using the data in the problem. This problem solving strategy is called **Make an Organized List.**

> Whenever Sherrie goes to her ice skating lesson, her father gives her one piece of fruit and one bag of nuts to take with her. How many different ways can she choose a snack?

If I choose the apple I can choose peanuts or cashews.

apple - peanuts
apple - cashews

If I choose the tangerine, I can choose peanuts or cashews.

tangerine - peanuts
tangerine - cashews

Sherrie has 4 different ways to choose a snack.

### TRY IT OUT

Read this problem and find the solution.

The ice hockey coach gave Luis a choice of brown, white, and black skates. Luis could also choose from two shirt numbers, 5 and 13. How many ways could Luis choose 1 pair of skates and 1 shirt number?

brown skates - shirt ⑤
brown skates - shirt ⑬
white skates - _____

_____ _____

_____ _____

_____ _____

1. How many different colors of skates can he choose?

2. How many different numbers can he choose?

3. Copy and complete the list to solve the problem.

78

1. The Angels, Barons, Colts, and Demons are ice hockey teams. Each team has to play each other team 1 time. How many games will be played?

2. For ice skating contests, Pat has a white sweater and a gold sweater. She also has a black skirt and a red skirt. How many different outfits can she make?

3. When the third-grade ice hockey teams played, the Hawks came in after the Eagles. The Lions came in before the Bears. The Eagles came in between the Bears and the Hawks. Who came in first and last?

4. Julie had a ten-dollar bill. She paid $3.75 for a ticket at the ice skating rink. How much change did she get back?

5. Janelle reserved the rink for 1 hour and 45 minutes for her birthday party. The party started at 3:00. When was it over?

| Problem Solving Strategies |
| --- |
| Act It Out |
| Use Objects |
| Choose an Operation |
| Draw a Picture |
| Make an Organized List |

6. Masako can take ice skating lessons on Tuesdays or Thursdays at 4 p.m., 5 p.m. or 8 p.m. How many choices does she have for a time to take her lesson?

7. Tickets for Bill's family to go skating cost $26.00. Their drinks at the rink cost about $3.00. How much does a day of skating cost for all of Bill's family?

8. The Wong family wants to skate for 1 hour and 15 minutes on Friday night. If they get to the rink at 8:30 p.m., what time will they finish? Will their plan work? Use this schedule.

| Evening Schedule, Winter Lodge | | | | |
| --- | --- | --- | --- | --- |
| Tuesday Lessons only | Wednesday Adults 8:00-10:00 | Thursday Lessons only | Friday Family 7:30-9:30 | Saturday Family 7:30-9:30 |

# Applied Problem Solving
## Group Decision Making

| |
|---|
| UNDERSTAND |
| FIND DATA |
| PLAN |
| ESTIMATE |
| SOLVE |
| CHECK |

**Group Skill:**

Check for Understanding

The junior museum has a real bee hive, an Indian canoe, and other things that children like to see. You and your group are going to help decide what hours the junior museum should be open.

### Facts to Consider

- Parents with preschoolers want to visit the museum in the morning.

- School children get out of school at different times between 2:30 and 4:00 p.m.

- The woman in charge only works from 4 to 6 hours each day. She works 5 days a week.

- Working parents can only take their children to the museum on weekends.

80

1. What is the earliest time on a weekday morning that you would want to open the museum?

2. Would it be a good idea to plan to have the same opening hours every day?

3. Suppose you opened the museum at 10 a.m. and closed it 4 hours later. What time would you close?

4. Suppose you opened the museum at 10 a.m. for 6 hours. Would students be able to come after school to visit?

Make a schedule for the museum hours that you could hang on the museum door. Show what hours the museum will be open each day of the week.

# Wrap Up

## Time, Date, and Money Match

Match each phrase on the left with a time, a date,
or an amount on the right. Explain your choices.

| | | | |
|---|---|---|---|
| **1.** | quarter past ten | **a.** | 3/1/91 |
| **2.** | twelve minutes past midnight | **b.** | 12:12 p.m. |
| **3.** | twenty minutes before seven | **c.** | 6¢ |
| **4.** | twelve minutes past noon | **d.** | $1.00 |
| **5.** | first of March, nineteen ninety-one | **e.** | 10:15 |
| **6.** | July fourth, nineteen ninety-six | **f.** | $1.05 |
| **7.** | half dollar | **g.** | 6:40 |
| **8.** | a nickel and a penny | **h.** | 7/4/96 |
| **9.** | three quarters, two dimes, one nickel | **i.** | $0.50 |
| **10.** | one dollar five cents | **j.** | 12:12 a.m. |

## Sometimes, Always, Never

Decide which word should go in the blank,
sometimes, always, or never. Explain your choices.

**11.** It _____ takes one hour for a hand on a clock to go around one time.

**12.** You _____ need 3 numbers and 5 digits to write a date in the month/day/year form like 7/20/93.

**13.** If you pay three quarters for a 64¢ pen, you _____ will receive a dime in your change.

**14.** If you have 1 quarter, 1 dime, 2 nickels, and 3 pennies, you _____ can buy a magazine that costs more than 50¢.

## Project

Keep a homework journal for a week. For each day, write the day of the week and the date. Then record the time you start your homework, the time you finish, and the amount of time you spend. On which day did you spend the most time doing your homework?

82

# CHAPTER REVIEW/TEST

## Part 1    Understanding

**1.** What are two ways to say this time?

**2.** Art class begins at 11:45. Look at Tim's clock. Decide if he arrived early or late.

**3.** How long does it take the hour hand to move around a clock? the minute hand?

**4.** Tell how you would figure out how much time passes between 2:00 and 3:15.

**5.** What does each number in this date mean? 8/31/92

## Part 2    Skills

**6.** At which of these times is it usually dark?

1 a.m.    11 a.m.    12 noon
1 p.m.    11 p.m.    12 midnight

Write the time in numbers. Write a.m. or p.m.

**8.** four thirty in the afternoon

**7.** How much time has passed?

**9.** quarter to ten in the morning

## Part 3    Applications

**10.** Anita had 3 comic books. Could she buy a 92¢ comic with 3 quarters and 3 nickels? What data is extra?

**11.** Elise gave the store owner $1 for an 86¢ notebook. How much change did she get back?

**12.** **Challenge** The Cub Scouts are planning a picnic. They can go on Friday, Saturday, or Sunday. They can go to the park or to the Scout Camp. How many choices do the Scouts have for a picnic?

# ENRICHMENT
## Clock Arithmetic

When you add or subtract numbers on a clock, some strange things can happen. These examples show how you can use the hour hand to find sums and differences.

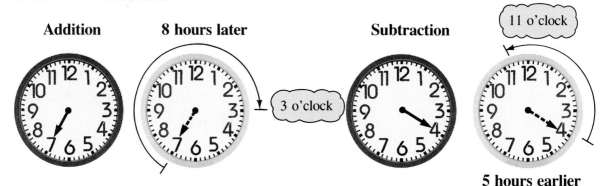

**Addition**  **8 hours later**    **Subtraction**    11 o'clock

3 o'clock

5 hours earlier

$$7 + 8 = 3$$

$$4 - 5 = 11$$

### Other Examples

$$9 + 8 = 5 \qquad 2 - 8 = 6$$

Solve these clock equations. Use the hour hand on a clock if you need help.

**1.** 8 + 4 = |||||

**2.** 10 + 6 = |||||

**3.** 12 + 5 = |||||

**4.** 6 + 4 = |||||

**5.** 5 − 8 = |||||

**6.** 2 − 10 = |||||

**7.** 1 − 6 = |||||

**8.** 11 − 9 = |||||

**9.** 8 + 6 = |||||

**10.** 4 − 10 = |||||

**11.** 7 − 12 = |||||

**12.** 7 + 12 = |||||

**13.** What clock number acts like zero? HINT: Look at exercises 11 and 12.

**14.** Does this statement work for all clock numbers? Show an example.

Adding 6 to a number gives the same answer as subtracting 6 from the number.

# CUMULATIVE REVIEW

Add or subtract.

**1.** $3 + 5$

    **A.** 2     **B.** 8

    **C.** 4     **D.** 9

**2.** $\begin{array}{r} 17 \\ -\ 8 \\ \hline \end{array}$

    **A.** 11     **B.** 9

    **C.** 10     **D.** 8

**3.** What is a fact for the addends 1 and 3?

    **A.** $1 + 1 = 2$   **B.** $3 + 3 = 6$

    **C.** $4 = 3 + 1$   **D.** $3 - 3 = 0$

**4.** What number is 2 hundreds, 5 tens, and 8 ones?

    **A.** 15     **B.** 582

    **C.** 825     **D.** 258

**5.** What is the next number in 30, 35, 40, 45, _____?

    **A.** 46     **B.** 50

    **C.** 45     **D.** 55

**6.** What time might you wake up for school?

    **A.** 1:15 a.m.     **B.** 7:06 a.m.

    **C.** 1:18 p.m.     **D.** 7:10 p.m.

**7.** How much time has passed?

    **A.** 1 hour     **B.** half hour

    **C.** 45 minutes   **D.** 15 minutes

**8.** Which is the tenth month?

    **A.** April     **B.** January

    **C.** August     **D.** October

**9.** You have 2 quarters, 2 dimes, 6 nickels, and 7 pennies. How much money do you have?

    **A.** $1.37     **B.** $1.07

    **C.** $1.57     **D.** $0.97

**10.** You are with 2 friends. You have 1 quarter, 3 dimes, 4 nickels. You want to buy juice for 60¢. What data is extra?

    **A.** 2 friends     **B.** 4 nickels

    **C.** 1 quarter     **D.** juice for 60¢

**11.** Terry gives a clerk $1.00 for an orange that costs $0.67. Find her change.

    **A.** $0.43     **B.** $0.63

    **C.** $0.33     **D.** $1.67

# 4

**M**ATH AND
LANGUAGE ARTS

DATA BANK

Use the Language
Arts Data Bank on
page 478 to answer
the questions.

# **A**DDITION

1 In "The Story of the
Four Little Children
Who Went Round th
World," the children saw re
parrots. Round that number
of parrots to the nearest ten

**2** The parrots in Lear's story lost many blue tail feathers. Give the total number of lost tail feathers in hundreds, tens, and ones.

**3** The Four Little Children saw a very tall tree on their trip. Which is taller, a tree 530 feet high or the tree that the children saw?

**4** **Use Critical Thinking** Can you tell whether the children saw more oranges than kangaroos and cranes? Explain your thinking.

87

# Mental Math
## Special Sums

LEARN ABOUT IT

**EXPLORE** **Use a Place Value Model**

The picture shows how to use 7 of your tens blocks to show $30 + 40$. How many other sums can you show using exactly 7 of your tens blocks?

3 tens  4 tens  7 tens in all

70 ones

$$30 + 40 = n$$
$$n = 70$$

**TALK ABOUT IT**

1. How is finding $30 + 40$ like finding $3 + 4$?

2. How is finding $300 + 400$ like finding $3 + 4$? Use your hundreds blocks to show $300 + 400$.

3. What number sentences can you write using exactly 5 hundreds blocks?

Finding special sums is very much like finding basic facts.

To find $90 + 50$, first add $9 + 5$ to get the number of tens.

$$90 + 50 = 140$$

To find $600 + 700$, first add $6 + 7$ to get the number of hundreds.

$$600 + 700 = 1300$$

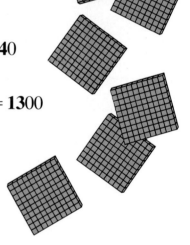

TRY IT OUT

Copy and solve.

1. 5 tens and 2 tens are 7 tens, so $50 + 20 = |||||$.

2. 6 hundreds and 2 hundreds are 8 hundreds, so $600 + 200 = |||||$.

3. Since $9 + 6 = 15$, we know that $90 + 60 = |||||$.

4. Since $8 + 2 = 10$, we know that $800 + 200 = |||||$.

5. $60 + 90$    6. $80 + 30$    7. $50 + 50$    8. $300 + 700$    9. $400 + 700$

Find the sums using mental math. Write answers only.

| 4 tens + 5 tens | 7 tens + 6 tens | 8 tens + 4 tens |
|---|---|---|
| **1.** 40 + 50 | **2.** 70 + 60 | **3.** 80 + 40 |

| 3 hundreds + 4 hundreds | 5 hundreds + 7 hundreds | 6 hundreds + 8 hundreds |
|---|---|---|
| **4.** 300 + 400 | **5.** 500 + 700 | **6.** 600 + 800 |

**7.** 50 + 20     **8.** 80 + 60     **9.** 80 + 30     **10.** 30 + 50

**11.** 50 + 50     **12.** 70 + 30     **13.** 60 + 10     **14.** 70 + 60

**15.** 50 + 60     **16.** 60 + 30     **17.** 80 + 40     **18.** 50 + 40

**19.** 900 + 200     **20.** 800 + 700     **21.** 400 + 400     **22.** 400 + 900

**APPLY**

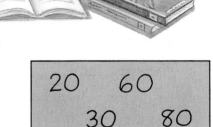

**MATH REASONING**

**23.** Pete said, ''The sum of any two numbers in the box has exactly two zeros.'' Prove that Pete made a mistake.

20    60
   30    80
40    70

**PROBLEM SOLVING**

**24.** Kimberly set a goal to read 70 pages of her book over the weekend. She liked the book and read 60 pages over her goal. How many pages did she read?

**MIXED REVIEW**

**25.** How many days are in a week?

**26.** What month and year are shown by the date 9/3/64?

**27.** What is the name and number of the last month of the year?

**28.** Which month is the tenth month?

# Estimating Sums Using Rounding

**LEARN ABOUT IT**

**EXPLORE  Analyze the Problem**

Phil worked in a bicycle shop. He used a calculator to find the total number of bikes sold in a year. This time Phil hit the wrong key and added 569 and 437 instead of 269 and 437.

| Bicycles Sold | |
|---|---|
| Jan - June | 269 |
| July - Dec | 437 |
| Total | ? |

**TALK ABOUT IT**

1. Look at Phil's answer on the calculator. How do you think he was able to discover that he had made this mistake?

2. Without using a calculator or pencil and paper, can you guess at a reasonable answer to Phil's problem?

You can estimate sums by rounding.

| Nearest ten | Nearest hundred | Nearest dollar |
|---|---|---|
| 148 → 150 | 589 → 600 | $7.95 → $8.00 |
| + 25 → + 30 | + 317 → + 300 | + 6.29 → + 6.00 |
| 180 | 900 | $14.00 |

**TRY IT OUT**

Estimate by rounding to the nearest ten.

**1.** 38 + 53          **2.** 45 + 82          **3.** 137 + 24          **4.** 132 + 49

Estimate by rounding to the nearest hundred.

**5.** 289 + 426          **6.** 515 + 196          **7.** 304 + 879          **8.** 750 + 138

Estimate by rounding to the nearest dollar.

**9.** $3.95 + $1.98  **10.** $2.15 + $3.99  **11.** $4.16 + $2.95  **12.** $7.90 + $0.97

90

Estimate by rounding to the nearest ten.

| **1.** 28<br>+ 29 | **2.** 48<br>+ 23 | **3.** 57<br>+ 24 | **4.** 69<br>+ 88 | **5.** 76<br>+ 32 | **6.** 55<br>+ 89 |
|---|---|---|---|---|---|

Estimate by rounding to the nearest hundred or dollar.

| **7.** 289<br>+ 304 | **8.** $6.13<br>+ 7.97 | **9.** 398<br>+ 242 | **10.** $4.79<br>+ 1.12 | **11.** 516<br>+ 296 |
|---|---|---|---|---|

Use estimation. Which displays are reasonable answers?

**12.** 487 + 296     **13.** 79 + 83        **14.** 885 + 709        **15.** 198 + 179

| 783 | 112 | 1294 | 377 |
|---|---|---|---|

**MATH REASONING**  Choose **less** or **greater** to complete the sentence.

**16.** When you round both addends **down,** the estimate is always _____ than the sum.

**17.** When you round both addends **up,** the estimate is always _____ than the sum.

**PROBLEM SOLVING**

**18.** During a holiday period, the bicycle shop sold 14 bicycles on Thursday, 28 on Friday and 39 on Saturday. About how many bicycles were sold on Friday and Saturday?

**19.** Write and answer an estimation question about the bicycles in stock at the store.

Single Speed         124
10-Speed              278
20-Speed Racing        69

▶ **USING CRITICAL THINKING  Support Your Conclusion**

**20.** Can you show examples to prove this statement is true? When you round one addend up and the other down, the estimate could be greater than, less than, or equal to the actual sum.

*More Practice, page 502, set B*

**91**

# Trading 10 Ones for 1 Ten

Trading blocks will help you understand addition of larger numbers. Make trades so that you show the same amount with the fewest possible pieces.

**EXPLORE  Use a Place Value Model**

Work in groups. Bev has 15 ones. If she trades 10 ones for 1 ten, she will still have the same amount, but with the fewest possible pieces. Try Bev's trade in your group.

**TALK ABOUT IT**

1.  How many tens and ones will Bev have after the trade?

2.  Can Bev make another trade of ones for a ten? Why or why not?

3.  What if Bev had started with 23 ones blocks? Explain the trading she could do.

Here is how to record the results after trading with 26 ones. Show this with your own place value blocks.

| Before the trade | After the trade | Record |
|---|---|---|

**TRY IT OUT**

In your group, take turns picking up a handful of ones blocks. Trade 10 ones for 1 ten until you have the fewest possible pieces. Record your results. Repeat until each student has recorded three examples.

92

Trade 10 ones for 1 ten until there are fewer than
10 ones. Record your results. Use blocks if you
need help.

**1.** 23 ones

| Tens | Ones |
|------|------|
|      |      |

**2.** 14 ones

| Tens | Ones |
|------|------|
|      |      |

**3.** 19 ones

| Tens | Ones |
|------|------|
|      |      |

**4.** 20 ones

| Tens | Ones |
|------|------|
|      |      |

Trade 10 ones for 1 ten. Give the missing numbers.

**5.** 12 ones ||||| ten ||||| ones

**6.** 27 ones ||||| tens ||||| ones

**7.** 18 ones ||||| ten ||||| ones

**8.** 22 ones ||||| tens ||||| ones

**9.** 40 ones ||||| tens ||||| ones

**10.** 11 ones ||||| ten ||||| ones

**MATH REASONING**

**11.** Suppose you pick up some ones blocks.
After all trades, can you have 0 tens? 0 ones?

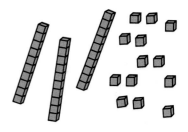

**PROBLEM SOLVING**

**12. Use Place Value Blocks**
Ann picked up 16 ones and Jack
picked up 18 ones. How many
ones did they pick up in all?

**13.** After making one trade, suppose
you still have 26 ones. How
many ones blocks did you start
with?

▶ **ALGEBRA**

Give the missing numbers.

**14.**
$$\begin{array}{r} 7 \\ + ||||| \\ \hline 12 \end{array}$$

**15.**
$$\begin{array}{r} ||||| \\ + 6 \\ \hline 15 \end{array}$$

**16.**
$$\begin{array}{r} ||||| \\ - 4 \\ \hline 9 \end{array}$$

**17.**
$$\begin{array}{r} 6 \\ + ||||| \\ \hline 14 \end{array}$$

# Adding 2-Digit Numbers
## Making the Connection

EXPLORE **Use a Place Value Model**

Work with a partner. You and your partner each choose a number from the number bank. Show your numbers with dimes and pennies. Record your numbers in a table like the one below.

Push the two piles of coins together. Trade 10 pennies for 1 dime if you can. Record the total in the table. Do this five times.

| Number Bank |
|:---:|
| 23¢  19¢  45¢  28¢ |
| 38¢  25¢  46¢  14¢ |
| 37¢  39¢  47¢  17¢ |

| | Dimes | Pennies |
|---|---|---|
| First partner | | |
| Second partner | | |
| Total | | |

**TALK ABOUT IT**

1. Did you have to trade each time?

2. Can you pick two numbers from the bank so that you would not have to trade? How?

3. Can you give a rule that tells when you will need to make a trade?

94

You have pushed coins together, traded, and figured out how many in all. You can also use place value blocks to show addition. Here is a way to record what you do. Use blocks to find 35 + 29.

**What You Do**          **What You Record**

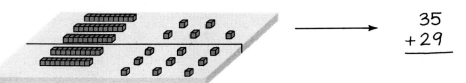

$$\begin{array}{r} 35 \\ +29 \\ \hline \end{array}$$

**1.** Are there enough ones on the table to make a trade?

**Trade**

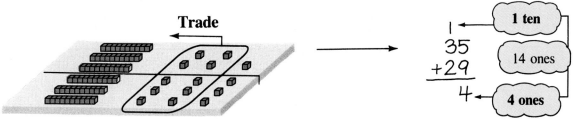

$$\begin{array}{r} 1 \\ 35 \\ +29 \\ \hline 4 \end{array}$$

1 ten
14 ones
4 ones

**2.** How many ones are there after the trade?

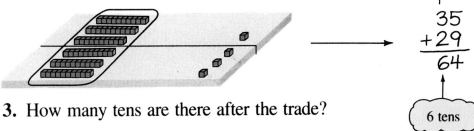

$$\begin{array}{r} 1 \\ 35 \\ +29 \\ \hline 64 \end{array}$$

6 tens

**3.** How many tens are there after the trade?

**4.** What is the sum of 35 and 29?

**5.** Suppose you use blocks to add 34 and 13. Would there be enough ones for a trade? Explain.

---

**TRY IT OUT**

Use place value blocks to add the numbers.
Record what you did.

**1.** Add 26 and 28.          **2.** Add 32 and 27.

**3.** Make up and solve your own addition problem using place value blocks.

# Adding 2-Digit Numbers

**EXPLORE Think About the Process**

If the Zebra went 29 miles and stopped to rest, then went 34 miles more to get to Jellibolee, how far did he go all together?

You add because you are putting together distances to find a total.

The Zigzag Zealous Zebra who carried five monkeys on his back all the way to Jellibolee

—Edward Lear

| Add the ones. | Trade if necessary. | Add the tens. |
|---|---|---|
| $\begin{array}{r} 29 \\ + 34 \\ \hline \end{array}$ (13) | $\begin{array}{r} \overset{1}{2}9 \\ + 34 \\ \hline 3 \end{array}$ (Trade) | $\begin{array}{r} \overset{1}{2}9 \\ + 34 \\ \hline 63 \end{array}$ |

**TALK ABOUT IT**

1. Why was it necessary to trade?

2. How would you have estimated the sum?

3. Use a complete sentence to answer the problem.

---

**Other Examples**

$\begin{array}{r} \overset{1}{2}7 \\ + 48 \\ \hline 75 \end{array}$ (Trade)  $\begin{array}{r} 53 \\ + 34 \\ \hline 87 \end{array}$ (No Trade)  $\begin{array}{r} \overset{1}{3}9 \\ + \ 4 \\ \hline 43 \end{array}$ (Trade)  $\begin{array}{r} 83 \\ + 62 \\ \hline 145 \end{array}$ (No Trade)

14 tens is 1 hundred and 4 tens

Find the sums.

**1.** $36 + 57$      **2.** $42 + 36$      **3.** $48 + 87$      **4.** $58 + 6$

Write **yes** if you need to trade and **no** if you do not need to trade. Then find the sum.

| **1.** 43 | **2.** 56 | **3.** 28 | **4.** 37 | **5.** 46 | **6.** 25 |
|---|---|---|---|---|---|
| + 16 | + 8 | + 19 | + 60 | + 39 | + 25 |

Find the sums.

| **7.** 83 | **8.** 28 | **9.** 65 | **10.** 16 | **11.** 28 | **12.** 63 |
|---|---|---|---|---|---|
| + 64 | + 47 | + 78 | + 95 | + 50 | + 88 |

**13.** $46 + 50$       **14.** $27 + 9$       **15.** $56 + 49$       **16.** $22 + 39$

**17.** $96 + 43$       **18.** $7 + 63$       **19.** $86 + 38$       **20.** $49 + 49$

**21.** Find the sum of 82 and 9.       **22.** What is 78 added to 47?

**MATH REASONING**

**23.** Write two addition problems that have a sum of 75. Make one of the problems use trading.

**PROBLEM SOLVING**

**24.** Mr. Lear's Piggiwiggia plants have 88 pigs. His Pollybirdia plants have 35 birds. How many pigs and birds are on the plants all together?

**25. Language Arts Data Bank**

If the Four Little Children saw 28 green parrots, how many parrots would they have seen all together? See page 478.

▶ **MENTAL MATH**

Use mental math. Think, "How many ones?" Then think, "How many tens?" Put the ones and tens together to find the sum. Write answers only.

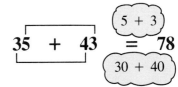

**26.** $27 + 41$       **27.** $32 + 53$       **28.** $41 + 28$       **29.** $16 + 70$

*More Practice, page 502, set C*

# Problem Solving
## Guess and Check

| UNDERSTAND |
| FIND DATA |
| PLAN |
| ESTIMATE |
| SOLVE |
| CHECK |

A problem like this one can be solved using the strategy called **Guess and Check.** You may want to use a calculator.

| Parts of Body | Number of Bones |
|---|---|
| Legs | 8 |
| Ankles | 14 |
| Feet | 10 |
| Toes | 28 |
| Spine | 26 |

Jeff and his mother glued together a lifesize plastic skeleton. The first day they glued together 42 bones. What two parts of the body did they do?

> I'll try the legs and ankles.

First guess: legs and ankles

8 $+$ 14 $=$ 22

Too few. They glued 42 bones.

> I'll try the toes and spine.

Second guess: toes and spine

28 $+$ 26 $=$ 54

Too many

> I'll try the toes and ankles.

Third guess: toes and ankles

28 $+$ 14 $=$ 42

It checks!

Jeff and his mother glued the toes and ankles first.

The next day, Jeff glued 20 bones in all. What two parts of the body did he do?

- How many bones did Jeff glue together?

- How many parts did he finish?

- Use Guess and Check to solve this problem.

| Parts of Body | Number of Bones |
|---|---|
| Skull | 8 |
| Face | 14 |
| Ears | 6 |
| Ribs | 24 |
| Spine | 26 |

First Guess: Skull + Face

Check: 8 + 14 = 22

98

Use the guess and check strategy to solve these problems.

**1.** Nina bought 3 kits to put together dinosaur bones. They cost $12 in all. Which kits did she buy?

| Tyrannosaurus Rex | $7 |
| Iguanodon | $4 |
| Brachiosaurus | $6 |
| Stegosaurus | $5 |
| Aristosaurus | $3 |

Solve. Choose a strategy from the list or use any other strategy you know.

**3.** Darlene needs to do a report on either bones or muscles. It can be handwritten or typed. How many different choices does she have?

**4.** How many bones are in the upper and lower limbs together?
Upper limbs   60 bones
Lower limbs   60 bones

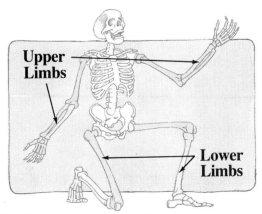

Upper Limbs

Lower Limbs

**2.** Use this list. Which two parts of the body together have 35 bones?

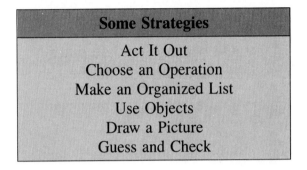

| Head | 29 | Fingers | 28 |
| Wrists | 16 | Arms | 6 |
| Hands | 10 | | |

| Some Strategies |
| --- |
| Act It Out |
| Choose an Operation |
| Make an Organized List |
| Use Objects |
| Draw a Picture |
| Guess and Check |

**5.** An adult has 206 bones. A baby can have as many as 144 more bones than an adult. How many bones can a baby have?

YOUR BONES $5   SONG BOOK $9   PET CARE $4
BUILD A MODEL $7   DINOSAURS $8

**6.** Brent has $13 to buy 2 books. What books can he buy? Is there more than one answer?

**7.** There are 26 bones in your spinal column. There are 28 bones in your fingers, and 28 bones in your toes. How many bones are in your fingers and toes together?

*More Practice, page 515, set C*

# Adding 3-Digit Numbers
## One Trade

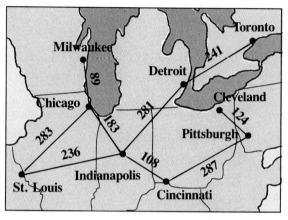

**Highway Distances in Miles**

**LEARN ABOUT IT**

**EXPLORE Think About the Process**

Look at the map. How far is it from
St. Louis to Detroit if you go through
Indianapolis?

You add because you are putting
together numbers to find the sum.

| Add the ones. Trade if necessary. | Add the tens. Trade if necessary. | Add the hundreds. |
|---|---|---|
| 236<br>+ 281<br>―――<br>7 | 1<br>236<br>+ 281<br>―――<br>17 | 1<br>236<br>+ 281<br>―――<br>517 |

**TALK ABOUT IT**

1. Was the trade made in the first or second step?
2. Would 400 be a good estimate for this sum?
3. Use a complete sentence to give a reasonable
   answer to the question.

### Other Examples

$$\begin{array}{r} 1 \\ 249 \\ + 38 \\ \hline 287 \end{array} \qquad \begin{array}{r} 1 \\ 346 \\ + 281 \\ \hline 627 \end{array} \qquad \begin{array}{r} 634 \\ + 720 \\ \hline 1,354 \end{array}$$

13 hundreds is 1 thousand
and 3 hundreds.

**TRY IT OUT**

Find the sum.

1. $327 + 45$
2. $472 + 264$
3. $826 + 432$
4. $27 + 258$

100

Find the sum.

| | | | | |
|---|---|---|---|---|
| **1.** 933 <br> +346 | **2.** 72 <br> +672 | **3.** 467 <br> +517 | **4.** 234 <br> +152 | **5.** 678 <br> +51 |
| **6.** 284 <br> +181 | **7.** 776 <br> +14 | **8.** 381 <br> +406 | **9.** 134 <br> +718 | **10.** 492 <br> +497 |
| **11.** 844 <br> +93 | **12.** 548 <br> +327 | **13.** 58 <br> +327 | **14.** 246 <br> +423 | **15.** 151 <br> +681 |

**16.** 575 + 802    **17.** 526 + 446    **18.** 166 + 253    **19.** 817 + 46

**20.** 173 + 183    **21.** 182 + 607    **22.** 216 + 635    **23.** 543 + 76

**24.** Add 236 and 145.          **25.** Find the sum of 221 and 90.

**APPLY**

**MATH REASONING**

**26.** Estimate this sum by rounding. Now find the sum. Why do you think your estimate is so far off?

$$\begin{array}{r} 352 \\ +\ 351 \\ \hline \end{array}$$

**PROBLEM SOLVING** Use the map on page 100.

**27. Write Your Own Problem** Aaron drove from Indianapolis to Detroit. He stayed overnight and then drove to Toronto.

**28.** How far is it from Chicago to Cincinnati if you go through Indianapolis?

▶ **COMMUNICATION Write to Learn**

**29.** Suppose you estimate a sum by rounding both addends up. Write a sentence that compares your estimate with the exact sum.

*More Practice, page 502, set D*

# Adding 3-Digit Numbers

**EXPLORE** **Think About the Process**

Read the chart. Suppose a full DC10 and a full
747 arrive at the airport. How many people get
off the two planes?

You add because you are putting together
numbers to find a total. Follow these steps.

| Aircraft | Number of Seats |
|---|---|
| McDonnell Douglas | |
| DC 9 | 155 |
| DC 10 | 278 |
| Boeing | |
| 747 | 486 |
| 757 | 185 |

| Add the ones. Trade if necessary. | Add the tens. Trade if necessary. | Add the hundreds. |
|---|---|---|
| $\begin{array}{r} 1 \\ 278 \\ + 486 \\ \hline 4 \end{array}$ | $\begin{array}{r} 1\ 1 \\ 278 \\ + 486 \\ \hline 64 \end{array}$ | $\begin{array}{r} 1\ 1 \\ 278 \\ + 486 \\ \hline 764 \end{array}$ |

**TALK ABOUT IT**

1. How many trades were made to find the sum?

2. How did you estimate the answer?

3. Use a complete sentence to give a reasonable
   answer to the question.

**Other Examples**

$$\begin{array}{r} 1 \\ 472 \\ + 856 \\ \hline 1,328 \end{array} \qquad \begin{array}{r} 1\ 1 \\ 765 \\ + 847 \\ \hline 1,612 \end{array} \qquad \begin{array}{r} 1\ 1 \\ 975 \\ + 96 \\ \hline 1,071 \end{array}$$

Find the sum.

1. $\begin{array}{r} 765 \\ + 840 \end{array}$     2. $\begin{array}{r} 256 \\ + 68 \end{array}$     3. $\begin{array}{r} 324 \\ + 435 \end{array}$

Find the sum.

| | | | | | | | | | |
|---|---|---|---|---|---|---|---|---|---|
| **1.** | 369 <br> + 543 | **2.** | 720 <br> + 393 | **3.** | 964 <br> + 375 | **4.** | 279 <br> + 61 | **5.** | 585 <br> + 840 |
| **6.** | 351 <br> + 284 | **7.** | 58 <br> + 756 | **8.** | 516 <br> + 230 | **9.** | 399 <br> + 208 | **10.** | 87 <br> + 371 |
| **11.** | 285 <br> + 199 | **12.** | 385 <br> + 76 | **13.** | 507 <br> + 758 | **14.** | 176 <br> + 556 | **15.** | 327 <br> + 610 |

**16.** 493 + 976     **17.** 686 + 921     **18.** 657 + 648     **19.** 483 + 462

**20.** 87 + 863     **21.** 231 + 456     **22.** 758 + 428     **23.** 742 + 682

**24.** Find the sum of 455 and 792.     **25.** Add 954 to 287.

**MATH REASONING** Find each sum using mental math. Write answers only.

**26.** If 385 + 579 = 964, then what is 379 + 585?

**27.** If 297 + 658 = 955, then what is 257 + 698?

**PROBLEM SOLVING**

**28.** How many people can ride on the two McDonnell-Douglas airplanes? Use the chart on page 102.

**29. Understand the Operation** At one airport, 12 planes took off and 7 planes landed. How many more planes took off than landed? What operation did you use? Why?

▶ **CALCULATOR**

**30.** Use a calculator to help you find the two numbers that add up to 762.

| 469 | 385 | 393 |
|---|---|---|
| 282 | 277 | 293 |

# Mental Math
## Breaking Apart Numbers

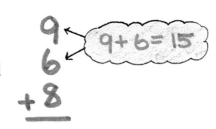

**LEARN ABOUT IT**

To add three numbers such as 9, 6 and 8, you need to add them two at a time. You will need to find sums such as $15 + 8$ in your head.

Can you figure out these sums in your head?

**TALK ABOUT IT**

1. How did you find the sum of 13 and 4?

2. Did you use a different method for $28 + 5$? Explain.

3. How did you find $17 + 8$?

$13 + 4$

$28 + 5$

$17 + 8$

Here are two ways of using mental math to find $15 + 8$. In each, you break apart numbers.

**Break Apart 15**

$10 + 5$

$15 + 8$

Since $5 + 8 = 13$, I know that $15 + 8 = 23$.

**Break Apart 8**

$5 + 3$

$15 + 8$

Since $15 + 5 = 20$, I know that $15 + 8 = 23$.

**PRACTICE**

Find the sums. Write answers only.

**1.** $16 + 5$

**2.** $16 + 4$

**3.** $12 + 6$

**4.** $11 + 7$

**5.**
8
6
$+7$
(14)
$(14 + 7)$

**6.**
5
6
$+9$
(11)
$(11 + 9)$

**7.**
6
7
$+4$
(13)
$(13 + 4)$

**8.**
8
9
$+7$
(17)
$(17 + 7)$

**9.**
6
9
$+7$

**10.**
4
7
$+8$

**11.**
7
7
$+7$

**12.**
9
8
$+6$

**13.**
6
5
$+8$

**14.**
8
8
$+8$

**104**

*More Practice, page 503, set A*

# MIDCHAPTER REVIEW/QUIZ

Find the sums using mental math.

**1.** $50 + 30$ **2.** $40 + 70$ **3.** $600 + 600$ **4.** $800 + 300$

Estimate by rounding to the nearest ten.

**5.** $13 + 54$ **6.** $66 + 47$ **7.** $227 + 61$ **8.** $484 + 25$

Estimate by rounding to the nearest hundred or dollar.

**9.**
$$\begin{array}{r} 216 \\ + 438 \\ \hline \end{array}$$

**10.**
$$\begin{array}{r} 376 \\ + 620 \\ \hline \end{array}$$

**11.**
$$\begin{array}{r} \$7.56 \\ + \ 3.75 \\ \hline \end{array}$$

**12.**
$$\begin{array}{r} \$1.42 \\ + \ 8.98 \\ \hline \end{array}$$

Trade 10 ones for 1 ten.

**13.** 26 ones → ____ tens ____ ones **14.** 60 ones → ____ tens ____ ones

Write yes if you will need to trade. Write no if
you will not need to trade. Then find the sums.

**15.**
$$\begin{array}{r} 23 \\ + 15 \\ \hline \end{array}$$

**16.**
$$\begin{array}{r} 35 \\ + \ 8 \\ \hline \end{array}$$

**17.**
$$\begin{array}{r} 48 \\ + 22 \\ \hline \end{array}$$

**18.**
$$\begin{array}{r} 9 \\ + 73 \\ \hline \end{array}$$

**19.**
$$\begin{array}{r} 57 \\ + 53 \\ \hline \end{array}$$

**20.**
$$\begin{array}{r} 60 \\ + 65 \\ \hline \end{array}$$

Find the sums.

**21.** $43 + 26$ **22.** $35 + 45$ **23.** $63 + 72$ **24.** $77 + 28$

## PROBLEM SOLVING

**25.** Ray used place value blocks to show these three problems. For which problems did he trade 10 ones for 1 ten? For which did he trade 10 tens for 1 hundred?

**A.** $27 + 18$ **B.** $46 + 70$
**C.** $38 + 64$

**26.** Carole earned the highest score on three of these five quizzes. She had 50 high score points. For which quizzes was Carole the high scorer?

| Quiz | High Score |
|------|-----------|
| 1 | 16 |
| 2 | 18 |
| 3 | 20 |
| 4 | 15 |
| 5 | 17 |

# Column Addition

| October Reading for Fun |
|---|
| **Mrs. Lim's Class** |
| Week 1     284 pages |
| Week 2     96 pages |
| Week 3     309 pages |
| Week 4     59 pages |

**LEARN ABOUT IT**

**EXPLORE  Think About the Process**

How many pages did the class read in October?

You add because you are putting together numbers to find a sum. Follow these steps.

| Line up the ones place. | Add the ones. Trade if necessary. | Add the tens. Trade if necessary. | Add the hundreds. |
|---|---|---|---|
| 284<br>96<br>309<br>+ 59 | 2<br>284<br>96<br>309<br>+ 59<br>8 | 2 2<br>284<br>96<br>309<br>+ 59<br>48 | 2 2<br>284<br>96<br>309<br>+ 59<br>748 |

**TALK ABOUT IT**

**1.** How many tens did you trade?

**2.** Was your estimate more or less than the total?

**3.** Use a sentence to give a reasonable answer.

### Other Examples

$$
\begin{array}{r} 2 \\ 328 \\ 7 \\ +\ 39 \\ \hline 374 \end{array}
\qquad
\begin{array}{r} 1 \\ 46 \\ 37 \\ 52 \\ +\ 21 \\ \hline 156 \end{array}
\qquad
\begin{array}{r} 1\ \ 1 \\ \$3.98 \\ 0.59 \\ +\ 1.30 \\ \hline \$5.87 \end{array}
\qquad
\begin{array}{r} 2\ \ 3 \\ \$2.76 \\ 0.09 \\ 4.77 \\ +\ 0.89 \\ \hline \$8.51 \end{array}
$$

*Line up the dollars and cents.*

*Show dollars and cents.*

**TRY IT OUT**

Add.

**1.** $19 + 8 + 17$      **2.** $48 + 36 + 155 + 32$      **3.** $\$2.84 + \$0.59 + \$3.09$

Find the sums.

| **1.** | 67 | **2.** | 156 | **3.** | 27 | **4.** | 175 | **5.** | $8.43 | **6.** | 283 |
| | 28 | | 95 | | 6 | | 96 | | 6.57 | | 456 |
| | + 9 | | + 9 | | + 8 | | + 9 | | +7.25 | | +329 |

**7.** $28 + 36 + 75 + 86$

**8.** $7 + 324 + 67 + 485$

**9.** $277 + 48 + 386$

**10.** $289 + 924 + 87 + 652$

**11.** $3.29 + $0.24 + $2.75

**12.** $0.67 + $0.38 + $0.26 + $0.95

**13.** $3.80 + $0.26 + $0.89

**14.** $1.75 + $0.89 + $0.08 + $1.26

**15.** Find the sum of 38, 9 and 56.

**16.** Add $3.95, $2.87 and $5.26.

**APPLY**

**MATH REASONING** Without finding the sum, tell if it is more or less than $5.00.

**17.** $1.98 + $0.79 + $0.87

**18.** $2.15 + $2.29 + $2.08

**PROBLEM SOLVING**

**19.** Edward Lear's *Complete Nonsense Book* has 156 alphabet poems, 11 poems that he wrote in letters, 212 limericks, and 29 other poems. How many poems is this all together?

**DATA BANK**

**20. Language Arts Data Bank**
Add together the number of red parrots with blue tails the Four Little Children saw, the number of tailfeathers the parrots lost, and the number of sons of the Old Person of Sparta.
See page 478.

▶ **MENTAL MATH**

Use the easy sums to find these sums using mental math.

**21.** $50 + 37 + 50$

**22.** $25 + 68 + 75$

**23.** $93 + 40 + 60$

**24.** $80 + 43 + 20$

**25.** $75 + 87 + 25$

**26.** $42 + 50 + 50$

| **Easy Sums** |
| --- |
| $25 + 75 = 100$ |
| $60 + 40 = 100$ |
| $50 + 50 = 100$ |
| $80 + 20 = 100$ |

# Adding 4-Digit Numbers

LEARN ABOUT IT

**EXPLORE** **Think About the Process**

Look at the table. How many people attended the Overlake and Hillside games?

You add because you are putting together numbers to find a total. Follow the steps.

| North Park School Basketball Games and Attendance | | |
|---|---|---|
| School | Score | Attendance |
| Central | Won 56-49 | 1,722 |
| South Side | Won 73-68 | 1,835 |
| Overlake | Won 47-45 | 1,946 |
| Hillside | Lost 62-63 | 2,408 |
| Parkview | Won 75-46 | 2,467 |

| Add the ones. Trade if necessary. | Add the tens. Trade if necessary. | Add the hundreds. Trade if necessary. | Add the thousands. |
|---|---|---|---|
| 1<br>1,946<br>+ 2,408<br>—————<br>4 | 1<br>1,946<br>+ 2,408<br>—————<br>54 | 1  1<br>1,946<br>+ 2,408<br>—————<br>354 | 1  1<br>1,946<br>+ 2,408<br>—————<br>4,354 |

**TALK ABOUT IT**

1. What step is new in finding this sum?
2. If you round both addends to 2,000, what is your estimate of the total?
3. Is the total more or less than your estimate?
4. Use a sentence to give a reasonable answer.

**Other Examples**

$$\begin{array}{r} \overset{1\,1}{3{,}287} \\ +\ 4{,}165 \\ \hline 7{,}452 \end{array} \qquad \begin{array}{r} \overset{1\,1}{5{,}672} \\ +\ 495 \\ \hline 6{,}167 \end{array} \qquad \begin{array}{r} \overset{1\,1\,1}{5{,}875} \\ +\ 1{,}659 \\ \hline 7{,}534 \end{array} \qquad \begin{array}{r} \overset{1\,1}{\$36.95} \\ +\ 28.50 \\ \hline \$65.45 \end{array}$$

TRY IT OUT

Add.

**1.** $3{,}876 + 2{,}540$   **2.** $6{,}275 + 178$   **3.** $\$39.95 + \$3.89$   **4.** $\$22.48 + \$98.75$

Add.

**1.**  1,374
  + 6,211

**2.**  6,784
  + 1,173

**3.**  2,392
  + 6,474

**4.**  $28.47
  + 16.12

**5.**  4,891
  + 709

**6.** 7,921 + 1,803

**7.** 3,928 + 5,063

**8.** 4,569 + 3,986

**9.** Add 4,291 and 306.

**10.** Find the sum of 2,760 and 1,404.

**APPLY**

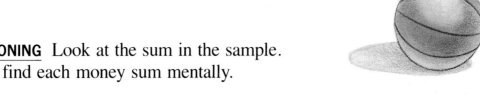

**MATH REASONING** Look at the sum in the sample.
Use it to find each money sum mentally.

**Sample:**    4,950
   + 2,890
   _____
    7,840

**11.**  $49.50
   + 28.90

**12.**  $4,950
   + 2,890

**13.**  $4.95
   + 2.89

**PROBLEM SOLVING**

**14.** If 14 games were played in all,
how many were played after the
Parkview game? Use the table
on page 108.

**15.** What was the total number of
points scored in the first two
games? Use the table on page
108.

**MIXED REVIEW**

Write the number for each.

**16.** next after 99

**17.** just before 23

**18.** between 378 and 380

Which students can buy a notepad that costs $1.69?

**19.**

**20.**

**21.**

# Problem Solving
## Deciding When to Estimate

| |
|---|
| UNDERSTAND |
| FIND DATA |
| PLAN |
| ESTIMATE |
| SOLVE |
| CHECK |

### LEARN ABOUT IT

When you solve problems, you must decide if
you need an exact answer or an estimate. When
you estimate, you are finding **about how many.**

> Suppose you start a lemonade stand. You need
> to compute your cost per glass, how much
> profit you want, and how much to charge per
> glass. How much will people buy?

You need to find out how much it
will cost you to make 1 glass of
lemonade so that you will know how
much to charge.

> I can estimate the cost per glass
> since the size of each glass will not
> be exact.

You need to charge a customer for 2
large glasses and 3 small glasses.

> I need an exact answer when I
> charge my customers.

You must understand the situation in the problem
before you can decide if you need an exact answer
or if an estimate will be enough to answer the question.

### TRY IT OUT

Think about the lemonade stand situation. Decide
if you need an exact answer or an estimate.

**1.** You figure the change from a
$5.00 bill after a purchase.

**2.** You figure how many glasses of
lemonade you can get from 3 full
pitchers of lemonade.

**3.** You tell a friend how much
lemonade you sold.

**4.** You want to figure a fair price for
a large glass of lemonade.

110

Decide if you need an exact answer or an estimate.

1. You have figured how much it costs you to make a glass of lemonade. You know how much profit you want to make for each glass. How much will you charge?

2. Someone asks you if a dollar is enough to buy two small glasses of lemonade.

3. Your father asks you about how much you made in profit.

4. It cost Holly 9 cents to make each glass of her lemonade. She decided to make 16 cents on each glass she sold. How much should she charge for 1 glass?

5. A customer asks you how much more a large glass of lemonade costs than a small glass.

6. Did you sell more or less than a gallon of lemonade?

7. You want to know how many lemons to buy for three batches of lemonade.

8. You want to compare the cost of large and small packages of paper cups.

9. A large glass of lemonade costs 50¢. A small glass costs 30¢. Gary has one dollar. Does he have enough to buy 1 small and 1 large glass of lemonade?

# Problem Solving
## Using Data from a Catalog

| UNDERSTAND |
| :---: |
| FIND DATA |
| PLAN |
| ESTIMATE |
| SOLVE |
| CHECK |

## LEARN ABOUT IT

To solve some problems, you will
need to use data from a catalog.

> Arnie was getting ready to do a magic show at
> his little brother's birthday party. At the House of
> Magic, he bought the Hidden Coin Trick and the
> Paperclip Tangle. How much did he spend in all?

I need
this. →

I need
this, too! ←

**Tricks**

| | | | |
| --- | --- | --- | --- |
| Ball and Vase | $1.12 | Vanishing Quarter | $1.00 |
| Coil and Ring | $0.98 | Guess the Card Trick | $1.87 |
| Magic Smoke | $1.20 | Disappearing Rabbit | $1.45 |
| **Hidden Coin Trick** | **$1.19** | Mind Reading Trick | $1.25 |
| Colored Scarves Trick | $2.00 | **Paperclip Tangle** | **$0.95** |
| Chinese Linking Rings | $6.00 | Magic Kit with 10 tricks | $12.98 |

**Costume**

| | | | |
| --- | --- | --- | --- |
| Cape | $11.00 | Wand | $1.00 |
| Top Hat | $3.00 | | |

I need to find the data in the catalog.

Hidden Coin Trick is $1.19.
Paperclip Tangle is $0.95.

Now I can solve the problem.

$$\begin{array}{r} \$1.19 \\ + \ \ 0.95 \\ \hline \$2.14 \end{array}$$

Arnie spent $2.14.

## TRY IT OUT

1. How much does it cost to buy
   both the Mind Reading Trick and
   the Guess the Card Trick?

2. Arnie saved $10 to buy a top hat.
   How much change would he get
   if he bought the hat?

112

Solve. You may need to use the table on page 112.

1. Bonnie has $8.00. How much change will she get if she buys the Colored Scarves Trick?

2. How much do the Ball and Vase, Disappearing Rabbit, and Vanishing Quarter tricks cost all together?

3. Houdini, the magician, could escape from a steel safe. If he went in the locked safe at 10:35 a.m., he could step out 14 minutes later. At what time could he step out?

4. Bea takes $13 to the House of Magic. If she buys the Chinese Linking Rings, how much money will she have left?

5. If you sewed a cape at home, it would cost $4.00. How much more does it cost to buy a cape than to sew one?

6. What would it cost to buy the Magic Kit with 10 tricks and the complete costume?

7. Arnie learned 2 magic tricks the first week, 4 tricks the second week, and 6 tricks the third week. How many do you think he learned the sixth week?

8. It cost $2.50 per person for Barney's Magic Show. Julia and her mother watched Barney's Magic Show for 65 minutes. Then they ate lunch for 45 minutes. How much time did they spend at the show and at lunch?

9. **Finding Another Way** Verne solved this problem by using objects. Find another way to solve the problem.

In the Money Growing Trick, if you use 4 coins, you only show the audience 2 of them. If you use 6 coins, you show 3. If you use 8 coins, you show 4. If you use 12 coins, how many coins do you show the audience?

# Data Collection and Analysis
## Group Decision Making

UNDERSTAND
FIND DATA
PLAN
ESTIMATE
SOLVE
CHECK

**Doing an Investigation**

**Group Skill:**
Listen to Others

Your group will need three rubber bands and a tape measure for this investigation. The rubber bands might be long and fat or short and thin, as long as they are different from each other. First, each of you will predict which rubber band you think you can blow the farthest. Write down your predictions.

> Nina
> I think I can blow the red rubber band farthest.
> Red - 15 inches
> Blue -
> Tan -

**Collecting Data**

1. Take turns blowing one of the rubber bands along a smooth floor or table top. Measure how far you blow the rubber band. Write down the measurement.

2. Take turns blowing the other two rubber bands. Write down how far you blow each one.

114

3. Make a table like this one to record the information for each person in your group.

**Nina**

| Red rubber band | 15 inches |
| --- | --- |
| Blue rubber band | |
| Tan rubber band | |

4. Which rubber band was blown the farthest on any turn? Were your predictions right? Why do you think that rubber band went farthest?

5. For each person, check to see which rubber band went the least distance. Was it the same rubber band for everyone? Explain what you think happened.

# WRAP UP

## Find the Missing Word

Choose one of these words to complete each sentence correctly: **estimated, rounded, traded.** Explain your choices.

1. To check his calculator sum, Jason _____ the addends to the nearest hundred and said the sum would be close to 800.

2. Alicia _____ 13 ones when she solved $18 + 25$.

3. Maxine _____she would need about 15 sandwiches for a party with 10 guests.

4. Juan _____ the prices of some model kits to the nearest dollar to see if they would cost more than $10.

## Sometimes, Always, Never

Decide which word should go in the blank, <u>sometimes</u>, <u>always</u>, or <u>never</u>. Explain your choices.

5. To find a sum, you _____ trade 10 tens for 1 hundred.

6. 842 _____ rounds to 840.

7. To estimate sums, you _____ round addends to the nearest ten.

8. 523 _____ rounds up to 600.

## Project

You have $10 to spend on pet supplies. Find groups of three items that would cost less than $5, more than $10, and about $10.

116

# CHAPTER REVIEW/TEST

## Part 1    Understanding

Explain how to use mental math to find these sums.

**1.** $60 + 60$

**2.** $800 + 400$

**3.** $17 + 7$

**4.** $8 + 5 + 9$

**5.** How many ones do you trade to make 1 ten and 9 ones? to make 3 tens?

**6.** Is 1,000 a reasonable estimate for the sum of $744 + 252$? Explain.

Would you estimate or find the exact amount? Why?

**7.** You need 3 cups of flour for 1 loaf of bread. How many cups do you need for 2 loaves?

**8.** You want to serve apple juice at a party for 15 friends. How much should you buy?

## Part 2    Skills

Estimate the sums by rounding.

**9.** $461 + 348$

**10.** $\$2.28 + \$5.47$

**11.** $217 + 68$

**12.** $\$7.09 + \$3.25$

Find the sums.

**13.**
$$\begin{array}{r} 96 \\ + 23 \\ \hline \end{array}$$

**14.**
$$\begin{array}{r} 672 \\ + 158 \\ \hline \end{array}$$

**15.**
$$\begin{array}{r} 4,384 \\ + 2,536 \\ \hline \end{array}$$

**16.**
$$\begin{array}{r} 53 \\ + 29 \\ \hline \end{array}$$

**17.**
$$\begin{array}{r} 572 \\ + 257 \\ \hline \end{array}$$

**18.** $256 + 175$ **19.** $\$3.62 + \$0.78 + \$4.35 + \$1.46$ **20.** $\$3.04 + \$2.73 + \$1.14$

## Part 3    Applications

**21.** Amy used 22 parts to complete two sections of a model airplane. What two sections did she complete?

| Sections | Parts |
|---|---|
| wings | 6 |
| landing gear | 8 |
| propeller | 17 |
| body | 14 |

**22.** **Challenge** Use the catalog. Find the sums of the prices of the tool and seeds that cost the most and those that cost the least.

| Tools | | Seeds | |
|---|---|---|---|
| shovel | $13.94 | tomato | $0.95 |
| hoe | $8.74 | peas | $0.79 |
| hose | $11.67 | squash | $1.07 |

**117**

# ENRICHMENT
## Front-End Estimation

This poster won a school contest. The judges gave points for artwork, ideas, and poetry.

You can estimate the total score using front-end estimation.

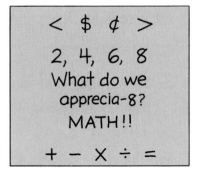

|     | Points | Estimate | Adjust |
|-----|--------|----------|--------|
| Art | 33 | 33 | |
| Idea | 35 | 35 | 3 + 5 + 1 |
| Poem | 11 | + 11 | about **10** more |
|     |    | 70 | 80 |

**First estimate**
Add the front numbers.

**Improved estimate**
Use the other places to adjust the first estimate.
The poster had a score of about 80.

Use front-end estimation to estimate these sums.

| **1.** | **2.** | **3.** | **4.** | **5.** |
|--------|--------|--------|--------|--------|
| 43 | 15 | 41 | 32 | 62 |
| 12 | 52 | 11 | 41 | 43 |
| + 23 | + 32 | + 32 | + 46 | + 21 |

Andrew earns points for special projects. Here is his estimate for each half of the school year.

|     | Points | Estimate | Adjust |
|-----|--------|----------|--------|
| First half | 448 | 448 | 48 + 51 |
| Second half | 451 | + 451 | about **100** more |
|     |     | 800 | 900 |
|     |     | **First estimate** | **Improved estimate** |

Use front-end estimation to estimate these sums.

| **6.** | **7.** | **8.** | **9.** | **10.** |
|--------|--------|--------|--------|---------|
| 235 | 319 | 642 | 429 | 118 |
| + 448 | + 550 | + 518 | + 317 | + 719 |

**11.** Linda bought a pencil for 23¢, a pen for 54¢, and paper or 42¢. About how much did she spend?

# CUMULATIVE REVIEW

1. Which number sentence belongs to the fact family for $7 - 4 = 3$?

    **A.** $7 + 4 = 11$    **B.** $7 + 3 = 10$

    **C.** $11 - 7 = 4$    **D.** $7 - 3 = 4$

2. Add. $9 + 5 + 7$

    **A.** 14    **B.** 21

    **C.** 20    **D.** 23

3. What is the next number in 655, 660, 665, _____?

    **A.** 666    **B.** 655

    **C.** 675    **D.** 670

4. Which seat is between the twelfth and fourteenth seats?

    **A.** third    **B.** thirteenth

    **C.** sixteenth    **D.** fifteenth

5. Which amount of money rounded to the nearest dollar is $5.00?

    **A.** $4.53    **B.** $5.54

    **C.** $4.48    **D.** $5.60

6. Which digit is in the thousands place in 7,853?

    **A.** 7    **B.** 8

    **C.** 5    **D.** 3

7. Which is nine hundred one?

    **A.** 91    **B.** 910

    **C.** 901    **D.** 1,910

8. In which place is the digit 6 in 763,400?

    **A.** hundred thousands    **B.** ones

    **C.** thousands    **D.** ten thousands

9. How much time has passed?

    **A.** 3 hours    **B.** 2 hours, 45 minutes

    **C.** 2 hours, 30 minutes    **D.** 4 hours

10. Which shows the date August fifth, nineteen ninety-three?

    **A.** 5/8/93    **B.** 93/5/8

    **C.** 8/5/93    **D.** 9/5/93

11. Maria has 1 quarter, 2 dimes, and 3 nickels. How much money does she have in all?

    **A.** $0.50    **B.** $0.60

    **C.** $0.59    **D.** $1.00

# 5

## MATH AND SOCIAL STUDIES

### DATA BANK

Use the Social Studies Data Bank on page 475 to answer the questions.

**1** Who flew kites first, Benjamin Franklin or the Wright Brothers?

# SUBTRACTION

**2** A string of 10 kites set the first record for height in 1910. Find the sum of these kites and the kites that broke the world's record in 1969.

**3** How many years after a kite was used for a radio was it used for an airplane?

**4** **Use Critical Thinking** Is it reasonable for someone to say, "My great-grandfather flew a plane more than 100 years ago?" Explain your answer.

# Mental Math
## Special Differences

**EXPLORE** **Use a Place Value Model**

Work in groups. This picture shows how to use 7 hundreds blocks to find $700 - 300$. The 3 hundreds that are marked with a ring have been taken away. Show your group some other differences starting with 7 of your hundreds blocks.

**TALK ABOUT IT**

1. How is finding $700 - 300$ like finding $7 - 3$?

2. Use your tens blocks to show $70 - 30$. How is this like finding $7 - 3$?

3. What differences can you show using 5 of your hundreds blocks?

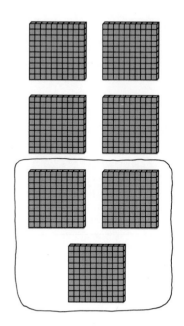

$$700 - 300 = 400$$

You can use basic subtraction facts to find special differences.

14 hundreds $-$ 5 hundreds $=$ 9 hundreds

$$1,400 - 500 = 900$$

8 tens $-$ 3 tens $=$ 5 tens

$$80 - 30 = 50$$

**TRY IT OUT**

Find the differences. Write only the answers.

1. $150 - 60$
2. $800 - 200$
3. $50 - 40$
4. $1,000 - 300$
5. $1,100 - 500$
6. $200 - 20$

Find the differences. Write only the answers.

**1.** 90 − 60 **2.** 130 − 70 **3.** 800 − 300 **4.** 1,300 − 400

**5.** 120 − 50 **6.** 1,100 − 700 **7.** 140 − 80 **8.** 70 − 50

**9.** 120 − 80 **10.** 170 − 80 **11.** 150 − 90 **12.** 2,000 − 500

**13.** Subtract 200 from 600.

**14.** Subtract 90 from 160.

**15.** Find the difference between 1,200 and 800.

**16.** What is 1,300 minus 500?

**APPLY**

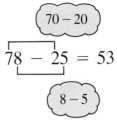

**MATH REASONING** Sometimes you can use your mental math skills to find other differences. Write only the answers.

**17.** 46 − 34 **18.** 94 − 22 **19.** 68 − 43 **20.** 59 − 36

**21.** 67 − 41 **22.** 98 − 72 **23.** 85 − 33 **24.** 76 − 24

**PROBLEM SOLVING**

**25.** Ted scored 700 points in a ring toss game. Sandra scored 1,300 points on her turn. How much more did Sandra score than Ted?

**26.** Bonnie told Ted that she scored 300 points more than his score of 700. She scored 300 points less than Sandra's 1,300. What was Bonnie's score?

▶ **ALGEBRA**

The ☐ and △ are hiding numbers.

**27.** ☐ + △ = 800

If the ☐ is hiding 300, what is the △ hiding?

**28.** ☐ − △ = 800

If the ☐ is hiding 1,200, what is the △ hiding?

# Estimating Differences Using Rounding

## LEARN ABOUT IT

**EXPLORE** **Compare the data**

Mr. Soto is in San Antonio. He wants to know which city is farther away, Amarillo or Houston. He also wants to know about how much farther.

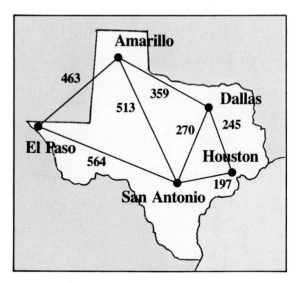

**Distances in Miles**

### TALK ABOUT IT

1. How far is it from San Antonio to Houston rounded to the nearest hundred miles? from San Antonio to Amarillo?

2. How can you use your answers to estimate the difference between the two distances?

| Round to nearest ten | Round to nearest hundred | Round to nearest dollar |
|---|---|---|
| 134 → 130 | 564 → 600 | $6.89 → $7.00 |
| − 69 → − 70 | − 359 → − 400 | − 2.25 → − 2.00 |
| Estimate: 60 | Estimate: 200 | Estimate: $5.00 |

## TRY IT OUT

Estimate by rounding to the nearest ten.

1.  79
   − 24

2.  52
   − 19

3.  123
   − 44

Estimate by rounding to the nearest hundred.

4.  495
   − 198

5.  712
   − 395

6.  832
   − 189

Estimate by rounding to the nearest dollar.

7.  $5.10
   − 3.96

8.  $9.80
   − 2.34

9.  $7.12
   − 1.79

Estimate by rounding to the nearest ten.

**1.** 58
  − 29

**2.** 48
  − 21

**3.** 127
  − 83

**4.** 156
  − 98

**5.** 92
  − 43

Estimate by rounding to the nearest hundred.

**6.** 817
  − 595

**7.** 795
  − 319

**8.** 921
  − 196

**9.** 1,716
  − 384

**10.** 921
  − 332

Estimate by rounding to the nearest dollar.

**11.** $5.10
  − 3.10

**12.** $8.05
  − 3.15

**13.** $2.90
  − 0.90

**14.** $5.98
  − 2.79

**15.** $8.88
  − 1.75

**16.** Round to the nearest dollar and solve. About how much more than $3.33 is $6.65?

**APPLY**

**MATH REASONING** Use estimation to choose the correct answer.

**17.** 811 − 203 = |||||
608 or 508?

**18.** 697 − 288 = |||||
409 or 309?

**19.** 82 − 39 = |||||
43 or 53?

**PROBLEM SOLVING**

**20.** About how much farther is it from San Antonio to El Paso than from San Antonio to Dallas? Use the Texas map.

**21. Write Your Own Problem**
Write and solve an estimation problem using the data on the map on page 124.

▶ **MENTAL MATH**

Use counting to find these differences. Write only the answers.

**22.** 500 − 498

**23.** 600 − 596

**24.** 501 − 498

**25.** 700 − 697

**26.** 900 − 895

**27.** 402 − 398

# Trading 1 Ten for 10 Ones

**EXPLORE  Use a Place Value Model**

Work in groups. Mark has 4 tens and 3 ones. If he trades 1 ten for 10 ones, he will still have the same amount but with 1 less ten and 10 more ones. Show this in your group with place value blocks.

Pick up some tens and ones blocks. Write the number they make. Trade 1 ten for 10 ones. Record the results.

**TALK ABOUT IT**

1. What are the different numbers of ones you might have after the trade?
2. Will you ever have more than 19 ones after a trade? Explain.
3. How many ones does Mark have before trading if he has 10 ones after a trade?

You can show Mark's trading this way.

| Before the trade | After the trade | | What you record | |
|---|---|---|---|---|
| | | tens | ones | tens | ones |
| | | | | 3 | 13 |
| | | 4 | 3 | 4 | 3 |

Trade 1 ten for 10 ones using blocks. Record the results.

| 1. | tens | ones |
|---|---|---|
| | 2 | 2 |

| 2. | tens | ones |
|---|---|---|
| | 5 | 6 |

| 3. | tens | ones |
|---|---|---|
| | 3 | 0 |

Trade 1 ten for 10 ones. Record the results.

| 1. | tens | ones |
|----|------|------|
|    | □    | □    |
|    | 3    | 7    |

| 2. | tens | ones |
|----|------|------|
|    | □    | □    |
|    | 2    | 0    |

| 3. | tens | ones |
|----|------|------|
|    | □    | □    |
|    | 5    | 2    |

Trade 1 ten for 10 ones. Write your answer above the number.

**4.** 76   **5.** 38   **6.** 47   **7.** 60   **8.** 25

**9.** 92   **10.** 67   **11.** 80   **12.** 73   **13.** 29

**14.** 55   **15.** 31   **16.** 54   **17.** 71   **18.** 10

## APPLY

**MATH REASONING**   What was the number before trading?

**19.** 4 tens 13 ones   **20.** 7 tens 18 ones   **21.** 5 tens 10 ones

### PROBLEM SOLVING

**22.** Vanessa has 6 dimes and 4 pennies. She gives 17 cents to her little sister. Did she have to trade 1 dime for 10 pennies? What coins does she have left?

**23. Problems With More Than One Answer** Mai has $34. What bills could she have if she has only tens and ones?

▶ **USING CRITICAL THINKING  Sort into Groups**

**24.** Suppose you have these blocks on a table. You can pick up 23 without a trade. You cannot pick up 17 without a trade. Can you explain why?

**25.** Sort these numbers into two groups.

21   32   6   15
  9   29   8   14

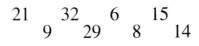

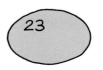

**Can Pick Up**   **Cannot Pick Up**

*More Practice, page 516, set B*

# Subtracting 2-Digit Numbers
## Making the Connection

**EXPLORE  Think About the Process**

Use play money to play a trading game.

- Show 75¢ by placing 7 dimes and 5 pennies on a table.

- Have a partner tell you a number from the bank. Take that number of cents away from 75¢. You may have to trade 1 dime for 10 pennies.

- Record your results in a table. Fill in the number of cents you took away. Then, fill in the amount left over.

- Do this four times, starting with 75¢ each time. Record your results.

- Give the play money to your partner and repeat.

**Trade**
**1 dime = 10 pennies**

|  | Dimes | Pennies |
|---|---|---|
| First number | 7 | 5 |
| Take away number | ___ | ___ |
| Difference | ___ | ___ |

| Number Bank |
|---|
| 19¢    23¢    46¢ |
| 47¢    38¢    25¢ |
| 28¢    45¢    14¢ |
| 37¢    17¢    39¢ |

**TALK ABOUT IT**

1. How many times did you need to make a trade when you played the game?

2. Taking away some numbers in the Number Bank did not need a trade. How can you tell what those numbers are?

3. Can you make a rule for when you need to make a trade?

128

In the trading game, you took away different numbers of cents from a given amount. Sometimes you traded and sometimes you did not. Now you will learn a way to record these results. Use place value blocks to show $42 - 15$.

**What You Do**　　　　　**What You Record**

$$\begin{array}{r} 42 \\ -\ 15 \end{array}$$

1. Can you take away 5 ones without trading? Why?

Trade

$$\begin{array}{r} \overset{3\ 12}{\cancel{42}} \\ -15 \end{array}$$

2. How many tens and ones are there after the trade?

$$\begin{array}{r} \overset{3\ 12}{\cancel{42}} \\ -\ 15 \\ \hline 27 \end{array}$$

3. How many blocks did you take away?

4. What is 42 minus 15?

5. Suppose you had 37 blocks on a table. Could you take away 15 without trading first? Explain.

---

**TRY IT OUT**

Use place value blocks to subtract these numbers. Record what you do using the example above.

1. Subtract 17 from 53.　　　　2. Subtract 17 from 48.

3. Write and solve your own subtraction problem using blocks.

# Subtracting 2-Digit Numbers

LEARN ABOUT IT

EXPLORE **Think About the Process**
Joanna and Krista are learning to fly kites. Joanna's kite flew for 50 minutes. Krista's stayed up 38 minutes. How much longer did Joanna's kite fly?

You subtract because you are comparing numbers.

| If you need more ones, trade a ten. | Subtract the ones. | Subtract the tens. |
|---|---|---|
| 4 10<br>5̶0̶ ( Trade )<br>− 38 | 4 10<br>5̶0̶<br>− 38<br>―――<br>2 | 4 10<br>5̶0̶<br>− 38<br>―――<br>12 |

**TALK ABOUT IT**

1. Why did you have to trade before you could subtract the ones?

2. Round 38 to the nearest ten. Then estimate the difference.

3. Answer the question in a complete sentence.

**Other Examples**

A    5 15
   6̶5̶ ( Trade )
   − 28
   ―――
    37

B    6 14
   7̶4̶ ( Trade )
   − 69
   ―――
     5

C    58 ( No Trade )
   − 27
   ―――
    31

D    7 15
   8̶5̶ ( Trade )
   −  7
   ―――
    78

TRY IT OUT

Find the difference. Trade if necessary.

**1.** 82 − 38          **2.** 79 − 20          **3.** 60 − 19          **4.** 67 − 9

Find the difference. Trade if necessary.

**1.** 51
−18

**2.** 62
−13

**3.** 83
− 4

**4.** 71
−47

**5.** 76
−55

**6.** 97 − 27   **7.** 50 − 18   **8.** 76 − 29   **9.** 88 − 36   **10.** 66 − 57

**11.** 54 − 17   **12.** 92 − 75   **13.** 62 − 53   **14.** 34 − 19   **15.** 71 − 32

**16.** What is 65 minus 47?       **17.** Subtract 29 from 96.

**APPLY**

**MATH REASONING** Find the missing numbers.

**18.** 72
− ||||
 45

**19.** 91
− ||||
 69

**20.** 33
− ||||
 15

**21.** 50
− ||||
 11

**PROBLEM SOLVING**

**22. Social Studies Data Bank** A huge kite called
Levitor lifted a man about 12 feet in the air.

 **DATA BANK** How much higher did Alexander Graham Bell's
kite Frost King lift somebody? See page 475.

**MIXED REVIEW**

Give the digital clock time shown by each clock face.

**23.**    **24.**    **25.**    **26.**

Match these times with the four clocks. Write the
times using a.m. or p.m.

**27.** catch morning bus to school       **28.** catch afternoon bus home

**29.** practice piano before dinner       **30.** eat lunch at school

# Problem Solving
## Make a Table

| UNDERSTAND |
| FIND DATA |
| PLAN |
| ESTIMATE |
| SOLVE |
| CHECK |

To solve some problems, it may help to put the data in a table first. This problem solving strategy is called **Make a Table.** You may want to use a calculator.

> I'll make a table to show what I already know.

> Now I'll fill in the table to find the answer.

> Juanita can buy 25 stones for $20.

Juanita collects rare and unusual rocks. She can buy a bag of 5 special stones for $4. How much would 25 special stones cost?

| Number of Stones | 5 |
| Cost | $4 |

ON/AC $5 + 5 = = = \ldots$

| Number of Stones | 5 | 10 | 15 | 20 | 25 |
|---|---|---|---|---|---|
| Cost | $4 | $8 | $12 | $16 | $20 |

ON/AC $4 + 4 = = = \ldots$

If Marshall collects 6 bonus stickers, he can get 5 free baseball cards. He sends in 30 bonus stickers. How many free baseball cards does he get?

- How many bonus stickers does Marshall need to collect to get free baseball cards?

- Then how many free baseball cards does he get?

- Copy and complete the table to solve.

| Bonus stickers | 6 | 12 | 18 | ___ | ___ |
|---|---|---|---|---|---|
| Free baseball cards | 5 | 10 | 15 | ___ | ___ |

132

Make a table to solve these problems.

1. Every time Tony adds 3 fossils to his collection, his mother gives him 2 from her old collection. If he collects 18 fossils, how many will his mother give him?

| Tony's fossils | 3 | 6 | 9 | ___ | ___ | ___ |
|---|---|---|---|---|---|---|
| Fossils from Mom | 2 | 4 | 6 | ___ | ___ | ___ |

**MIXED PRACTICE**

Solve.

3. Jody's grandfather offered her a shell from his collection. She has a choice of mussel, limpet, or conch shells. She can also choose a large shell or a small shell. How many different choices does Jody have?

4. Penny had 132 bottle caps in her collection. She gave away 13 to her brother. She lost 25 when her family moved. How many bottle caps does she have left?

5. Jeff kept his postcard collection in shoeboxes. He had 112 cards in one box, 98 in another and 83 in another. About how many cards does Jeff have?

2. Lin has the same number of Circus stamps as First Man on the Moon Stamps. Circus stamps cost 5¢, and First Man on the Moon stamps cost 10¢. If she has 30¢ worth of Circus stamps, how many First Man on the Moon stamps does she have?

| Some Strategies |
|---|
| Act It Out |
| Choose an Operation |
| Make an Organized List |
| Use Objects |
| Draw a Picture |
| Guess and Check |
| Make a Table |

6. Ekwa strings her bead collection in a pattern of 4 African beads and 2 Indian beads. She has 20 African beads and wants to continue this pattern. How many Indian beads does Ekwa need?

7. Jim and his father looked through a coin catalog for Spanish dubloons. One sold for $365, another for $286, and a third for $316. What is the difference between the most expensive dubloon and the least expensive dubloon?

*More Practice, page 516, set C*

# Subtracting 3-Digit Numbers
## One Trade

One year    365 days
Average School
    year    36 weeks
One School
    week    5 days
School Days
    in a year  180 days

**LEARN ABOUT IT**

**EXPLORE** **Think About the Process**
How many days in a year are there when you do not go to school?

You subtract because you are taking away one number from another. Follow the steps.

| Subtract the ones. Trade if necessary. | Subtract the tens. Trade if necessary. | Subtract the hundreds. |
|---|---|---|
| $\begin{array}{r} 365 \\ -180 \\ \hline 5 \end{array}$ | $\begin{array}{r} \overset{2\ 16}{3\cancel{6}5} \\ -180 \\ \hline 85 \end{array}$ | $\begin{array}{r} \overset{2\ 16}{3\cancel{6}5} \\ -180 \\ \hline 185 \end{array}$ |

**TALK ABOUT IT**

1. Did you trade in the first step or in the second step of this problem? Why?

2. Estimate the difference using rounding.

3. Use a sentence to give a reasonable answer.

**Other Examples**

A $\begin{array}{r} 138 \\ -\ 53 \\ \hline 85 \end{array}$

B $\begin{array}{r} \overset{7\ 12}{48\cancel{2}} \\ -135 \\ \hline 347 \end{array}$

C $\begin{array}{r} 675 \\ -243 \\ \hline 432 \end{array}$

D $\begin{array}{r} \overset{2\ 15}{3\cancel{5}6} \\ -\ 92 \\ \hline 264 \end{array}$

**TRY IT OUT**

Subtract.

1. $\begin{array}{r} 835 \\ -362 \end{array}$

2. $\begin{array}{r} 152 \\ -\ 85 \end{array}$

3. $\begin{array}{r} 378 \\ -\ 85 \end{array}$

4. $\begin{array}{r} 610 \\ -301 \end{array}$

134

Subtract. What do you notice about your answers?

| **1.** 563 | **2.** 627 | **3.** 689 | **4.** 517 | **5.** 444 | **6.** 916 |
|---|---|---|---|---|---|
| −128 | −192 | −254 | − 82 | − 9 | −481 |

Find the differences.

| **7.** 671 | **8.** 143 | **9.** 926 | **10.** 764 | **11.** 134 | **12.** 824 |
|---|---|---|---|---|---|
| −134 | − 82 | −432 | −132 | − 74 | −194 |

| **13.** 773 | **14.** 628 | **15.** 128 | **16.** 692 | **17.** 892 | **18.** 725 |
|---|---|---|---|---|---|
| −444 | −575 | − 95 | − 38 | −152 | − 92 |

**19.** 683 − 456    **20.** 425 − 180    **21.** 126 − 92    **22.** 821 − 340

**23.** Subtract 293 from 866.    **24.** Find the difference between 375 and 892.

**MATH REASONING** Use estimation to choose the correct answer.

| **25.** 685 | **26.** 685 | **27.** 528 | **28.** 528 | **29.** 419 |
|---|---|---|---|---|
| + 137 | − 137 | + 296 | − 296 | − 292 |
| 822 or 722? | 448 or 548? | 724 or 824? | 232 or 332? | 127 or 27? |

**PROBLEM SOLVING**

**30.** Mrs. Lee worked 239 days last year. Because it was a leap year, there were 366 days in all. How many days was Mrs. Lee not at work?

**31. Missing Data** If a year has 52 weeks, how many weeks are not spent in school? What data do you need to solve this problem? Where can you find it?

▶ **CALCULATOR**

Show the first number on your calculator. Find what you must subtract to get the second number.

**32.** 582 → 502    **33.** 693 → 93    **34.** 4786 → 4780

*More Practice, page 504, set B*    **135**

# Subtracting 3-Digit Numbers

EXPLORE **Think About the Process**

David's cocker spaniel eats 365 grams of dry dog food each day. Mariko's poodle eats 198 grams each day. How much more dog food does David's dog eat than Mariko's dog?

You subtract because you are comparing.

| Subtract the ones. Trade if necessary. | Subtract the tens. Trade if necessary. | Subtract the hundreds. |
|---|---|---|
| 5 15<br>3̸6̸5̸<br>− 198<br>———<br>7 | 2 15 15<br>3̸6̸5̸<br>− 198<br>———<br>67 | 2 15 15<br>3̸6̸5̸<br>− 198<br>———<br>167 |

**TALK ABOUT IT**

1. How many trades did you need? Why?

2. Estimate the difference by rounding.

3. Answer the question in a complete sentence.

|  | 6 10 14 |  | 8 13 13 |  | 13 12 |  | 5 12 14 |
|---|---|---|---|---|---|---|---|
| **Other Examples   A** | 7̸1̸4̸<br>− 189<br>———<br>525 | **B** | 9̸4̸3̸<br>− 878<br>———<br>65 | **C** | 1 4̸2̸<br>− 68<br>———<br>74 | **D** | 6̸3̸4̸<br>− 89<br>———<br>545 |

Find the differences.

| 1. 835<br>− 289 | 2. 973<br>− 272 | 3. 346<br>− 98 | 4. 440<br>− 177 |
|---|---|---|---|

136

Find the differences.

| | | | | | | | | | |
|---|---|---|---|---|---|---|---|---|---|
| **1.** | 643<br>− 578 | **2.** | 135<br>− 77 | **3.** | 516<br>− 97 | **4.** | 437<br>− 359 | **5.** | 228<br>− 79 |
| **6.** | 927<br>− 258 | **7.** | 875<br>− 154 | **8.** | 712<br>− 638 | **9.** | 483<br>− 126 | **10.** | 324<br>− 167 |
| **11.** | 462<br>− 325 | **12.** | 312<br>− 172 | **13.** | 693<br>− 596 | **14.** | 840<br>− 652 | **15.** | 898<br>− 199 |

**16.** 324 − 68  **17.** 123 − 87  **18.** 654 − 590  **19.** 743 − 555

**20.** 327 − 213  **21.** 530 − 364  **22.** 483 − 45  **23.** 815 − 337

**24.** How much less is 356 than 824?  **25.** How much more is 634 than 281?

**APPLY**

**MATH REASONING** Do not subtract. Compare to find the greater difference.

**26.** 654<br>− 289  or  654<br>− 298  **27.** 617<br>− 286  or  617<br>− 298  **28.** 728<br>− 379  or  782<br>− 379

**PROBLEM SOLVING**

**29.** Mariko's poodle eats 198 grams and her collie eats 596 grams of dry dog food each day. How much dry dog food does Mariko feed her two dogs each day?

**30.** Beads are 3 for 5 cents. How many can you get if you have 40 cents?

| Beads | 3 | 6 | 9 |
|---|---|---|---|
| Cents | 5 | 10 | 15 |

▶ **ESTIMATION**

What basic fact helps you estimate each difference?

**31.** 784<br>− 503  **32.** 219<br>− 98  **33.** 841<br>− 286  **34.** 310<br>− 194

*More Practice, page 504, set C*

# Using Critical Thinking

Which number in the cloud is the correct difference? Remember what you know about the relationship between addition and subtraction. See if you can use addition, not subtraction, to find the answer.

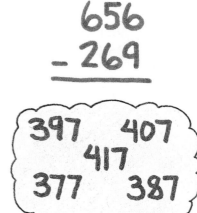

**TALK ABOUT IT**

1. If you know that $7 + 5 = 12$, what two subtraction facts do you know?

2. Which number in the cloud adds to 269 to give 656?

3. Can you give a rule for using addition to check answers to subtraction problems?

Check your answers after subtracting.

| Subtract | Check | | Subtract | Check |
|----------|-------|---|----------|-------|
| 976 | 634 | | 624 | 471 |
| − 342 | + 342 | | − 153 | + 153 |
| 634 | 976 | | 471 | 624 |

**TRY IT OUT**

Subtract. Check each answer by adding.

1. 588
   − 257

2. 690
   − 78

3. 654
   − 121

4. 723
   − 380

5. 362
   − 148

6. 735
   − 364

7. 483
   − 195

8. 921
   − 286

9. 854
   − 767

10. 776
    − 309

**138**

# MIDCHAPTER REVIEW/QUIZ

Find the differences. Write answers only.

**1.** $90 - 30$ **2.** $150 - 60$ **3.** $1,100 - 800$ **4.** $1,000 - 400$

Estimate by rounding to the nearest ten, hundred, or dollar.

| **5.** | **6.** | **7.** | **8.** | **9.** | **10.** |
|---|---|---|---|---|---|
| 53 | 166 | 836 | 550 | \$7.39 | \$6.22 |
| $-27$ | $-84$ | $-421$ | $-425$ | $-2.65$ | $-3.98$ |

Write **yes** if you will need to trade. Write **no** if you will not need to trade. Then find the difference.

| **11.** | **12.** | **13.** | **14.** | **15.** | **16.** |
|---|---|---|---|---|---|
| 67 | 65 | 25 | 72 | 54 | 41 |
| $-25$ | $-27$ | $-16$ | $-64$ | $-24$ | $-3$ |

Subtract.

| **17.** | **18.** | **19.** | **20.** | **21.** | **22.** |
|---|---|---|---|---|---|
| 96 | 129 | 856 | 27 | 368 | 241 |
| $-56$ | $-89$ | $-279$ | $-18$ | $-271$ | $-75$ |

| **23.** | **24.** | **25.** | **26.** | **27.** | **28.** |
|---|---|---|---|---|---|
| 137 | 263 | 570 | 63 | 324 | 483 |
| $-29$ | $-126$ | $-172$ | $-48$ | $-315$ | $-396$ |

**29.** $643 - 55$ **30.** $231 - 158$ **31.** $730 - 554$ **32.** $818 - 609$

## PROBLEM SOLVING

**33.** For which of the three problems would you trade 1 ten for 10 ones? For which would you trade 1 hundred for 10 tens?

    **A.** $156 - 78$
    **B.** $345 - 254$
    **C.** $657 - 649$

**34.** At Peter's school, 5 students walk to school for every 3 students who ride the bus. If 30 students walk, how many ride the bus?

| walkers | 5 | 10 | 15 | |
|---|---|---|---|---|
| **bus riders** | 3 | 6 | | |

139

# Subtracting Across a Middle Zero

**EXPLORE** **Think About the Process**

Rosa measured her arm span. It is 129 cm. How does Rosa's arm span compare to the wingspread of the King Vulture?

You subtract because you are comparing.

| World's Largest Flying Birds | |
|---|---|
| Bird | Wingspread |
| Albatross | 366 cm |
| Condor | 335 cm |
| King Vulture | 305 cm |
| White Pelican | 276 cm |
| Bald Eagle | 244 cm |

| Subtract the ones. Trade if necessary. | Subtract the tens. Trade if necessary. | Subtract the hundreds. |
|---|---|---|
| $\begin{array}{r} {}^{2\,9\,15}\\ \cancel{305}\\ -129\\ \hline 6 \end{array}$  30 tens minus 1 ten equals 29 tens. | $\begin{array}{r} {}^{2\,9\,15}\\ \cancel{305}\\ -129\\ \hline 76 \end{array}$ | $\begin{array}{r} {}^{2\,9\,15}\\ \cancel{305}\\ -129\\ \hline 176 \end{array}$ |

**TALK ABOUT IT**

1. After you used 1 of the 30 tens for a trade, how many tens are left?

2. Estimate the answer by rounding.

3. Use a sentence to give a reasonable answer.

**Other Examples**

A  $\begin{array}{r} {}^{6\,9\,10}\\ \cancel{700}\\ -367\\ \hline 333 \end{array}$   B  $\begin{array}{r} {}^{5\,9\,15}\\ \cancel{605}\\ -\ 88\\ \hline 517 \end{array}$   C  $\begin{array}{r} {}^{9\,13}\\ \cancel{103}\\ -\ 76\\ \hline 27 \end{array}$   D  $\begin{array}{r} {}^{6\,10}\\ \cancel{705}\\ -382\\ \hline 323 \end{array}$

Find the differences.

1.  $\begin{array}{r} 704\\ -256 \end{array}$   2.  $\begin{array}{r} 602\\ -\ 37 \end{array}$   3.  $\begin{array}{r} 104\\ -\ 85 \end{array}$   4.  $\begin{array}{r} 903\\ -774 \end{array}$   5.  $\begin{array}{r} 409\\ -122 \end{array}$

140

Find the differences.

| | | | | |
|---|---|---|---|---|
| **1.** 605 −416 | **2.** 524 −392 | **3.** 703 − 59 | **4.** 502 −287 | **5.** 104 − 36 |
| **6.** 703 −634 | **7.** 800 −342 | **8.** 401 − 89 | **9.** 906 −528 | **10.** 628 −273 |

**11.** 309 − 142     **12.** 600 − 536     **13.** 904 − 627     **14.** 504 − 146

**15.** Subtract 627 from 705.     **16.** What is 503 minus 188?

APPLY

**MATH REASONING** Write <u>sometimes</u>, <u>never</u>, or <u>always</u> to complete the sentence.

**17.** The answer to a subtraction problem is _____ less than the number that was subtracted.

**18.** When you subtract a number greater than zero, the answer is _____ less than the number you subtracted from.

**PROBLEM SOLVING**

**19.** How much greater is the wingspread of the King Vulture than the White Pelican? Use the chart on page 140.

**20. Write Your Own Problem** Write a subtraction problem using the data in the chart on page 140.

MIXED REVIEW

Find the sums.

| | | | | | |
|---|---|---|---|---|---|
| **21.** 25 + 63 | **22.** 604 + 327 | **23.** 237 + 396 | **24.** 345 + 34 | **25.** $46.38 + 27.19 | **26.** 131 + 576 |

**27.** 2,604 + 3,396     **28.** 527 + 405     **29.** 7,424 + 1,588     **30.** 459 + 276

*More Practice, page 504, set D*

# Estimating Differences with Money

**EXPLORE  Solve to Understand**

Brian wants to buy a U.S. map puzzle, but he can save money by buying an atlas instead. How much money can he save, not counting tax?

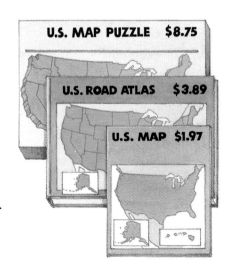

U.S. MAP PUZZLE  $8.75

U.S. ROAD ATLAS  $3.89

U.S. MAP  $1.97

**TALK ABOUT IT**

1. Can you tell by estimating whether the answer will be more or less than $5.00?

2. How is subtracting money like subtracting whole numbers?

Since you want to know how much less, you subtract.

| **Estimating the Difference** | **Finding the Difference** |
|---|---|

Round to the nearest dollar.

Line up the dollars and cents.

$$
\begin{array}{r}
\$8.75 \longrightarrow \$9.00 \\
-\ 3.89 \longrightarrow -\ 4.00 \\
\hline
\$5.00
\end{array}
$$

$$
\begin{array}{r}
\$8.75 \\
-\ 3.89 \\
\hline
\$4.86
\end{array}
$$

Show dollars and cents.

## Other Examples

| **Exact** | **Estimate** | | **Exact** | **Estimate** |
|---|---|---|---|---|
| $6.19 ⟶ | $6.00 | | $9.74 ⟶ | $10.00 |
| − 1.85 ⟶ | − 2.00 | | − 2.59 ⟶ | − 3.00 |
| $4.34 | $4.00 | | $7.15 | $7.00 |

**TRY IT OUT**

Estimate each difference. Then find the exact difference.

| | | | | | | | | | |
|---|---|---|---|---|---|---|---|---|---|
| **1.** | $7.79 | **2.** | $8.70 | **3.** | $6.24 | **4.** | $5.00 | **5.** | $3.06 |
| | − 2.65 | | − 1.29 | | − 1.98 | | − 2.79 | | − 1.88 |

142

Estimate each difference. Then find the exact difference.

| | | | | |
|---|---|---|---|---|
| **1.** $3.69 <br> − 1.25 | **2.** $6.15 <br> − 1.30 | **3.** $7.25 <br> − 2.50 | **4.** $5.32 <br> − 2.47 | **5.** $5.00 <br> − 1.87 |
| **6.** $5.00 <br> − 2.75 | **7.** $8.50 <br> − 1.75 | **8.** $1.00 <br> − 0.45 | **9.** $7.52 <br> − 2.75 | **10.** $8.00 <br> − 6.25 |

**11.** $6.32 − $4.44     **12.** $3.09 − $1.99     **13.** $1.36 − $1.27

**14.** How much is $750 minus $6.98?     **15.** Subtract $2.98 from $5.00.

**MATH REASONING** Use the first difference
for finding the other differences in your head.

| | | | |
|---|---|---|---|
| $1.00 <br> − 0.68 <br> $0.32 | **16.** $3.00 <br> − 0.68 | **17.** $3.00 <br> − 1.68 | **18.** $2.00 <br> − 1.68 |

**PROBLEM SOLVING**

**19. Determining Reasonable Answers** How much less does the map cost than the road atlas? Look at the prices on page 142. Use estimation to choose the correct answer. Is it $2.92, $1.92, or $0.92?

**20.** Reggie thinks he can buy a road atlas and a map for less than the puzzle alone. Look at the prices on page 142. Is he right? Use estimation to help you decide.

▶ **CALCULATOR**

**21.** Estimate to find amounts that add up to $11.56. Check with a calculator. Write an equation to show each way.

$4.20     $5.22     $3.12

$3.22     $4.60

$6.96     $6.34

# Subtracting 4-Digit Numbers

LEARN ABOUT IT

EXPLORE **Think About the Process**

Today Mary's kite went 2,903 feet in the air.
Jeb's kite went up 4,132 feet. How much higher
did Jeb's kite go?

You subtract because you are comparing.

| Subtract the ones. Trade if necessary. | Subtract the tens. Trade if necessary. | Subtract the hundreds. Trade if necessary. | Subtract the thousands. |
|---|---|---|---|
| 2 12<br>4,13̸2̸<br>− 2,903<br>**9** | 2 12<br>4,13̸2̸<br>− 2,903<br>**29** | 3 11 2 12<br>4,13̸2̸<br>− 2,903<br>**229** | 3 11 2 12<br>4,13̸2̸<br>− 2,903<br>**1,229** |

**TALK ABOUT IT**

1. How many trades do you make in finding this difference?

2. How can you estimate the difference using rounding to the nearest thousand?

3. Give a complete sentence to answer the question.

|  | | 6 11 14 13 | | 8 11 14 | | 4 12 8 16 | | 8 13 9 13 |
|---|---|---|---|---|---|---|---|---|
| **Other Examples** | **A** | 7,25̸3̸<br>− 2,568<br>4,685 | **B** | 6,92̸4̸<br>−   78<br>6,846 | **C** | $5̸2.9̸6̸<br>− 48.37<br>$4.59 | **D** | 9,40̸3̸<br>− 1,628<br>7,775 |

TRY IT OUT

Find the differences.

| 1. | 7,816<br>−   79 | 2. | 4,374<br>−  891 | 3. | 6,234<br>− 1,569 | 4. | 1,723<br>− 1,644 | 5. | 8,304<br>− 2,758 |
|---|---|---|---|---|---|---|---|---|---|

Find the difference.

| | | | |
|---|---|---|---|
| **1.** 8,746<br> − 578 | **2.** 9,538<br> − 5,675 | **3.** 9,327<br> − 58 | **4.** 4,313<br> − 2,765 |
| **5.** 7,620<br> − 6,341 | **6.** 9,314<br> − 8,487 | **7.** 5,759<br> − 4,287 | **8.** 7,287<br> − 1,532 |

**9.** $71.86 − $5.49   **10.** 6,473 − 3,284   **11.** $68.00 − $23.57

**12.** 7,620 − 307   **13.** 7,204 − 6,487   **14.** 8,683 − 4,362

**15.** 6,247 − 768   **16.** $58.32 − $51.60   **17.** 4,130 − 1,586

**18.** How much more is $76.28 than $54.03?

**19.** How much less is $17.65 than $63.04?

**APPLY**

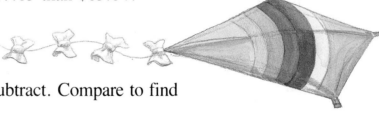

**MATH REASONING**  Do not subtract. Compare to find the greater difference.

**20.** 7,326   7,326   **21.** 8,562   8,652   **22.** 3,002   3,002
      − 4,158 or − 4,518        − 3,879 or − 3,879        − 1,099 or − 1,009

**PROBLEM SOLVING**

**23.** Jeff's kite went up 1,307 feet in the air. Barry's kite went up 2,139 feet. How much higher did Barry's kite go?

**24. Social Studies Data Bank**
The Cygnet and the Frost King were two huge kites built by Alexander Graham Bell. How many more cells did the Cygnet have than the Frost King? See page 475.

▶ **COMMUNICATION  Write to Learn**

**25.** Suppose you estimate a sum by rounding both numbers down. Write a sentence that compares your estimate with the exact sum.

*More Practice, page 504, set F*

# Problem Solving
## Choosing a Calculation Method

| UNDERSTAND |
| FIND DATA |
| PLAN |
| ESTIMATE |
| SOLVE |
| CHECK |

When you solve a problem you need to choose the calculation method that works best.

This table shows how much tax is added to sales between $78.90 and $80.49. For each dollar spent, 5 cents tax is added.

Do this when you choose a calculation method.

■ First try mental math. Look for numbers you can work with easily in your head.

■ If you cannot use mental math, choose between paper and pencil and a calculator. Some problems have many steps and trades. You may want to use a calculator for these.

| Sales Tax Table (5¢ per dollar) | |
|---|---|
| $78.90–79.09 | $3.95 |
| 79.10–79.29 | 3.96 |
| 79.30–79.49 | 3.97 |
| 79.50–79.69 | 3.98 |
| 79.70–79.89 | 3.99 |
| 79.90–80.09 | 4.00 |
| 80.10–80.29 | 4.01 |
| 80.30–80.49 | 4.02 |

How much does an $80.00 sale cost with tax?

$80 plus $4 tax.
That's easy. I'll do it in my head.

How much does a sale of $78.95 cost with tax?

$78.95 plus $3.95 tax.
I'll need a pencil and some paper, or maybe a calculator would be better.

**TRY IT OUT**

Choose a calculation method for each problem. Tell why you chose it. Then solve. Use each method at least twice.

**1.** $70 + 80$ **2.** $195 - $28 **3.** $69.95 - $2.89 **4.** $65 - $20

**5.** $75 - 39$ **6.** $60.00 + $15.00 **7.** $90.25 + $17.89 **8.** $6.95 + $28.98

Choose a calculation method. Use each method at least twice.

**1.** $182 - 28$

**2.** $150 - 70$

**3.** $59.98 - $26.75

**4.** $875 + 698 + 89$

**5.** $7.80 + $0.25

**6.** $50.00 + $80.00

**7.** $600 + 800$

**8.** $56 + $28

**9.** $170 - $90

**10.** $726 - 192$

**11.** $75 + 25$

**12.** $40.25 - $17.89

Solve. Use each method at least once. Use data from the tax table on page 146.

**13.** How much more is the tax on $80 than on $79?

**14.** How much would a sale of $79.90 cost with tax?

**15.** Suppose a city adds 2 cents sales tax for every $5 spent. What would be the tax on a sale of $40?

| Sale | $5 | $10 | $15 |
|------|-----|------|------|
| Tax | 2¢ | 4¢ | 6¢ |

**16.** Lucas bought a shirt for $19.85 and slacks for $59.98. How much did these two items cost in all? What was the tax? How much did Lucas spend for the clothes and the tax?

**17.** Mrs. Tyler bought a suit on which she paid $3.95 in tax. Use the tax table. Make up a price that Mrs. Tyler could have paid for the suit. Give the total cost for your price.

**18.** Stacy paid $85.03 with tax for some school clothes. The tax was $4.05. What did the clothes cost without tax?

**19.** Write a paragraph to tell why sometimes it is easier to use pencil and paper than to get out your calculator.

# Problem Solving
## Multiple-Step Problems

| UNDERSTAND |
| FIND DATA |
| PLAN |
| ESTIMATE |
| SOLVE |
| CHECK |

### LEARN ABOUT IT

Some problems can be solved by using both addition and subtraction. These problems are called Multiple-Step Problems.

> Brett had $35 to spend on his new puppy. He spent $4 on a collar and leash. He also spent $12 for a fancy dog dish. How much money did he have left?

First I'll add to find out how much money Brett spent in all.

$$\$4 + \$12 = \$16$$

Then I'll subtract to find out how much money he has left.

$$\$35 - \$16 = \$19$$

Brett has $19 left.

### TRY IT OUT

Solve.

1. Ali had $16 saved. He earned $5 more walking Mrs. Trent's dog. He spent $9 on a haircut. How much money does he have left?

2. It cost Louis $2 for cat litter and $6 for a litter box. He gave the clerk a $10 bill. How much change did he get back?

3. Delia earned $26 washing cars and $12 raking leaves. She spent $25. A basket for her kitten costs $15. How much more money does she need to buy the basket?

4. A video store had 12 videos about animals. There were 6 about dogs and 3 about monkeys. The rest were about cats. How many were about cats?

Solve.

1. Bill added up how much he spends on pet food in a week. What did he find out?

   $3.78 for Trixie the cat
   $5.11 for Bilbo the mutt
   $5.78 for Bob the collie

2. At the beginning of April there were 16 dogs at the animal shelter. The next week 15 more dogs were brought in. At the end of the month 4 dogs were adopted. How many dogs were left at the animal shelter?

3. Dana took her dog Rocky to the vet. It cost $27 in all for a rabies shot and for a checkup. The shot cost $7 more than the checkup. What did the checkup cost?

4. When Fernando's dog got lost, his parents advertised in the newspaper for 3 days. They paid $7.11. They also paid $4.22 to have posters made. How much did they pay in all?

5. There were 11 animal posters at the pet store. 5 were kitten posters and 6 were dog posters. The rest were of horses. How many posters were of horses?

6. Katie wants to buy a flea collar and a brush for her dog. About how many dollars does she need?

Collar $2.89     Brush $3.29

7. Ling had to take her cat Tiger to the emergency vet for a broken leg. It cost $79.50 for the cast and $15 for the emergency visit. How much did Ling pay?

8. Winston's family wanted to make their backyard safe for their dog. How much did they pay in all?

| Yard and Garden Center | Neighborhood Pet Store |
| --- | --- |
| fence . . . . $268 | doghouse . . . $125 |

9. **Think About Your Solution**
   Dog collars come in 3 colors, red, black, and white. Dog name tags come in 2 sizes, small and large. How many choices do you have if you want to buy a collar and a name tag?

   Write your answer in a complete sentence. Name the strategy and show the steps you used to solve the problem.

# Applied Problem Solving
## Group Decision Making

UNDERSTAND
FIND DATA
PLAN
ESTIMATE
SOLVE
CHECK

### Group Skill:
Encourage and Respect Others

Your group is in charge of the balloon booth at the school festival. You need to decide what kinds of balloons to buy for the booth. You also need to decide what price to charge for each balloon.

### Facts to Consider

1. You hope to sell about 125 balloons.

2. You want to make 50¢ to 75¢ profit on each balloon. You will probably sell more balloons if you charge less.

3. The large shiny balloons cost you $1.19 each.

4. The small plain-colored balloons cost you $0.29 each.

5. It costs about 5¢ to fill a balloon with helium.

1. Do you think more people will buy the shiny balloons or the plain balloons?

2. Should you try to make the same amount of profit on both kinds of balloons?

3. How much do you think people will pay for the shiny balloons?

4. How much do you think people will pay for the plain balloons?

Write down how many of each kind of balloon your group would buy for the booth. Tell how much you would charge for each kind of balloon. Tell how much money you will make on each balloon.

# WRAP UP

## Tell All About It

Solve. Tell how you would follow the directions
for each problem.

1. $1{,}800 - 900$
   Use mental math.

2. $105 - 32$
   Make trades.

3. $4{,}981 - 3{,}692$
   Make trades.

4. $999 - 372$
   Use addition to check the answer.

5. $2{,}443 - 1{,}944$
   Make trades.

6. $\$790.38 - \$525.59$
   Estimate answer to the nearest
   $10.

## Sometimes, Always, Never

Decide which word should go in the blank,
<u>sometimes</u>, <u>always</u>, or <u>never</u>. Explain your choices.

7. When a number in a subtraction problem
   contains a zero, you _____ will need to trade.

8. You _____ can subtract money the same way
   that you subtract whole numbers.

9. When you subtract, you _____ start with the
   numbers on the left, such as the thousands
   in numbers with four digits.

10. When you subtract one 3-digit number from
    another 3-digit number, you _____ make
    trades.

## Project

Plan a day at an amusement park. Make up
names for 5 different rides. Set a price for each
ride. Decide on a bargain price for riding all 5
rides one time. Tell how you would spend $20
and why.

152

# CHAPTER REVIEW/TEST

## Part 1 Understanding

1. Explain how to use mental math to find this difference. $1,100 - 400$

2. Do you trade to subtract 35 from 53? Explain how to find the difference.

3. Tell how you would estimate this difference. $53 - 34$.

4. Tell how you would estimate $6.25 - $1.96$.

5. What calculation method would you use to solve $15.41 - $6.59$? Tell why.

6. Draw a picture to show your trading to subtract 19 from 31.

## Part 2 Skills

Estimate the differences.

7. $459 - 103$
8. $57 - 21$
9. $6.59 - $5.17$
10. $372 - 34$

Find the differences.

11. $\begin{array}{r} 75 \\ -48 \\ \hline \end{array}$
12. $\begin{array}{r} 81 \\ -33 \\ \hline \end{array}$
13. $\begin{array}{r} \$7.24 \\ -\$2.51 \\ \hline \end{array}$
14. $\begin{array}{r} \$52.52 \\ -\ 18.73 \\ \hline \end{array}$
15. $\begin{array}{r} 904 \\ -613 \\ \hline \end{array}$

16. $452 - 295$
17. $706 - 144$
18. $7,408 - 2,659$
19. $836 - 208$

## Part 3 Applications

20. Mindy bought 2 sweaters. One cost $15.47 and the other cost $30.03. How much did she spend? What calculation method would you use to solve the problem? Tell why.

21. Gene bought 2 sheets of drawing board for $3. How much would he have to pay for 8 sheets of drawing board?

22. **Challenge** Use the table. Karen bought a birdcage and a bird bell. Her brother Leon bought a wood perch and one box of bird food. Who spent more? How much more?

| Bird Supplies | |
|---|---|
| Box of food | $4.43 |
| Birdcage | $10.93 |
| Bird bell | $3.52 |
| Wood perch | $2.04 |

# ENRICHMENT
## Using a Calculator

**Cal's Cool Car Company**

| Fire Power | Luxor |
|---|---|

$23,595          $56,879

| Dragster | Road King |
|---|---|

$15,925          $24,789

| Super Dupe | Ulty Mate |
|---|---|

$19,969          $37,798

You can use a calculator to help you solve these problems.

**A.** Diane sells cars at Cal's. One month she sold four cars, a Fire Power, a Dragster, a Road King, and an Ulty Mate. What was Diane's sales total that month?

**B.** One month, Diane set a sales goal of $150,000. After selling a Luxor, a Super Dupe, and a Road King, how much does Diane have to sell to reach her goal?

Use the memory keys.

$\boxed{M+}$ adds the displayed number to the memory.

$\boxed{M-}$ subtracts the displayed number from the memory.

$\boxed{MRC}$ recalls the number stored in the memory.

**Solving Problem A**
1. $\boxed{ON/AC}$ Enter $23,595.
2. Press $\boxed{M+}$.
3. Enter $15,925.
4. Press $\boxed{M+}$.
5. Enter $24,789.
6. Press $\boxed{M+}$.
7. Enter $37,798.
8. Press $\boxed{M+}$.
9. Press $\boxed{MRC}$.

**Solving Problem B**
1. $\boxed{ON/AC}$ Enter $150,000.
2. Press $\boxed{M+}$.
3. Enter $56,879.
4. Press $\boxed{M-}$.
5. Enter $19,969.
6. Press $\boxed{M-}$.
7. Enter $24,789.
8. Press $\boxed{M-}$.
9. Press $\boxed{MRC}$.

Make up and solve some addition and subtraction problems about the cars. Use a calculator.

# CUMULATIVE REVIEW

**1.** Choose the symbol that belongs in the ⫸.  798 ⫸ 789.

 **A.** >  **B.** =

 **C.** <  **D.** $\overline{)}$

**2.** Which number is three hundred six thousand, one hundred two?

 **A.** 360,120  **B.** 306,120

 **C.** 306,102  **D.** 360,102

**3.** Which date is the fourth of June, nineteen hundred ninety?

 **A.** 6/4/09  **B.** 4/5/09

 **C.** 4/6/90  **D.** 6/4/90

**4.** Elise has 2 quarters, 3 dimes, 2 nickels, and 5 pennies. How much money does she have?

 **A.** $0.95  **B.** $1.15

 **C.** $0.97  **D.** $1.07

**5.** You give the clerk 75¢ to pay for an eraser that costs 64¢. Which shows the change that you should get back?

 **A.**   **B.**

 **C.**   **D.**

**6.** Estimate the sum to the nearest hundred. 142 + 661.

 **A.** 500  **B.** 600

 **C.** 700  **D.** 800

**7.** Add. 46 + 25

 **A.** 61  **B.** 71

 **C.** 70  **D.** 81

**8.** Add. 356 + 199

 **A.** 445  **B.** 665

 **C.** 555  **D.** 655

**9.** Add. 1,537 + 3,664

 **A.** 5,193  **B.** 5,201

 **C.** 5,211  **D.** 5,191

**10.** Mari has $5.90, Mike has $6.85, and Morgan has $4.77. How much money do they have in all?

 **A.** $17.52  **B.** $16.42

 **C.** $17.42  **D.** $15.52

**11.** Manuel gave the clerk $5.00 to pay $2.57. How should the clerk figure the change?

 **A.** Estimate. **B.** Add $5.00 and $2.57.

 **C.** Round $2.57 to $3.00 **D.** Subtract to get the exact amount.

# 6

# DATA, GRAPHS, AND PROBABILITY

**M**ATH AND SCIENCE

**DATA BANK**

Use the Science Data Bank on page 472 to answer the questions.

1 The length of the shortest dinosaur we know of was about 2 feet. What is the length of the shortest dinosaur that is listed in the table of lengths?

**2** The length of the longest dinosaur we know of was about 100 feet. What is the length of the longest dinosaur in the table of lengths?

**3** Which of the dinosaurs listed in the table were the same length as Triceratops? Which of the dinosaurs were longer than Triceratops?

**4** Use Critical Thinking Look at the table of heights. Which dinosaur could eat the top leaves of trees 35 feet high? Explain your thinking.

157

# Reading Bar and Picture Graphs

LEARN ABOUT IT

**EXPLORE Study the Data**

Susie's class made a **bar graph** to show how they help at home. Can you tell what the most students named?

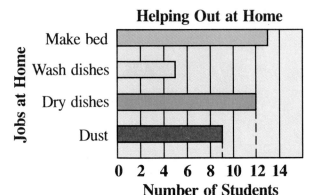

**Helping Out at Home**

**TALK ABOUT IT**

1. Susie always makes her bed and often dries dishes. What colors show her choices?

2. Which job was chosen by the fewest students?

Look at the dashed lines from the ends of the bars to the numbers along the bottom of the graph. They help you tell how many students chose each job.

Curtis made this **picture graph.** It shows how many hours some students spend helping and doing homework. Count the number of pictures to find the hours. You can see that Karen works 4 hours in a week.

**Working at Home**

Curtis  ⊘ ⊘ ⊘ ⊘ ⊘
Karen   ⊘ ⊘ ⊘ ⊘
Peter   ⊘ ⊘ ⊘ ⊘ ⊘
Molly   ⊘ ⊘ ⊘ ⊘ ⊘ ⊘

Each ⊘ means 1 hour each week.

TRY IT OUT

1. In the bar graph, how many students said they make the bed?

2. In the picture graph, who works the greatest number of hours each week? the least?

158

Read the bar graph.

**1.** What is the most common job?

**2.** How many students do outside cleaning?

**3.** How many students in all have summer jobs?

Read the picture graph.

**4.** Whose class has fewest helpers?

**5.** Are there more helpers in Brent's or Tanya's class?

**6.** Mark helps cook. He moved from Brent's to Tanya's class. How should the graph look now?

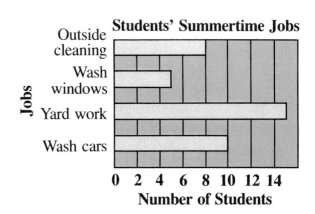

**Students' Summertime Jobs**

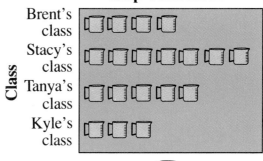

**Help Cook Meals**

Each [🍶] means 1 cooking helper.

**MATH REASONING**

**7.** If each [🍶] means 2 helpers instead of 1, how many students in each class help cook?

**PROBLEM SOLVING**

**8.** Use the picture graph. In Stacy's class, 4 of the helpers are boys. How many are girls?

**9. Write Your Own Problem** Use data from the bar graph to write a problem. Solve.

▶ **ESTIMATION**

To read a bar graph, you estimate where the end of the bar falls. Estimate each of these.

**10.**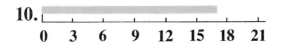

| 0 | 3 | 6 | 9 | 12 | 15 | 18 | 21 |

**11.**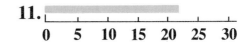

| 0 | 5 | 10 | 15 | 20 | 25 | 30 |

# Collecting and Organizing Data

Before you make graphs, you must collect and organize the data.

**EXPLORE  Make a Tally Chart**
Work in groups. Bradley, Lauren, and Drew collected data on the length of the first names of their classmates. They organized the data in a tally chart.

Make a tally chart for names in your class. Show 5 tallies like this ⁙ so that they are easy to count.

| Letters in First Names | | | |
|---|---|---|---|
| 1 |  | 6 | I |
| 2 |  | 7 | I |
| 3 |  | 8 |  |
| 4 | I | 9 |  |
| 5 |  | 10 |  |

**TALK ABOUT IT**

1. Which tally mark in this chart stands for Lauren's name?

2. What would you do if Christopher were in your class?

The chart shows data for the number of letters in last names. The bar graph shows the same data.

How can the tally chart help you make a graph?

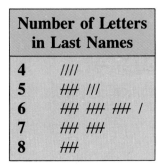

| Number of Letters in Last Names | |
|---|---|
| 4 | //// |
| 5 | ⁙ /// |
| 6 | ⁙ ⁙ ⁙ / |
| 7 | ⁙ ⁙ |
| 8 | ⁙ |

1. How many people had 6 letters in their last names?

2. What color is the bar showing that 10 people had 7 letters in their last names?

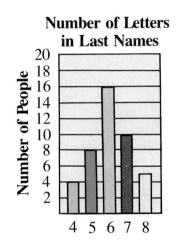

**Number of Letters in Last Names**

160

| Washington School Science Club | | |
|---|---|---|
| Evan | Jean | Gary |
| Julie | Jonathan | Jeremy |
| Alicia | Benjamin | Courtney |
| Ricardo | Allison | Angela |
| Samuel | Susan | Amber |
| Alejandro | Felicia | Danny |
| Caitlin | Amanda | Jacqueline |
| Victoria | Christopher | Patrick |
| Hannah | Leah | Timothy |
| Dana | Tamara | Zachary |
| Kimberly | Mallory | Dawn |
| Kenneth | Russell | Andy |

1. Make a tally chart. Which name length is most common?

2. Make a tally chart that sorts the names by first letter. How many names start with A?

3. Look at the list. Try to find how many names start with C. Then look at your tally chart. Which way shows the data more clearly?

4. Match the tally chart with the correct graph.

**A**

| Name Length | |
|---|---|
| 5 | /// |
| 6 | ### /// |
| 7 | ### |
| 8 | // |

**B**

| Name Length | |
|---|---|
| 5 | //// |
| 6 | ### |
| 7 | ### // |
| 8 | // |

**C**

Name Length

Number of People — 8, 6, 4, 2

Number of Letters — 5 6 7 8

**D**

Name Length

Number of People — 8, 6, 4, 2

Number of Letters — 5 6 7 8

**MATH REASONING**

5. A new girl's name is shorter than Crystal and longer than Marie. Is her name Nichole or Masako?

**PROBLEM SOLVING**

6. **Science Data Bank** Make a tally chart for the number of letters in the names of all the dinosaurs. What do you find? See page 472.

DATA BANK

▶ **USING CRITICAL THINKING Making Predictions**

7. Suppose a new student joined the science club listed above. What is your guess for the most likely length of her name, within 1 letter?

# Making Bar Graphs

The table shows the number of dinosaur bones some scientists discovered in a river bank.

**EXPLORE  Use Graph Paper**

You can make a graph to compare the numbers of bones they found. Begin by finding the greatest number you will need to show.

**TALK ABOUT IT**

1. What numbers will you write along the side? Do you need to write every number?

2. Do you have to estimate to draw the bars? Explain.

3. How does a bar graph help you compare data?

Here is how to make a bar graph.

**A.** Choose numbers for the spaces so that all the data will fit.

**B.** Label the side and bottom.

**C.** Draw the bars.

**D.** Name the graph.

| Kinds of Dinosaur Bones Found in the River Bank | |
|---|---|
| Neck bones | 5 |
| Tail bones | 13 |
| Leg bones | 8 |
| Jaw bones | 3 |
| Back bones | 10 |

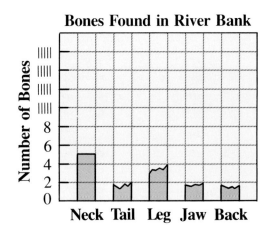

**Bones Found in River Bank**

1. Copy and complete the graph about dinosaur bones.

2. Use data from this table to make a bar graph.

| Favorite Prehistoric Animal Still Living | |
|---|---|
| Turtle | ﬀ //// |
| Lizard | ﬀ ﬀ / |

1. Big River School District put on a science fair. The table shows how many projects students showed at the fair. Copy and complete the graph to show the data in the table.

| Kind of project | Number of projects |
|---|---|
| Stars | 5 |
| Health | 14 |
| Animals | 31 |
| Plants | 20 |
| Dinosaurs | 10 |

2. How many bars did you draw in the science fair graph? Why?

3. When did you have to estimate to draw the bars in the graph?

4. Make a list of 4 or 5 common pets. Have members of your class vote on their favorite. Record their votes. Make a graph of favorite pets for your class. What will you name your graph?

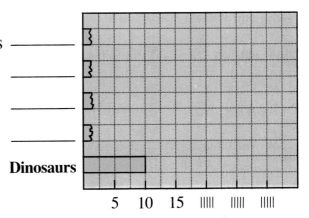

## MATH REASONING

5. The science club had 6 more members than the math club. Sara changed from science to math. What is the difference now?

## PROBLEM SOLVING

6. **Data Bank** Suppose you were making a bar graph of the dinosaur heights. What is the greatest number your graph would show? See page 472.

See page 472.

▶ **USING CRITICAL THINKING  Analyze the Data**

7. Write two or three sentences about the story the graph tells.

**Science Club Members**

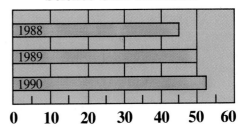

*More Practice, page 517, set B*

# Making Picture Graphs

**EXPLORE** **Study the Data**

A science teacher asked students to name a wild animal they think is most like a pet cat. The numbers in the table tell how many students chose each animal. Alex started a picture graph to show the results.

Do the pictures help you understand the data?

**TALK ABOUT IT**

1. How many more faces should Alex draw in the row for Tiger?

2. What is the greatest number of faces Alex will draw in one row?

3. What name should Alex give the graph?

Here is how to make a picture graph.

**A.** Choose a picture. Tell what it means.

**B.** Find how many pictures you will need for each row.

**C.** Draw the pictures.

**D.** Write the labels. Give the graph a title.

| Wild Animals Most Like Pet Cats | |
|---|---|
| Tiger | 7 |
| Lion | 9 |
| Jaguar | 2 |
| Leopard | 4 |

**How Students Voted**

Each [cat] means 1 vote.

**TRY IT OUT**

1. Copy and complete Alex's graph.

2. Copy and complete the picture graph for this data.

| Wild Animals Most Like Pet Dogs | |
|---|---|
| Wolf | ⁂ / |
| Fox | ⁂ ⁂ |
| Coyote | /// |

Each [wolf] means 1 vote.

164

| Science Club Books | |
|---|---|
| Rain forest animals | 4 |
| Desert plants | 2 |
| Animals from long ago | 8 |
| Space | 6 |
| The human body | 3 |

1. Make a picture graph to show the number of books that the science club has. Give your graph a title. The graph has been started to help you.

2. Choose 5 classmates. Find out about how many hours of TV each one watches on a Saturday. Make a picture graph to show your findings. What picture will you use to show hours?

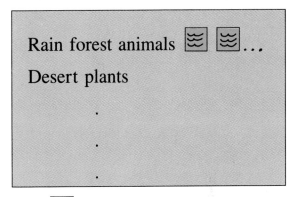

Rain forest animals ⊠ ⊠ ...

Desert plants

Each ⊠ means _____

**MATH REASONING**

3. Suppose you used 🮂 to show 2 TV sets for a picture graph. How many would you draw to show 14 TV sets? HINT: Think of addition doubles.

**PROBLEM SOLVING**

4. Look at the data for books. How many more animal books than plant books are there?

5. **Suppose . . .** What if ⊠ means 2 books? Then how might you show 5 books in a picture graph?

6. Estimate. Do you have more or less than 79¢?

7. You give the clerk 50¢ for an apple that costs 37¢. Count the change.

*More Practice, page 517, set C*

# Exploring Algebra

Troy used this chart to figure out that the number for his name is 78.

**T    R    O    Y**

Troy = 20 + 18 + 15 + 25 = 78

| | | |
|---|---|---|
| **A** = 1 | **B** = 2 | **C** = 3 |
| **D** = 4 | **E** = 5 | **F** = 6 |
| **G** = 7 | **H** = 8 | **I** = 9 |
| **J** = 10 | **K** = 11 | **L** = 12 |
| **M** = 13 | **N** = 14 | **O** = 15 |
| **P** = 16 | **Q** = 17 | **R** = 18 |
| **S** = 19 | **T** = 20 | **U** = 21 |
| **V** = 22 | **W** = 23 | **X** = 24 |
| **Y** = 25 | **Z** = 26 | |

## TALK ABOUT IT

1. Why do you think Troy's number is so large?

2. Could someone have a name with more than 4 letters and still have a number much lower than Troy's number?

3. What number do you get for your name?

4. How does your name number compare to Troy's number?

## TRY IT OUT

Find the number for each word. You may want to use a calculator.

1. Pony    2. Eagle    3. Frog    4. Poodle    5. Multiply

6. Find 3 words with a number greater than 75.

7. What is the least number for a 3-letter word that you can find? What it the greatest?

8. Make up a problem of your own about name numbers. Have a classmate solve it.

# MIDCHAPTER REVIEW/QUIZ

Students answered three questions about their pets. The picture graph and the bar graph show the results for two of the questions.

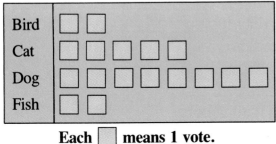

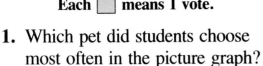

Each ☐ means 1 vote.

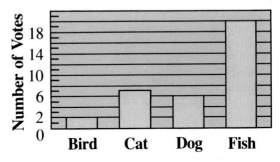

**1.** Which pet did students choose most often in the picture graph? in the bar graph?

**2.** Which tally chart matches the picture graph? the bar graph?

Two of these tally charts match the graphs.

| A | **Which would you rather be?** | B | **How many do you have at home?** | C | **Which makes you afraid?** |
|---|---|---|---|---|---|
| | Bird    // | | Bird    // | | Bird    // |
| | Cat    ##  | | Cat    ## // | | Cat    ## |
| | Dog    ## / | | Dog    ## / | | Dog    ## /// |
| | Fish    /// | | Fish    ## ## ## ## | | Fish    // |

**3.** Make a picture graph for the data that is not graphed above.

**4.** Give a title to each graph.

## PROBLEM SOLVING

In a story, a princess had to choose the magic door. Tilly's class made these guesses of door A, B, C, or D.

| | | | | | |
|---|---|---|---|---|---|
| C | A | A | C | D | D |
| C | B | B | B | C | C |
| D | A | B | B | B | C |
| A | C | D | C | B | B |
| B | C | C | B | D | B |
| C | A | B | D | D | B |

**5.** Make a tally chart of the results.

**6.** Which letter appears most often?    Least often?

**7.** Make a bar graph of the results.

167

# Problem Solving
## Look for a Pattern

| UNDERSTAND |
| FIND DATA |
| PLAN |
| ESTIMATE |
| SOLVE |
| CHECK |

### LEARN ABOUT IT

You may need to discover a pattern in data in order to solve some problems. This problem solving strategy is called Look for a Pattern.

> The lion tamer, Petey Doyle, started with 2 lions his first year at the circus. The second year, he had 4 lions, and the third year he had 6. How many lions were in his act the sixth year?

I'll start a table using the data in the problem.

| Year | 1 | 2 | 3 |
|---|---|---|---|
| Lions | 2 | 4 | 6 |

There's a pattern in the table! Every year there are 2 more lions than the year before.

| Year | 1 | 2 | 3 | 4 |
|---|---|---|---|---|
| Lions | 2 | 4 | 6 | 8 |

Now I can complete the table.

| Year | 1 | 2 | 3 | 4 | 5 | 6 |
|---|---|---|---|---|---|---|
| Lions | 2 | 4 | 6 | 8 | 10 | 12 |

There were 12 lions in the act in the sixth year.

### TRY IT OUT

After going to the circus, Marja decided to practice riding her unicycle. The first day she rode past 5 houses. The second day she rode past 6 houses. The third day she rode past 7 houses. How many houses did she ride past on the eighth day?

- How many more houses did Marja ride past on the second day than on the first day?

- Copy and complete the table.

| Day | 1 | 2 | 3 |
|---|---|---|---|
| Houses | 5 | 6 | |

Look for a pattern to help you solve these problems.

1. The Jokicki family does teeterboard tricks. The first trick ends with 3 people standing on each other's shoulders. The second trick ends with 4 people. The third trick ends with 5 people. How many people are standing on shoulders at the end of the seventh trick?

2. After seeing the circus, Julia started practicing her flips. The first day she practiced 5 minutes. The second day she practiced 10 minutes. The third day, Julia practiced 15 minutes. How long did she practice the ninth day?

**MIXED PRACTICE**

Solve. Choose a strategy from the strategies list or use other strategies that you know.

3. The Ringling Brothers Circus started in 1884 as a tiny wagon circus. How old is the circus?

4. The Bengal tiger in the circus weighs 510 pounds. His trainer, Ursula, weighs 104 pounds. About how much more does the tiger weigh?

5. Bigtop Circus has 31 clowns. They each went to clown school for 8 weeks. 19 are men. How many are women?

6. Kai brought $12 to the circus. He bought a flashlight and a patch. How much money did he have left?

| Some Strategies |
|---|
| Act It Out |
| Use Objects |
| Choose an Operation |
| Draw a Picture |
| Make an Organized List |
| Guess and Check |
| Make a Table |
| Look for a Pattern |

7. Mervin the Magnificent balances on one hand on blocks on top of a board on top of a ball. For the first trick, he uses 1 block. In the second trick he uses 3 blocks. In the third trick he uses 5 blocks. How many blocks does Mervin use for the fifth trick?

| Souvenirs - Price List | | | |
|---|---|---|---|
| Program | $2 | Flashlight | $4 |
| Badge | $5 | Poster | $6 |
| Patch | $2 | T-shirt | $8 |

*More Practice, page 518, set A*

# Analyzing Data

## EXPLORE  Study the Data

The list shows data from a survey. Can you tell which bedtime parents and teachers chose most often?

Look at the table under the list. Does it help you compare how parents and teachers answered the question?

| What bedtime is best for third graders? | | |
|---|---|---|
| Parents | Children | Teachers |
| 8:30 | 9:00 | 9:00 |
| 9:00 | 9:00 | 8:30 |
| 8:30 | 9:30 | 9:00 |
| 8:30 | 10:00 | 9:00 |
| 9:00 | 9:00 | 9:00 |
| 9:00 | 9:30 | 8:30 |
| 8:30 | 9:00 | 8:00 |
| 8:30 | 9:00 | 8:30 |
| 9:00 | 9:30 | 9:00 |
| 8:30 | 8:30 | 9:30 |
| 8:30 | 10:00 | 9:00 |
| 8:00 | 9:00 | 9:00 |

## TALK ABOUT IT

1. How do you think the table helps you analyze the data?

2. Who sets an earlier bedtime, parents or teachers?

You can use data to compare answers and draw conclusions. Organizing the data into tables helps you do this.

| Best Bedtime | | | |
|---|---|---|---|
| | 8:00 | 8:30 | 9:00 | 9:30 |
| Parents | 1 | 7 | 4 | 0 |
| Teachers | 1 | 3 | 7 | 1 |

Here is how to make a table.

- Draw the table.
- Label the parts.
- Put data into the table.
- Choose a title.

Copy and complete this table for children and teachers. Who sets a later bedtime?

| Children | 8:00 | 8:30 | ___ | ___ |
|---|---|---|---|---|
| ___ | | | | |

Use the data on page 170.

**Parents** 8:00 8:30 ___ ___

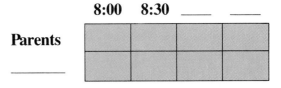

1. Copy and complete this table for parents and children. What do you conclude about their bedtime choices?

2. Copy and complete the bar graph that shows how parents feel about bedtimes. What can you conclude about the parents' choices?

3. Make a table to compare the answers from children, parents and teachers. Draw at least two conclusions from your table.

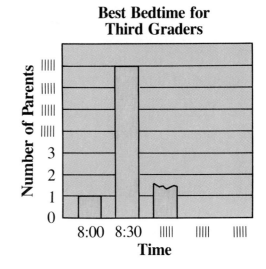

**Best Bedtime for Third Graders**

4. Think about the table on page 170. Suppose 120 parents and 120 teachers had voted instead of 12 of each. Estimate what the numbers would be.

**PROBLEM SOLVING**

5. Suppose that grandparents voted this way. 4 voted 8:00, 7 voted 8:30, and 3 voted 9:00. How many grandparents voted?

6. How much later was the most popular bedtime set by children than the most popular bedtime set by parents?

▶ **CALCULATOR**

Use your calculator to help you fill in numbers for the bottom of these graphs.

7.

8.

# Probability
## Fair and Unfair Games

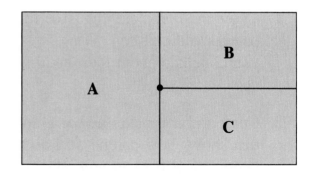

**LEARN ABOUT IT**

**EXPLORE** **Play a Game**

Work in groups. Use a sheet of paper or cardboard to make a game board like this one. Play the game in groups of three.

- Decide on a way to give each player a different letter, A, B, or C.

- Make a tally table. Score a point each time a counter is completely inside the space with your letter.

- Drop 10 counters or pebbles. Aim for the center dot.

- Each person gets a turn. High score wins.

**TALK ABOUT IT**

1. Do you think this is a fair game? Why?

2. If you could choose your letter, which one would you pick?

3. If A wins, who is more likely to be second, B or C? Explain.

A game is fair when everyone has an equal chance to win. You can use tally tables to help you decide whether or not a game is fair.

**TRY IT OUT**

1. Is this a fair game?

2. Predict who will have the highest score.

3. Predict who will have the lowest score.

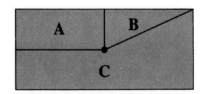

172

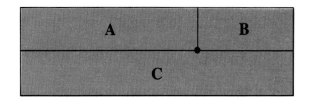

1. Which letter on this board would you like to have?

2. Which would you least want?

Write *fair* or *unfair* for each game board.

3. 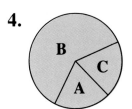  4. 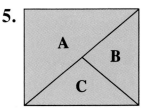  5.   6.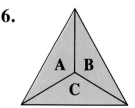

7. These are tallies for two different game boards. Which game do you think is fair? Why?

| A | ＃＃ ＃＃ // |
| B | ＃＃ ＃＃ ＃＃ ＃＃ ＃＃ ＃＃ // |
| C | ＃＃ ＃＃ ＃＃ /// |

**Brad's Game**

| A | ＃＃ ＃＃ //// |
| B | ＃＃ ＃＃ ＃＃ / |
| C | ＃＃ ＃＃ /// |

**Bonnie's Game**

**MATH REASONING**

8. You and a friend use this game board. You flip a coin to see who gets A. What is fair and what is unfair about the coin flip and the game?

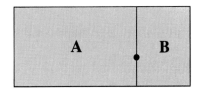

**PROBLEM SOLVING**

9. Look at the tallies for Brad's and Bonnie's games. What is the difference in the number of turns? How many turns were there in all?

▶ **COMMUNICATION** Write to Learn

10. Explain why we use tallies this way ＃＃ ＃＃ ＃＃ instead of this way //////////////.

# Using Critical Thinking
## Making Predictions from Line Graphs

**EXPLORE** **Study the Data**

Nathan's class planted a grain of corn. They made a bar graph to show how it grew. Nathan connected the tops of the bars with a red line. He said, "The red line will help us predict how tall the corn will be the fourth week." Do you agree?

**TALK ABOUT IT**

1. Explain how you might use the red line to help you predict the height for the fourth week.

2. What is your prediction?

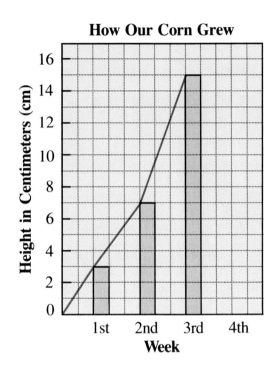

**How Our Corn Grew**

The red line on the graph about corn is an example of a **line graph.** You can read a line graph just as you would read a bar graph. Think of the points as tops of the bars.

Notice if the changes in the line go up or down. Sometimes this helps you to predict where the next point will be.

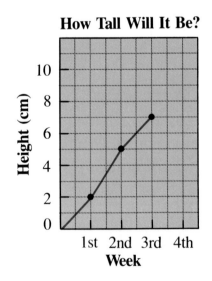

**How Tall Will It Be?**

**TRY IT OUT**

1. This graph shows how a soy bean plant grew. How does its growth compare to the corn?

2. What do you predict for the soy bean's height in the fourth week?

**174**

These plants grow at different rates. Predict the height of each plant in the fourth week

**1.**

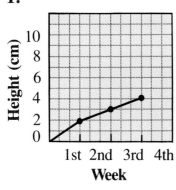

**2.**

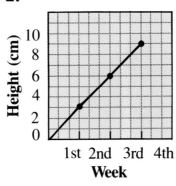

**3.**

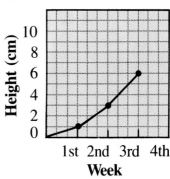

**MATH REASONING**

**Ken's Cup of Popcorn**

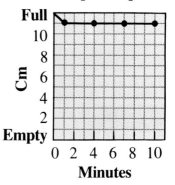

4. Write a sentence to tell what might have happened to the person eating this cup of popcorn.

**PROBLEM SOLVING**

5. Julie's corn plant is taller than Keith's. April's plant is shorter than Keith's. Hank's plant is taller than Keith's but shorter than Julie's. List the plants in order from tallest to shortest.

**MIXED REVIEW**

Add or subtract. Write answers only.

**6.** 70 + 60    **7.** 800 − 300    **8.** 1,300 − 400    **9.** 500 + 700    **10.** 120 − 50

Round to the nearest ten to estimate the sum or difference.

**11.** 37 + 26    **12.** 154 + 78    **13.** 84 − 72    **14.** 166 + 97    **15.** 385 − 22

**16.** 234 − 29    **17.** 94 + 51    **18.** 816 + 25    **19.** 78 − 67    **20.** 35 + 51

# Problem Solving
## Using a Calculator

| UNDERSTAND |
| FIND DATA |
| PLAN |
| ESTIMATE |
| SOLVE |
| CHECK |

### LEARN ABOUT IT

The Wheels Galore Company wants to make 7,500 skateboards by the end of the year. How many more do they have to make to meet the goal?

**Skateboards Made**

| January | 1,079 |
|---------|-------|
| February | 894 |
| March | 1,186 |
| April | 749 |

Jean and Peter both used calculators, but they solved the problem in different ways.

**Jean's Work**

Jean added first and then subtracted.

ON/AC 1079 + 894 + 1186 + 749 = *3908 skateboards made so far.*

Then ON/AC 7500 − 3908 =

They need to make 3592 more skateboards.

**Peter's Work**

Peter just used subtraction.

ON/AC 7500 − 1079 − 894 − 1186 − 749 =

*3592*

The company has to make 3592 more skateboards.

### TRY IT OUT

Use a calculator.

1. Wheels Galore made 2,083 scooters in January and 1,809 in February. How many more scooters should they make to meet a goal of 9,500?

2. Suppose you have $45.00. If you buy a skateboard for $23.95, elbow pads for $7.99, and knee pads for $5.49, how much money would you have left?

3. A truck delivered skateboards to stores in nearby states. How many miles did the truck travel in the first three months of the year? Use the bar graph.

**Wheels Galore Trucks**

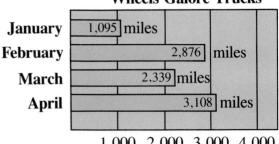

Use the table for Exercises 1 to 3.

| Sales at Wheels and Deals Shop | | |
|---|---|---|
| Year | Bikes | Skateboards |
| 1985 | 1,123 | 362 |
| 1986 | 1,256 | 652 |
| 1987 | 1,347 | 578 |
| 1988 | 1,312 | 492 |

1. How many more skateboards did Wheels and Deals sell in 1987 than in 1985?

2. The goal was to sell 5,000 bikes in 4 years at Wheels and Deals. Did they make their goal?

3. How many more bikes than skateboards did Wheels and Deals sell?

4. José is making a skateboard ramp. Plywood sheets cost $6.98 each. If he needs 2 sheets of plywood, how much will it cost all together?

5. Wheels and Deals put on a bike safety fair last spring. There were 53 riders in the Slalom contest and 15 in Freestyle Riding. Riders came from 12 different schools. How many riders were in the 2 contests?

6. For every 3 skateboard stickers Jean buys, she gets 2 free. How many free stickers will Jean have if she buys 12 stickers?

7. The most expensive mountain bike at Hamada's Bike Shop is $780. The least expensive is $209. What is the difference in price?

8. Fancy scooters cost $170. Jan has saved $90 by helping with her sister's paper route. How much more money must she save?

9. **Talk About Your Solution**
   Jimmy started with $25.00. He needed to repair his bicycle. The flat tire cost $9.89. Repairing the brakes cost $8.95. A new kickstand cost $4.00. How much did he have left?

   ■ Explain your solution to a classmate.
   ■ Compare your solutions.

*More Practice, page 518, set D*

# Data Collection and Analysis
## Group Decision Making

UNDERSTAND
FIND DATA
PLAN
ESTIMATE
SOLVE
CHECK

**Doing a Questionnaire**

**Group Skill:**
Check for Understanding

Computers can be used to keep records, to make graphs, to publish books, or to play games. They can be helpful at work, at home, or at school. Who uses computers? Do people like to use computers? You can write down questions to ask people. This is called a questionnaire.

### Collecting Data

1. Work with your group to make a questionnaire about using computers. You can make up questions like these.

   How often do you use a computer?

   ☐ never ☐ a little ☐ a lot

   Do you like using a computer?

   ☐ yes ☐ no ☐ I don't know

2. Make at least 10 copies of the questionnaire. Have 10 or more people each mark answers.

3. Make a table to record the information from your questionnaires. Make a tally for each person's answer.

| Question 1 | Question 2 |
| --- | --- |
| never | yes |
| a little | no |
| a lot | I don't know |

4. How many people answered your questionnaire?

5. How many people use the computer a lot?

6. Did more people like the computer than did not like it?

7. Write two or three sentences to tell what you found out.

179

# WRAP UP

## Choose a Graph

Would you use a bar graph or a picture graph to show the data? Explain your choice.

1. You want to show how many library books your classmates have read in one year.

2. You want to show how fast each of 6 friends can run a mile.

3. You want to show the favorite school subjects of 100 third graders.

4. You want to show the temperature every hour during the school day.

## Sometimes, Always, Never

Decide which word should go in the blank, *sometimes, always,* or *never.* Explain your choices.

5. Bar graphs _____ use a number scale to help the reader understand them.

6. The name of a graph _____ may be omitted.

7. If the number scale on a graph is labeled with even numbers 0, 2, 4, 6, 8, then the number halfway between 6 and 8 is _____ 7.

8. When each player has the same chance to win, the game is _____ fair.

## Project

Work in groups. On the playground, take turns trying to bounce a ball or jump rope as many times as you can without missing. Use a tally chart to record the results for your group.

Make a bar graph showing the data in the tally chart. Skip count by fives or tens to choose numbers to label the side of the graph. Write a good title for your graph.

Write two questions about the data in the graph.

# Chapter Review/Test

## Part 1  Understanding

1. How do you find how much a bar in a bar graph stands for?

2. Explain what a picture graph tells you.

3. Is this spinner fair? Explain.

4. | A = 7 | B = 4 | T = 2 | U = 3 |
   |---|---|---|---|

   Change the letters to these number values to decide if TUB or BAT gives a greater sum. How did you decide?

## Part 2  Skills

5. Make a tally chart to find how many names in the box start with the letters A, B, C, and D.

6. Use your tally chart. Did more names start with A or B?

| Duran | Ahn | Beck |
|---|---|---|
| Barr | Boyd | Dakin |
| Conti | Daly | Borja |
| Allen | Dixon | Davis |
| Acosta | Dal | Chao |

Use the table.

7. What subject do these 4th graders like best?

8. Make a bar graph for math as a favorite subject for 3rd, 4th, and 5th graders.

9. Make a picture graph for 4th grade favorite subjects. Use a stick figure to show 5 students.

| Favorite Subjects | | | |
|---|---|---|---|
| Grade | 3rd | 4th | 5th |
| Math | 10 | 5 | 15 |
| Reading | 10 | 15 | 9 |
| Art | 10 | 10 | 6 |

## Part 3  Applications

10. Naomi swam 2,358 meters on Monday and 1,998 meters on Tuesday. How many meters must she swim on Wednesday to swim a total of 6,280 meters in 3 days?

11. **Challenge**  Carl rowed 2,555 meters one day, 3,110 meters the next, and 3,665 meters the next. At this rate, in how many more days will he row 6,440 meters in one day?

# ENRICHMENT
## A Probability Game

Try this game with a partner. You will need a copy of the gameboard, a coin, and two markers.

- Put both markers on START.
- Take turns tossing the coin to find which direction to move your marker up the board.
- The player with the highest score at the top wins.

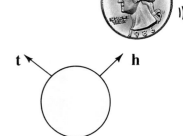

Toss heads,
move right.

Toss tails,
move left.

Look carefully at the board. Then answer the questions. After you play the game with your partner, you may want to revise your answers.

**1.** Why do you think the outer scores are greater?

**2.** Is it possible to score a 20? Is it very likely?

**3.** Estimate your total score for 10 games. Check your estimate by playing the game.

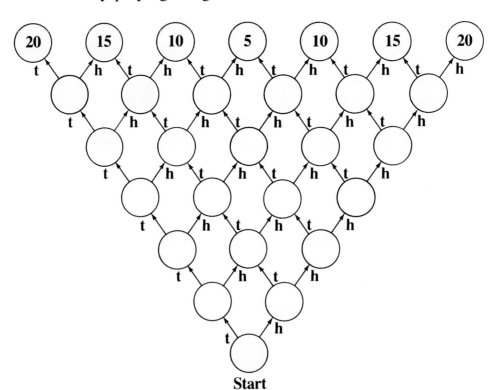

# CUMULATIVE REVIEW

1. Which is lunchtime?

   **A.** 3:00 a.m.  **B.** 12 noon

   **C.** 12 midnight  **D.** 8:00 p.m.

2. You have 3 quarters, 4 dimes, and 3 nickels. Which book costs too much for you to buy?

   **A.**   **B.**

   **C.**   **D.**

3. You must trade 10 ones for 1 ten when you add 13 and 29. How many ones and tens do you have after the trade?

   **A.** 3 tens, 2 ones
   **B.** 4 tens, 2 ones
   **C.** 3 tens, 3 ones
   **D.** 4 tens, 3 ones

4. Which two classes together have 52 students?

   | Class | Students |
   | --- | --- |
   | Math | 29 |
   | Gym | 28 |
   | Music | 23 |
   | Art | 14 |

   **A.** gym, art  **B.** math, music

   **C.** math, art  **D.** gym, music

5. 1,700 − 900

   **A.** 1,000  **B.** 1,200

   **C.** 900  **D.** 800

6. Estimate $3.49 − $1.72 to the nearest dollar.

   **A.** $1.00  **B.** $2.00

   **C.** $2.50  **D.** $3.00

7. Show the results of trading 1 ten for 10 ones in 65.

   **A.** 6 tens, 5 ones
   **B.** 5 tens, 15 ones
   **C.** 6 tens, 15 ones
   **D.** 5 tens, 5 ones

8. Mark can draw a picture with crayon or pen on either paper or plastic. Which items would appear on a list showing the ways he can make the picture?

   **A.** paper-crayon
   **B.** paper-picture
   **C.** pen-crayon
   **D.** paper-plastic

9. It cost Leona $17 for a bird cage and $4 for a bird feeder. How much change did she get back from $25?

   **A.** $1  **B.** $2

   **C.** $4  **D.** $5

# 7

# CUSTOMARY MEASUREMENT

**M**ATH AND FINE ARTS

DATA BANK

Use the Fine Arts Data Bank on page 469 to answer the questions.

1 The complete Statue of Liberty monument includes the statue, the pedestal, and the foundation. Which part is the tallest? The shortest?

**2** Visitors can climb steps inside the pedestal and the statue to the crown. Which has more steps, the statue or the pedestal? How many more?

**3** People can stand in the statue's crown and on the balcony around the torch. How many more people can fit in the crown than on the balcony?

**4** Use Critical Thinking Imagine that you are standing on the pedestal, next to the statue. How tall is the statue in comparison to you?

185

# Measuring Length in Inches

**LEARN ABOUT IT**

### EXPLORE  Use a Ruler

You can use nonstandard units, like a key, to measure length. You can also use standard units, like the **inch.**

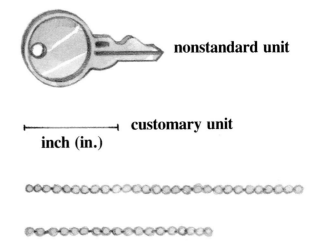

nonstandard unit

customary unit

inch (in.)

Estimate the length of each chain in keys and in inches. Measure the chains with a ruler to check your estimates.

### TALK ABOUT IT

**1.** Which was easier, measuring in keys or inches?

**2.** Suppose you estimated that a chain is 4 keys long. What is your estimate of its length in inches?

You can use a ruler to measure to the nearest inch.

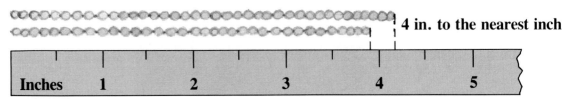

4 in. to the nearest inch

Inches   1        2        3        4        5

**TRY IT OUT**

Estimate each length to the nearest inch. Check your estimates by measuring to the nearest inch.

**1.**

**2.**

**3.**

**4.**

186

Estimate each bracelet's length to the nearest inch.
Check your estimates by measuring to the nearest inch.

**1.**

**2.**    **3.**

**4.**

APPLY

**MATH REASONING**

**5.** Should your estimate of the number of keys in a length be greater or less than the number of inches? Why?

**PROBLEM SOLVING**

**6.** Wilma is making a flag 8 inches long. She has one piece of felt that is a little over 3 in. long, and another that is 4 in. long. Does she have enough felt?

**7. Data Hunt** Measure two objects in your desk to the nearest inch. Write each object's name and length.

▶ **COMMUNICATION** **Write to Learn**

Many different words are used to describe measurement comparisons. Name an object in your classroom that makes these sentences true.

**8.** My desk is **shorter** than ____.   **9.** My pencil is **thinner** than ____.

**10.** My math book is **thicker** than ____.   **11.** My chair is **taller** than ____.

*More Practice, page 519, set A*

# Estimating Length
## Using a Benchmark

**LEARN ABOUT IT**

**EXPLORE** **Study the Information**

If you know the length of an object, you can use it as a **benchmark** to estimate the length, height, and width of other objects.

About 3 in. ⟶
long

**TALK ABOUT IT**

1. Why is the crayon a good object to use as a benchmark?

2. Explain how you can use the crayon as a benchmark to estimate the length of the pencil.

About ? in. ⟶
long

To use an object as a benchmark, you must know its length to the nearest inch. Then compare the benchmark to the object you are estimating.

The eraser is a little shorter than the crayon. It is about 2 inches long.

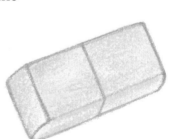

The pencil is not quite double the length of the crayon. It is about 5 in. long.

**TRY IT OUT**

Measure your pencil's length to the nearest inch.
Use it as a benchmark to estimate these measurements.
Check by measuring to the nearest inch.

1. Length and width of your desk top

2. Length and width of your chair seat

188

Use your pencil as a benchmark to make estimates
of your math book cover in inches. Then check
your estimates by measuring to the nearest inch.

**1.** Math book
width

**2.** Math book
length

**3.** Math book
diagonal

**4.** Math book
thickness

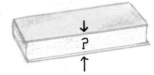

### MATH REASONING

**5.** Explain the steps for using your hand span as
a benchmark.

### PROBLEM SOLVING

**6.** Imagine using a crayon as a benchmark to
estimate the length of your pet snake. The
snake is about 3 crayons long. What is your
estimate of the snake's length?

Estimate if the difference is more or less than
$5.00. Then find the difference.

**7.** $14.88 − $9.59　　**8.** $8.25 − $3.32　　**9.** $12.00 − $6.95

Find each sum and difference.

**10.** $27.86 + $3.24　　**11.** $18.05 + $15.97　　**12.** $13.54 − $9.25

**13.** $4.08 − $3.75　　**14.** $32.57 − $19.48　　**15.** $46.35 + $18.95

# Foot, Yard, and Mile

The **foot** and the **yard** are customary
units used to measure length.

1 foot (ft) = 12 inches

1 yard (yd) = 36 inches, or 3 feet

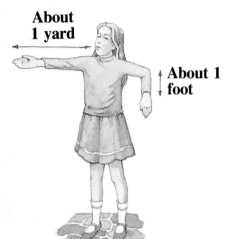

About
1 yard

About 1
foot

**EXPLORE** **Use a Ruler**

Work in groups. You can use benchmarks to
estimate lengths in feet and yards. Use the
benchmarks pictured here to list objects about 1 ft
long and about 1 yd long. Then measure each
object to the nearest inch to check your estimate.

**TALK ABOUT IT**

1. Is it easier to estimate and measure curved
   objects or straight objects? Explain why.

2. Is the distance around your head closer to
   a foot or a yard? How do you know?

Another customary unit of length is the **mile.**
1 mile (mi) = 5,280 feet or 1,760 yards.
A mile is a very long unit. It takes about 20
minutes to walk 1 mile.

Write <u>shorter</u> or <u>longer</u> to complete the sentences.

1. 15 inches is ____ than 1 foot.

2. 1,520 yards is ____ than 1 mile.

3. 28 inches is ____ than 1 yard.

4. 2,000 yards is ____ than 1 mile.

190

1. Which object is shorter than 1 ft?

2. Which object is 1 ft tall?

3. Which objects are shorter than 1 yd?

4. Which object is 1 yard tall?

**Table 12 in.**

**Lamp 36 in.**

**Clock 10 in.**

**APPLY**

**MATH REASONING** Is the distance more, less, or the same as 1 mile?

5. from Mike's house to the school

6. from Mike's house to the post office

7. from the post office to the school

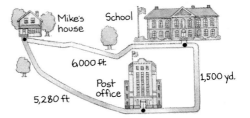

**PROBLEM SOLVING**

8. The Statue of Liberty's nose is 54 in. long. Its mouth is 36 in. wide. Which measurement is more than 1 yd?

9. **Fine Arts Data Bank** Which measurement of the Statue of Liberty is closest to 1 ft? Which is closest to 8 ft? See page 469.

▶ **ESTIMATION**

How many of these small squares will cover each figure? Choose the number that you think is the best estimate.

10.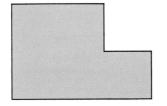

**A** 5 **B** 10 **C** 15

11.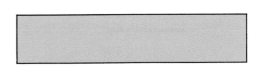

**A** 5 **B** 10 **C** 15

# Perimeter

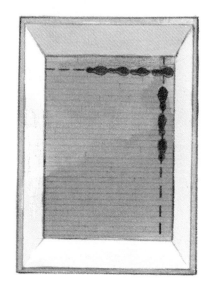

### EXPLORE  Use Objects

Work in groups. Find and record the length and width of your classroom using your shoes as units. What is the distance all the way around the room in shoe units? Use the two measurements you recorded to find out.

### TALK ABOUT IT

1. How did you use your measurements to figure the distance around the room?

2. What other units could you use to measure your classroom?

The distance around an object or figure is its **perimeter.** One way to find the perimeter is to measure the length of each side and then add the lengths to find the total.

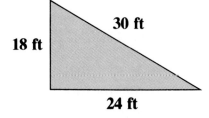

To find perimeter on a calculator, use this key code.

$\boxed{\text{ON/AC}}$ length of side $\boxed{+}$ length of side
$\boxed{+}$ length of side $\boxed{=}$

$30 + 18 + 24 = 72$
**The perimeter of the triangle is 72 feet.**

### TRY IT OUT

Use a calculator to find the perimeter of each figure.

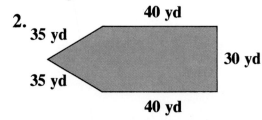

1.
25 in.
15 in.   15 in.
25 in.

2.
40 yd
35 yd
35 yd
30 yd
40 yd

192

Use a calculator to find the perimeters.

**1.**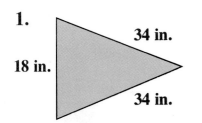

34 in.

18 in.

34 in.

**2.**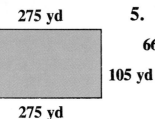

21 yd

14 yd       14 yd

21 yd

**3.**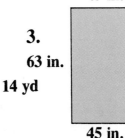

45 in.

63 in.       63 in.

45 in.

**4.**

275 yd

105 yd

275 yd

**5.**

660 ft

660 ft       660 ft

660 ft

105 yd

**6.**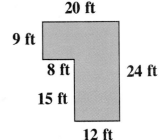

20 ft

9 ft

8 ft       24 ft

15 ft

12 ft

**MATH REASONING**

**7.** How can you use mental math to find the perimeter of the stop sign in feet? What is its perimeter?

12 in.

12 in.       12 in.

12 in.   STOP   12 in.

12 in.       12 in.

12 in.

**PROBLEM SOLVING**

**8.** The Statue of Liberty stands on a square pedestal. Each side of the pedestal is 65 ft long. What is its perimeter?

**DATA BANK**

**9. Fine Arts Data Bank** The Statue of Liberty has a square foundation. What is its perimeter? See page 469.

▶ **ALGEBRA**

Use a calculator. Find the unknown length in each figure.

**10.** The perimeter is 150 yd.

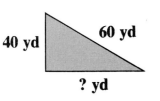

40 yd       60 yd

? yd

**11.** The perimeter is 1,200 yd.

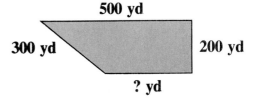

500 yd

300 yd       200 yd

? yd

*More Practice, page 519, set D*

**193**

# Problem Solving
## Use Logical Reasoning

| UNDERSTAND |
|---|
| FIND DATA |
| PLAN |
| ESTIMATE |
| SOLVE |
| CHECK |

**LEARN ABOUT IT**

To solve some problems, you cannot just quickly add or subtract. You need to organize your work in a different way. This problem solving strategy is called **Use Logical Reasoning.**

> Mark and Jean visit their grandparents every summer. Grandma told them her age this way, "I am between 70 and 80 years old. My age is less than 73. I have lived an odd number of years." How old is she?

I'll list the numbers between 70 and 80. )  70 71 72 73 74 75 76 77 78 79 80

I'll cross out all numbers that are not less than 73. )  70 71 72 7̶3̶ 7̶4̶ 7̶5̶ 7̶6̶ 7̶7̶ 7̶8̶ 7̶9̶ 8̶0̶

I'll cross out any even numbers left. )  7̶0̶ 71 7̶2̶ 7̶3̶ 7̶4̶ 7̶5̶ 7̶6̶ 7̶7̶ 7̶8̶ 7̶9̶ 8̶0̶

Grandma is 71 years old.

**TRY IT OUT**

Grandpa told Jean and Mark how many trout he and Grandma caught on their last fishing trip. "We caught less than 25 and more than $12 + 7$. The number of trout we caught is an even number. It is more than 22." How many trout did they catch?

■ The number of trout they caught is between what two numbers?

■ Is the number an even or an odd number?

■ Copy the numbers here. Use logical reasoning to solve the problem.

19  20  21  22
23  24  25

Use logical reasoning to solve each problem.

1. Grandpa gave Jean a hint about what her bedtime would be. "It is later than 7:00 and earlier than 11:00. The hour is not an even number." When is bedtime?

2. Grandma told how many tomatoes she had grown. "I grew an even number between 40 and 50. I grew more than 45, but I did not grow 46." How many tomatoes did she grow?

## MIXED PRACTICE

Solve. Choose a strategy from the list or other strategies you know.

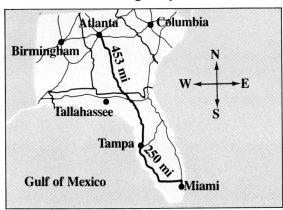

| Some Strategies |
|:---:|
| Act It Out |
| Choose an Operation |
| Make an Organized List |
| Make a Table |
| Use Objects |
| Draw a Picture |
| Guess and Check |
| Look for a Pattern |
| Use Logical Reasoning |

3. Jean and Mark's parents have to drive from Atlanta to Miami to get to Grandma and Grandpa's house. How far is it?

4. Jean and Mark's uncle lives in Tallahassee. About how far is that from Atlanta?

5. Since Grandpa is over 65 years old, he often can pay less for tickets. Grandpa only paid $21.45 for a $24.00 book of tickets to the swimming pool. How much did he save?

6. Mark took $12.85 to his grandparents' house. He spent $9.95 on a Miami Dolphins poster. Grandpa gave him $3.50 for raking leaves. Then how much money did Mark have?

7. Jean told how many hamsters she has. "I have more than 10 and less than 20. I have an odd number of hamsters. I have less than 9 + 6 and I do not have 11. How many do I have?"

*More Practice, page 519, set E*

# Problem Solving
## Deciding When to Estimate

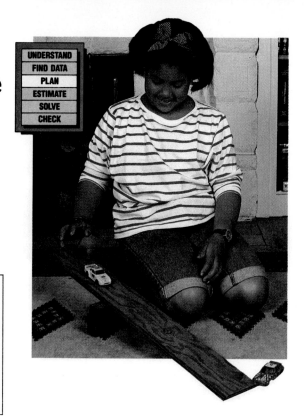

UNDERSTAND
FIND DATA
PLAN
ESTIMATE
SOLVE
CHECK

### LEARN ABOUT IT

To solve some problems, you need to decide if you need an estimate or an exact measurement.

> Marva wants to make a new shelf for her bookcase. She also wants to make a ramp for her model cars. Does Marva need to estimate or measure the length of the boards?

The board for the shelf needs to be cut an exact size so it fits into the bookcase.

Marva needs to measure the board for her bookcase.

The board for the car ramp does not need to be cut an exact size. Almost any length will work as a ramp.

Marva needs to estimate the length of the board for her car ramp.

### TRY IT OUT

1. Mr. Adams is replacing the broken glass in his window. Should he estimate or measure the window's size?

2. Rocio is cutting string to tie up balloons for a party. Should she estimate or measure the length of each string?

3. Tiko is making a picture frame. Should he estimate or measure the size of the picture?

4. You are shaping popcorn balls. Should you estimate or measure each amount of popcorn?

# MIDCHAPTER REVIEW/QUIZ

1. Is a 3-in. or a 30-in. benchmark better to use to measure a door? Why?

Measure to the nearest inch.

2. 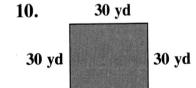    3. ▬▬▬▬▬

Choose, <u>inches</u>, <u>feet</u>, <u>yards</u>, or <u>miles</u> to make the sentence reasonable.

4. The ceiling is about 3 _____ high.

5. A bicycle tire is about 2 _____ across.

6. In 15 minutes you could ride your bike about 2 _____.

7. In 1 hour a car could go about 55 _____.

8. A chair seat is about 18 _____ high.

9. An airplane flying at 10,000 _____ is about 2 _____ high.

Find the perimeter of each.

10.
30 yd
30 yd    30 yd
30 yd

11.
12 in.
8 in.    8 in.
12 in.

12.
15 ft
7 ft    8 ft
12 ft    10 ft

## PROBLEM SOLVING

13. The football team ran around the outer edge of this field 5 times. Did each player run more or less than 1 mile?

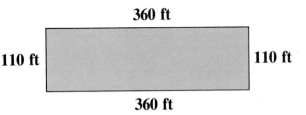

360 ft
110 ft    110 ft
360 ft

14. After Yoko and her 3 friends found seats on the bus, there were 2 empty seats left. If the bus holds 24 people, how many people were on the bus before Yoko and her friends got on?

15. Tony grew 2 inches during second grade. If he grows 3 inches during third grade he will be 54 inches tall. How tall was Tony at the beginning of second grade?

197

# Weight

The **pound** (lb) and the **ounce** (oz) are customary units used to measure weight.

**EXPLORE** **Use a Scale**

Work in groups. Choose five objects that you think each weigh about 1 pound. Check your estimate by measuring each object's weight in ounces.

**TALK ABOUT IT**

1. How did you estimate which objects weighed about 1 pound?

2. Did any of your objects weigh more than 1 pound? How do you know?

Thinking of objects that weigh 1 oz or 1 lb can help you estimate the weight of other objects.

A spoon
weighs about 1 ounce (oz).

A soccer ball
weighs about 1 pound (lb).
1 lb = 16 oz

**TRY IT OUT**

Estimate whether these objects weigh more or less than 1 pound.

1. Letter

2. Bananas

Choose the best estimate of weight.

**1.**

6 oz or 6 lb

**2.**

12 oz or 12 lb

**3.**

10 oz or 10 lb

**4.**

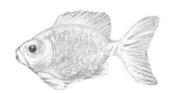

1 oz or 1 lb

**APPLY**

**MATH REASONING** Decide if the weight is more or less than 2 lb.

**5.** 36 oz          **6.** 40 oz          **7.** 30 oz          **8.** 24 oz

**PROBLEM SOLVING**

**9.** Mai needs to buy 24 oz of rice. The rice comes in bags that weigh 1 lb 12 oz. Is one bag of rice enough?

**10.** Mrs. Garcia bought 10 oz of peas for $1.15. How much would it cost if she bought 20 oz of peas?

**MIXED REVIEW**

Find the sums and differences.

**11.** $461 + 284 + 37$          **12.** $32 + 75 + 58 + 24$          **13.** $64 + 9 + 158 + 71$

**14.**
$$\begin{array}{r} 38 \\ -17 \\ \hline \end{array}$$

**15.**
$$\begin{array}{r} 275 \\ -\ 92 \\ \hline \end{array}$$

**16.**
$$\begin{array}{r} 324 \\ -175 \\ \hline \end{array}$$

**17.**
$$\begin{array}{r} 60 \\ -34 \\ \hline \end{array}$$

**18.**
$$\begin{array}{r} 407 \\ -128 \\ \hline \end{array}$$

**19.**
$$\begin{array}{r} 821 \\ -727 \\ \hline \end{array}$$

**20.**
$$\begin{array}{r} 700 \\ -238 \\ \hline \end{array}$$

**21.**
$$\begin{array}{r} 610 \\ -538 \\ \hline \end{array}$$

**22.**
$$\begin{array}{r} 804 \\ -595 \\ \hline \end{array}$$

**23.**
$$\begin{array}{r} 283 \\ -176 \\ \hline \end{array}$$

*More Practice, page 520, set A*

# Capacity

LEARN ABOUT IT

The **quart** (qt) is a customary unit used to measure liquids. You can measure amounts of liquid to find the **capacity** of the container.

**EXPLORE** **Estimate and Measure**

Sort some empty containers into three groups: those that you estimate hold about 1 quart, more than 1 quart, or less than 1 quart. Measure a quart of water. Use it to check your estimates.

**TALK ABOUT IT**

**1.** How did you compare the containers when you first sorted them?

**2.** Do containers have to have the same shape and size to hold the same amount?

Other customary units of capacity are the **gallon** (gal), **pint** (pt), and **cup** (c).

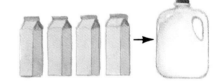

2 cups = 1 pint    4 cups = 2 pints = 1 quart    4 quarts = 1 gallon

**PRACTICE**

Is this more than, less than, or the same as 1 gallon?

**1.** 8 cups        **2.** 6 pints        **3.** 5 quarts        **4.** 8 pints

Is this more than, less than, or the same as 1 quart?

**5.** 3 pints        **6.** 4 cups        **7.** 1 gallon        **8.** 2 cups

**200**

*More Practice, page 520, set B*

# Temperature

**LEARN ABOUT IT**

**EXPLORE** **Study the Information**
The **degree Fahrenheit** (°F) is a
unit for measuring temperature. This
thermometer reads 63°F.

**TALK ABOUT IT**

1. What is the weather like on this
   day in March? What kind of
   clothes would you wear?

2. Is 30°F on a thermometer warmer
   or cooler than 70°F?

**July**

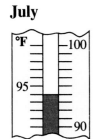

**December**

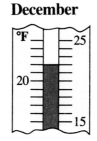

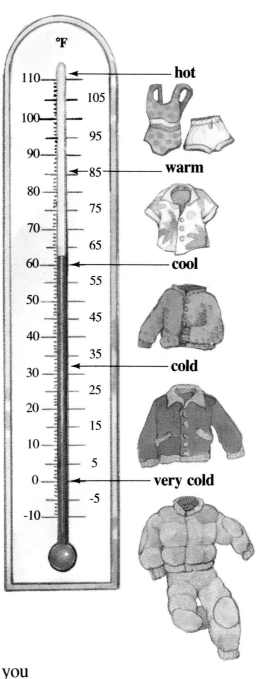

**PRACTICE**

Give the temperature for each of these
thermometers. Decide what kind of clothes you
would wear outdoors.

1. February

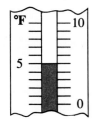

2. May

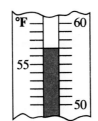

3. August

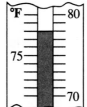

4. November

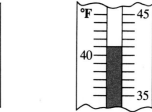

*More Practice, page 520, set C*

**201**

# Problem Solving
## Using Data from a Chart

| UNDERSTAND |
|------------|
| FIND DATA |
| PLAN |
| ESTIMATE |
| SOLVE |
| CHECK |

**LEARN ABOUT IT**

To solve some problems, you will need to read the numbers in a chart to find the data you need.

Gorillas are some of the largest living animals. Koko and Michael are two famous gorillas. How much more does Michael weigh than Koko?

| Facts About Koko and Michael | Koko | Michael |
|------------------------------|------|---------|
| Year of birth | 1971 | 1973 |
| Height | 61 in. | 72 in. |
| Weight | 250 lb | 425 lb |
| Vegetables and fruit eaten | 70 lb each week | 98 lb each week |
| Number of signs known | 600 | 250 |

*I'll find the data I need in the chart.*

Koko weighs 250 lb.
Michael weighs 425 lb.

*Now I'll solve the problem.*

$425 - 250 = 175$

Michael weighs 175 pounds more than Koko.

**TRY IT OUT**

1. Koko and Michael are famous because they know sign language. How many more signs does Koko know than Michael?

2. About how old is Koko? About how old is Michael?

3. How much do Koko and Michael weigh together?

4. How many pounds of food does Michael eat in 2 weeks?

202

| Number of Signs Koko Knew | |
| --- | --- |
| **Age in Months** | **Number of Signs** |
| 42 | 111 |
| 44 | 135 |
| 46 | 163 |
| 48 (4 years old) | 182 |
| 50 | 199 |
| 52 | 214 |

Solve.

1. At the time Michael was measured, he was 15 years old. An average 15-year-old boy weighs about 112 lb. How much heavier is Michael?

2. How many signs do Koko and Michael know all together?

3. How much taller is Michael than Koko?

4. A chimp named Washoe was also taught sign language. As a young chimp, Washoe knew 85 signs. At the same age, Koko knew 112 signs. How many more signs did Koko know?

5. When Michael's arm span was last measured, it was between 90 and 100 in. The number of inches was an even number. It was greater than 95, and it was not 98 or 100. How long was his arm span?

6. How many pounds of fruit and vegetables do Koko and Michael eat in 2 weeks?

7. How many more signs did Koko know at 50 months than at 48 months?

8. How many fewer signs did Koko know at 46 months than at 52 months?

9. **Write Your Own Problem** Write your own problem that can be solved using addition or subtraction. Use the data from one of the charts.

# Applied Problem Solving
## Group Decision Making

UNDERSTAND
FIND DATA
PLAN
ESTIMATE
SOLVE
CHECK

**Group Skill:**

Explain and Summarize

You bought a sweatshirt as a birthday gift. You need to decide whether you will wrap the gift yourself or have it done in the store.

### Facts To Consider

1. The store would charge $4.50 to wrap the sweatshirt in a gift box. This would include your choice of wrapping paper and bow.

2. If you wrap the box, you will need a sheet of paper at least 30 inches long and 18 inches wide.

3. One kind of paper comes on rolls that are 30 inches wide and 102 inches long. A roll costs $2.79.

4. Another kind of paper comes two sheets to a package. Each sheet is 20 inches wide and 30 inches long. A package costs $1.49.

5. A roll of ribbon costs $1.59. A large bow costs $1.19.

1. Is the paper that comes on a roll big enough to wrap the box?

2. Would you like to have some paper left over for another gift sometime? Would you have more of the sheet paper or more of the rolled paper left over?

3. Do you want to use some ribbon, a bow, or both?

4. What would it cost to buy a package of paper, a roll of ribbon, and a bow?

List the things that you would buy if you were going to wrap the package yourself. Show the total cost. Compare that cost to what the store would charge for wrapping the gift. Will you wrap the sweatshirt yourself or have it done at the store?

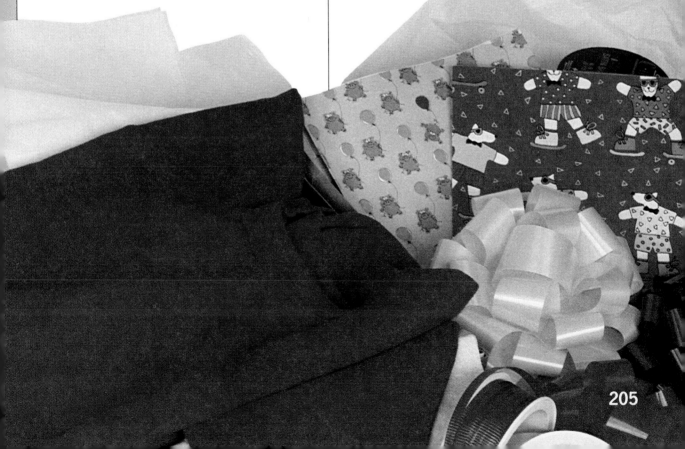

# WRAP UP

## What Would You Measure?

What would you measure with the given unit?

**1.** pound

   **a.** weight of a ring
   **b.** weight of a feather
   **c.** weight of a dog

**2.** foot

   **a.** length of a stamp
   **b.** height of a grasshopper
   **c.** height of a ladder

**3.** yard

   **a.** length of a pool
   **b.** width of a book
   **c.** height of a radio

**4.** gallon

   **a.** water in a fishtank
   **b.** soup in a cup
   **c.** juice in a glass

## Sometimes, Always, Never

Decide which word should go in the blank, *sometimes*, *always*, or *never*. Explain your choices.

**5.** Two pints of water _____ will fill an empty quart container.

**6.** A quart milk carton _____ holds five cups of milk.

**7.** You _____ measure weight in pounds.

## Project

Estimate the perimeter of your classroom. Write your estimate on a slip of paper. Work with a group or a partner to think of a way to measure the perimeter of the room using measuring tools at hand. Make the measurement. Compare the estimates your group made to the actual measurement.

Is your answer a reasonable one? What units did you use to give the perimeter?

Estimate the combined weight of the members of your group. Find the actual weight. Is the answer reasonable? How did you estimate?

# CHAPTER REVIEW/TEST

## Part 1  Understanding

Choose a good benchmark for estimating each unit.

**1.** inches          **2.** pounds          **3.** yards          **4.** ounces

Would you use pounds or ounces to find the weight?

**5.** a typewriter          **6.** a handful of beans          **7.** a chair

**8.** Give a likely temperature in degrees Fahrenheit for weather for building a snowman.

**9.** Do you need an estimate or an exact measurement to make a card that fits into an envelope?

## Part 2  Skills

Measure to the nearest inch.

**10.** ├───────┤          **11.** ├──────────────────────┤          **12.** ├──────────┤

Find the perimeter of each figure.

**13.**

4 in.
2 in.
4 in.

**14.**

2 ft
2 ft
3 ft
4 ft
2 ft
5 ft

Write <, >, or = in the ▦.

**15.** 3 cups ▦ 1 quart     **16.** 2 pints ▦ 1 quart     **17.** 1 gallon ▦ 1 quart

**18.** 5 feet ▦ 1 yard     **19.** 1 mile ▦ 3,000 yards     **20.** 2,000 feet ▦ 1 mile

## Part 3  Applications

**21. Challenge** Use the chart. At this school, 2,500 students each play at least one sport. No student plays more than two of the sports. How many students play two sports?

| Sport | Student Players |
|---|---|
| Basketball | 1,892 |
| Soccer | 759 |
| Tennis | 171 |
| Swimming | 172 |

# ENRICHMENT
## How Many Seats?

Jamie is looking at some ways that she can use square tables to seat people at a party. Here are some different ways that she can use 2 tables.

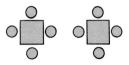

**2 tables, 8 people**

**2 tables, 6 people**

Here are some ways that Jamie can use 4 tables.

**A**  **B**  **C**

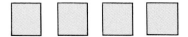

1. How many people can sit at 4 tables in Plan A?

2. How many people can Jamie seat when 4 tables are in a row as in Plan B?

3. How many people can sit at the tables in Plan C?

4. Use blocks to show different ways to arrange 6 tables. Tell how many people can sit at each of your arrangements of tables.

5. Use graph paper to show different arrangements of 12 tables. Tell how many people can sit at each arrangement.

Use blocks to find a pattern.

6. Jamie is borrowing folding tables for a big school picnic. She will arrange the square tables in a long row like Plan B. Write a rule to tell how many places Jamie adds each time she adds one table to the row.

# CUMULATIVE REVIEW

**1.** Find $300 + 500$.

   **A.** 200    **B.** 700

   **C.** 800    **D.** 900

**2.** Estimate to the nearest 10.
$35 + 54$

   **A.** 70    **B.** 80

   **C.** 85    **D.** 90

**3.** You traded 10 ones for 1 ten to get 32. You started with:

   **A.** 2 tens, 12 ones
   **B.** 2 tens, 23 ones
   **C.** 3 tens, 2 ones
   **D.** 3 tens, 12 ones

**4.** You trade 1 ten for 10 ones in 75. Now you have:

   **A.** 7 tens, 15 ones
   **B.** 7 tens, 5 ones
   **C.** 6 tens, 5 ones
   **D.** 6 tens, 15 ones

**5.** $426 - 248$

   **A.** 222    **B.** 178

   **C.** 278    **D.** 122

**6.** How many? ‖‖ ‖‖ ///

   **A.** 8    **B.** 18

   **C.** 13    **D.** 23

**7.** Which do you use to check $563 - 368 = 195$?

   **A.** $563 + 195$    **B.** $563 + 368$

   **C.** $368 - 195$    **D.** $195 + 368$

**8.** What do you draw on a bar graph to show how many?

   **A.** bars    **B.** labels

   **C.** dashed lines    **D.** title

**9.** If each picture shows 1, how many pictures do you draw on a picture graph to show 4?

   **A.** 1    **B.** 2

   **C.** 3    **D.** 4

**10.** Juanita is taller than Gary. Anna is shorter than Gary. Harry is taller than Gary but shorter than Juanita. If you list them from shortest to tallest, who is third?

   **A.** Juanita    **B.** Anna

   **C.** Harry    **D.** Gary

**11.** John has $55.00. If he buys a racket for $24.58 and 2 cans of tennis balls at $9.75 each, how much money is left?

   **A.** $10.92    **B.** $34.33

   **C.** $20.67    **D.** $5.08

8

# MULTIPLICATION CONCEPTS AND FACTS

**M**ATH AND HEALTH AND FITNESS

DATA BANK

Use the Health and Fitness Data Bank on page 482 to answer the questions.

1 Indians once lived in Yosemite Valley. If you rode your bike from Yosemite Lodge to the Indian Caves and back, how many miles would you ride

**2** Use the distance from the hotel to Indian Caves as a benchmark. What is your estimate for the distance from Indian Caves to Happy Isles?

**3** About how far do you think Upper River Campground is from Happy Isles? Find some distances on the map to help you.

**4** **Use Critical Thinking** Mirror Lake Loop takes about 2 hours. Predict how long it would take to hike from the Visitor Center to Happy Isles. Why?

# Problem Solving
## Understanding Multiplication

| UNDERSTAND |
|---|
| FIND DATA |
| PLAN |
| ESTIMATE |
| SOLVE |
| CHECK |

**LEARN ABOUT IT**

The action in a problem helps you decide what operation to use. Show the action with counters and cups. Complete a number sentence about the action.

| **Problem** | **Action** | **Operation** |
|---|---|---|

On a tandem bicycle, 2 riders pedal together. How many people can ride 3 tandem bicycles?

$3 \times 2 = ?$

Multiplication is like addition because you put together groups to find the total. But multiplication can be used only when the groups are the same size.

| **We see:** | **We write:** | **We read:** |
|---|---|---|
| 3 groups of 2 | $3 \times 2$ | "Three times two" |

**TRY IT OUT**

Use counters to show the action in each problem. Decide whether the problem suggests addition or multiplication. Tell why.

**1.** Teri bought 5 orange drinks and 2 grape drinks. How many drinks did she buy?

**2.** Juan bought 6 packages of Huggy Bear cards. Each package held 3 cards. How many cards did he buy?

Here is a list of the key actions you have learned so far. These actions in a problem tell you what operation to use to solve the problem.

Show the action in each problem with counters. Tell which action you are showing. Decide which operation you would use to solve the problem.

| Key Action | Operation |
|---|---|
| Put together | Add |
| Take away | |
| Compare | Subtract |
| Find a missing part | |
| Put together same-size sets | Multiply |

1. Katie bought a 6-pack of fruit juice. She and a friend drank 3 cans. How many cans did she have left?

2. Keisha had 7 dollars in her bank. She put 8 dollars more in the bank. How much does she have in the bank now?

3. Jessica wants 5 tickets for the play. The tickets are $2 each. How many dollars does Jessica need for the tickets?

4. Melissa sold 6 tickets, Shauna sold 8, and Mallory sold 7. How many tickets did the 3 girls sell?

5. Joshua has 4 cans of tennis balls. There are 3 balls in each can. How many tennis balls does he have?

6. Travis sold 4 tickets to his father, 4 tickets to his uncle, and 4 tickets to his grandfather. How many tickets did he sell?

7. Shane wants to buy a biking shirt that costs $14. He has $8. How much more does he need?

8. Nicholas rode his bike 12 miles from camp. Steven rode only 8 miles. How much farther did Nicholas ride than Steven?

*More Practice, page 520, set D*

**213**

# More About Multiplication

### EXPLORE  Work in Groups

Use counters, cups, and a number cube.

- Choose 2, 3, or 4 cups. Roll a number cube labeled 1 to 6 to choose a number. Put that number of counters in each cup.

- Write a multiplication equation for the cups and counters as in the example.

- Change jobs and repeat until you have 4 different equations from your group.

### TALK ABOUT IT

**3 cups, 4 in each cup**

**12 in all**
**3 × 4 = 12**

1. What equation would you write for "4 cups, 3 in each cup"? How is this equation like the one in the example? How is it different?

2. What addition equation could you write for "3 cups, 4 in each cup"?

This picture shows two important ideas about multiplication.

- Multiplication and addition are related.

- You can multiply in another order. The answer stays the same.

$3 + 3 = 6$      $2 + 2 + 2 = 6$

$2 × 3 = 6$      $3 × 2 = 6$

Use counters and cups to show each. Does changing the order change the total number of counters?

1. Two fours and four twos  **2.** $3 × 5$ and $5 × 3$  **3.** two ones and one two

214

Add and multiply.

**1.**

2 fives    $5 + 5 = |||||$
           $2 \times 5 = |||||$

**2.**

5 twos    $2 + 2 + 2 + 2 + 2 = |||||$
          $5 \times 2 = |||||$

You can look at the 12 dots in different ways.
Write a multiplication equation for each picture.

**3.**

**4.**

APPLY

**MATH REASONING**

**5.** If you know that $8 \times 9 = 72$, what is $9 \times 8$?

**PROBLEM SOLVING**

**6.** Josh put 3 counters in each of 5 cups. If he put the same number of counters 5 to a cup, how many cups would he need?

**7. Missing Data** What else do you need to know to answer the question? Kristen put 4 counters in each cup. How many counters does she have?

▶ **CALCULATOR**

Write the final calculator display for each key code.

**8.** 7 $+$ 7 $+$ 7 $+$ 7 $=$

**9.** 7 $+$ 7 $=$ $+$ 7 $=$ $+$ 7 $=$

**10.** $+$ 7 $=$ $=$ $=$ $=$

**11.** 7 $\times$ 4 $=$

**12.** Which key code would you choose to find the total of 4 sevens?

# 2 as a Factor
## Using Mental Math

**EXPLORE** **Think About the Picture**
Thinking about addition doubles can
help you multiply when 2 is a factor.

**TALK ABOUT IT**

1. Which picture reminds you of 2 fives?
   How much is $2 \times 5$?

2. Choose another of the pictures. Give an
   addition fact and a multiplication fact for your
   picture.

Here are some ways to find a product when 2 is a factor.

**Think:** $4 + 4 = 8$, so $\qquad$ $2 \times 4 = 8$ $\qquad$ $4 \leftarrow$ factor
$\qquad$ $2 \times 4 = 8$ $\qquad$ factor ) (factor) (product $\qquad$ $\underline{\times 2} \leftarrow$ factor
$\qquad\qquad\qquad\qquad\qquad\qquad\qquad\qquad$ $8 \leftarrow$ product

**Think:** 2 fours $\qquad\qquad$ The product of 2 and 4 is 8.
$\qquad$ $2 \times 4 = 8$ $\qquad\qquad\qquad\qquad\qquad$ 4 multiplied by 2 is 8.

**TRY IT OUT**

Find the products.

**1.** $2 \times 8$ $\qquad$ **2.** $8 \times 2$ $\qquad$ **3.** $2 \times 4$ $\qquad$ **4.** $4 \times 2$ $\qquad$ **5.** $2 \times 6$ $\qquad$ **6.** $6 \times 2$

**7.** $\begin{array}{r} 7 \\ \times 2 \\ \hline \end{array}$ $\quad$ **8.** $\begin{array}{r} 2 \\ \times 7 \\ \hline \end{array}$ $\quad$ **9.** $\begin{array}{r} 9 \\ \times 2 \\ \hline \end{array}$ $\quad$ **10.** $\begin{array}{r} 2 \\ \times 9 \\ \hline \end{array}$ $\quad$ **11.** $\begin{array}{r} 3 \\ \times 2 \\ \hline \end{array}$ $\quad$ **12.** $\begin{array}{r} 2 \\ \times 3 \\ \hline \end{array}$

Multiply.

| | | | | | | | | | | | |
|---|---|---|---|---|---|---|---|---|---|---|---|
| **1.** | 3<br>×2 | **2.** | 2<br>×6 | **3.** | 2<br>×2 | **4.** | 5<br>×2 | **5.** | 2<br>×7 | **6.** | 4<br>×2 |
| **7.** | 2<br>×5 | **8.** | 8<br>×2 | **9.** | 6<br>×2 | **10.** | 9<br>×2 | **11.** | 2<br>×4 | **12.** | 7<br>×2 |

**13.** $9 \times 2$    **14.** $4 \times 2$    **15.** $6 \times 2$    **16.** $2 \times 5$

**17.** $7 \times 2$    **18.** $2 \times 8$    **19.** $2 \times 7$    **20.** $3 \times 2$

**21.** Find the product of 2 and 9.    **22.** Multiply 8 by 2.

**MATH REASONING**

**23.** When 2 is a factor and 12 is the product, what is the other factor?

**24.** When 2 is a factor and 16 is the product, what is the other factor?

**PROBLEM SOLVING**

**25.** At a bicycle race, Alan saw 6 tandem bikes. How many riders were racing on tandem bikes?

**DATA BANK**

**26. Health and Fitness Data Bank** Kate's mother and father rode their tandem bike from Yosemite Lodge to Happy Isles Nature Center and back. How far did they ride? See page 482.

▶ **ALGEBRA**

Find the missing factors. Think about doubles.

What number doubled gives 14?

**27.** $2 \times \text{||||} = 14$    **28.** $2 \times \text{||||} = 12$    **29.** $2 \times \text{||||} = 10$

**30.** $2 \times \text{||||} = 18$    **31.** $2 \times \text{||||} = 16$    **32.** $2 \times \text{||||} = 8$

*More Practice, page 505, set A*

# 5 as a Factor
## Using Mental Math

### EXPLORE  Use Play Money

Thinking about nickels can help you find products when 5 is a factor. The figure shows $4 \times 5$. How many ways can you find the product $8 \times 5$?

5

10

15

20

$4 \times 5 = 20$

### TALK ABOUT IT

**1.** Can you count by fives to find $8 \times 5$?

**2.** Emily said "$8 \times 5$ is double $4 \times 5$." Is she right? Explain.

**3.** If you know $8 \times 5$, how can you find $9 \times 5$?

Here are some ways to find the product when 5 is a factor.

■ Skip count. Count in pairs to help you keep track.

    5, 10    15, 20    25, 30    35, 40    45

■ Use a known fact. Since 6 fives are 30, 7 fives are 35.

$$6 \times 5 = 30 \longrightarrow 7 \times 5 = 35$$

### TRY IT OUT

Find each product.

**1.** $7 \times 5$      **2.** $5 \times 3$      **3.** $9 \times 5$      **4.** $5 \times 6$

**5.**  $\begin{array}{r} 5 \\ \times 4 \\ \hline \end{array}$      **6.** $\begin{array}{r} 8 \\ \times 5 \\ \hline \end{array}$      **7.** $\begin{array}{r} 5 \\ \times 7 \\ \hline \end{array}$      **8.** $\begin{array}{r} 5 \\ \times 5 \\ \hline \end{array}$      **9.** $\begin{array}{r} 2 \\ \times 5 \\ \hline \end{array}$

Multiply.

| 1. | 5 | 2. | 7 | 3. | 8 | 4. | 5 | 5. | 5 | 6. | 2 |
|---|---|---|---|---|---|---|---|---|---|---|---|
| | ×6 | | ×5 | | ×2 | | ×5 | | ×3 | | ×7 |

| 7. | 8 | 8. | 4 | 9. | 9 | 10. | 5 | 11. | 2 | 12. | 4 |
|---|---|---|---|---|---|---|---|---|---|---|---|
| | ×5 | | ×5 | | ×5 | | ×2 | | ×6 | | ×2 |

**13.** $5 \times 2$     **14.** $5 \times 8$     **15.** $4 \times 5$     **16.** $3 \times 2$

**17.** $5 \times 6$     **18.** $9 \times 2$     **19.** $9 \times 5$     **20.** $5 \times 3$

**21.** Find the product of 7 and 5.     **22.** Multiply 8 by 5.

**APPLY**

**MATH REASONING**  Finish the sentence.

**23.** Whenever 5 is a factor, the ones digit of the product is either 0 or ‖‖‖.

**PROBLEM SOLVING**

**24.** **Missing Data** What data do you need to answer the question? Nicole bought some 5¢ stamps and some 8¢ stamps. How much did Nicole spend in all?

**25.** Erin has one row of 40 stamps and another row of 25 stamps. How many more stamps are in the longer row?

**MIXED REVIEW**

Choose the word that gives the better estimate.

**26.** A person's hand is about 3 _____ wide.     A. inches     B. feet

**27.** A table top is about 1 _____ high.     A. foot     B. yard

**28.** In one hour, you can bike about 6 _____.     A. miles     B. yards

**29.** A car is about 10 _____ long.     A. yards     B. feet

**30.** One mile is _____ than 2,000 yards.     A. more     B. less

*More Practice, page 505, set B*

# 9 as a Factor
## Using Mental Math

**LEARN ABOUT IT**

Katie likes to ride her mountain bike up 9-Mile Trail. She rode the trail 3 times last week. She multiplied to find how many miles she went.

**EXPLORE  Look for a Pattern**
Katie discovered a pattern for multiplying by 9. The arrows give you a hint about what she is thinking. Can you figure out the pattern?

**TALK ABOUT IT**

1. Using Katie's method, what subtraction would give you the product for $9 \times 5$? for $9 \times 7$?

2. Which product is less, $10 \times 3$ or $9 \times 3$? How much less? Can you explain why Katie's method works?

$$30 - 3$$
$$9 \times 3 = 27$$

$$40 - 4$$
$$9 \times 4 = 36$$

Here is a method that you can use to find the product when 9 is a factor.

■  Look at the other factor. Multiply by 10.

■  Subtract that factor.

$$9 \times 6 \text{ is } 60 - 6, \text{ or } 54.$$

**TRY IT OUT**

Multiply.

1. $9 \times 4$    2. $9 \times 6$    3. $3 \times 9$    4. $8 \times 9$    5. $9 \times 9$

6. $\begin{array}{r} 2 \\ \times 9 \\ \hline \end{array}$    7. $\begin{array}{r} 3 \\ \times 9 \\ \hline \end{array}$    8. $\begin{array}{r} 7 \\ \times 9 \\ \hline \end{array}$    9. $\begin{array}{r} 9 \\ \times 4 \\ \hline \end{array}$    10. $\begin{array}{r} 9 \\ \times 5 \\ \hline \end{array}$

Multiply.

| **1.** 5 | **2.** 9 | **3.** 3 | **4.** 8 | **5.** 9 | **6.** 4 |
|---|---|---|---|---|---|
| $\times 9$ | $\times 6$ | $\times 9$ | $\times 9$ | $\times 7$ | $\times 9$ |

| **7.** 5 | **8.** 9 | **9.** 9 | **10.** 7 | **11.** 9 | **12.** 9 |
|---|---|---|---|---|---|
| $\times 8$ | $\times 1$ | $\times 5$ | $\times 5$ | $\times 9$ | $\times 2$ |

**13.** $6 \times 9$ **14.** $4 \times 8$ **15.** $9 \times 4$ **16.** $9 \times 0$

**17.** Find the product of 9 and 6. **18.** Multiply 9 by 5.

Use Katie's pattern on page 220 to help you find
the missing number.

**19.** $9 \times \text{||||} = 54$ **20.** $9 \times \text{|||||} = 72$ **21.** $9 \times \text{|||||} = 63$

**MATH REASONING**

**22.** Brent said, "If one factor is odd, the product
is odd." Give an example to prove Brent is
not right.

**PROBLEM SOLVING**

**23. Health and Fitness Data Bank**
For 9 Sundays, a park ranger led a hike
around the Mirror Lake loop. How many
miles did she hike? See page 482.

▶ **MENTAL MATH**

Sometimes you need to remember a product long
enough to add a number to it. Use mental math to
find each answer. First multiply the numbers in
the box. Then add to the product.

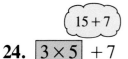

**24.** $\boxed{3 \times 5} + 7$ **25.** $\boxed{4 \times 5} + 8$ **26.** $\boxed{9 \times 4} + 5$

*More Practice, page 505, set C* **221**

# Using Critical Thinking

Jan's little brother Timmy knew his addition and subtraction facts, but he didn't know about multiplication.

When he looked at this list of multiplication 9 facts, he said "Oh, I see!" and quickly completed the list. Jan sure was surprised!

| 9 Facts | | |
|---|---|---|
| $9 \times 1 = 9$ | $9 \times 4 = 36$ | $9 \times 7 =$ |
| $9 \times 2 = 18$ | $9 \times 5 =$ | $9 \times 8 =$ |
| $9 \times 3 = 27$ | $9 \times 6 =$ | $9 \times 9 =$ |

## TALK ABOUT IT

1. What did Timmy do that surprised Jan?

2. How do you think that Timmy was planning to complete the list?

3. What patterns do you see in the list?

4. In a 9 fact, what is the relationship between the first digit of the product and the smaller factor?

5. What is the sum of the digits in the product?

1. Write a fact list for the 2 facts. Do you see a pattern? Explain.

2. Make a fact list for the 5 facts. What patterns do you see? Explain.

222

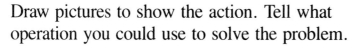
Draw pictures to show the action. Tell what operation you could use to solve the problem.

**1.** You have 3 fish bowls with 2 fish in each bowl. How many fish in all?

**2.** The calendar showed 2 cats and 3 dogs. How many more dogs than cats?

Copy the picture. Draw lines to show groups. Finish the equations.

**3.** 4 twos
$2 + 2 + 2 + 2 =$ ‖‖‖
$4 \times 2 =$ ‖‖‖

**4.** 2 fours
$4 + 4 =$ ‖‖‖
$2 \times 4 =$ ‖‖‖

Draw the objects. Write an addition equation and a multiplication equation for each drawing.

**5.** Draw hands to show 2 fives.

**6.** Draw shoe boxes to show 5 twos.

Complete the sentence.

**7.** If 3 twos are 6, then 2 threes are ‖‖‖.

**8.** If 2 fives are 10, then 3 fives are ‖‖‖.

**9.** If 10 fours are 40, then 9 fours are ‖‖‖.

**10.** If $8 + 8$ is 16, then 2 eights are ‖‖‖.

**11.** If 10 nines are 90, then 9 nines are ‖‖‖.

**12.** If 8 fives are 40, then 9 fives are ‖‖‖.

Multiply.

**13.** $2 \times 4$ **14.** $2 \times 8$ **15.** $8 \times 2$ **16.** $5 \times 7$ **17.** $5 \times 2$

**18.**
$\begin{array}{r} 4 \\ \times 5 \\ \hline \end{array}$
**19.**
$\begin{array}{r} 8 \\ \times 5 \\ \hline \end{array}$
**20.**
$\begin{array}{r} 5 \\ \times 9 \\ \hline \end{array}$
**21.**
$\begin{array}{r} 4 \\ \times 9 \\ \hline \end{array}$
**22.**
$\begin{array}{r} 2 \\ \times 9 \\ \hline \end{array}$
**23.**
$\begin{array}{r} 2 \\ \times 7 \\ \hline \end{array}$
**24.**
$\begin{array}{r} 6 \\ \times 2 \\ \hline \end{array}$

**25.**
$\begin{array}{r} 9 \\ \times 3 \\ \hline \end{array}$
**26.**
$\begin{array}{r} 6 \\ \times 9 \\ \hline \end{array}$
**27.**
$\begin{array}{r} 7 \\ \times 9 \\ \hline \end{array}$
**28.**
$\begin{array}{r} 9 \\ \times 5 \\ \hline \end{array}$
**29.**
$\begin{array}{r} 5 \\ \times 5 \\ \hline \end{array}$
**30.**
$\begin{array}{r} 9 \\ \times 8 \\ \hline \end{array}$
**31.**
$\begin{array}{r} 9 \\ \times 9 \\ \hline \end{array}$

# Problem Solving
## Work Backward

| UNDERSTAND |
|---|
| FIND DATA |
| PLAN |
| ESTIMATE |
| SOLVE |
| CHECK |

### LEARN ABOUT IT

To solve some problems, you need to undo the key actions in the problem. This problem solving strategy is called Work Backward. Sometimes, you may want to use a calculator.

> At carnival time in Peru, children throw water balloons filled with colored water. Maria had a bagful of balloons. She found 7 more. She threw away 5 that had holes in them. Maria ended with 16 balloons. How many did she have at first?

Maria ended with 16 balloons.　　16

Since she threw away 5 balloons, I'll add those back on.　　$16 + 5 = 21$　　ON/AC　16　+　5　=

Since she found 7 more, I'll subtract those.　　$21 - 7 = 14$　　−　7　=

Maria had 14 balloons at first.

### TRY IT OUT

On the 4th of July, the police department set off skyrockets. They lit 7 of them. The next 3 fizzled out. There were 12 left to light. How many skyrockets did they have to start?

- How many skyrockets did the police light?
- How many skyrockets fizzled out?
- How many skyrockets were left to light?
- Copy and complete the table to solve.

12 skyrockets left　　12

Lit 7. Add those back　　12 + 7

Threw away 3.

Work backward to help you solve each problem.
You may want to use a calculator.

1. In Italy, children celebrate the Festival of Crickets. In the morning, Dominic's family bought some crickets. 2 got away. They bought 5 more. Then they had 11 crickets. How many did they buy in the morning?

2. The day before a party Becka made a batch of popcorn balls. That night, she made 6 more balls. The next day, she made 12 more balls. Becka took all 23 balls to the party. How many balls were in the first batch?

**MIXED PRACTICE**

3. For Girls' Day, Reiko received 2 new dolls. Reiko keeps her dolls on 4 shelves. There are 7 dolls on each shelf. How many dolls does she have in all?

4. Carlos saw 3 rows of piñatas for sale. There were 8 piñatas in each row. All together, how many piñatas were for sale?

5. In October, the Costume Bank sold 5 times as many clown hats as they sold in September. How many did they sell in October?

| Some Strategies |
| --- |
| Act It Out |
| Choose an Operation |
| Make an Organized List |
| Make a Table |
| Use Logical Reasoning |
| Use Objects |
| Draw a Picture |
| Guess and Check |
| Look for a Pattern |
| Work Backward |

6. On Valentine's Day, Sam bought a box of cards. His mother gave him 5 blank cards. He found 7 more from last year. Sam sent 22 cards to friends. He had 4 cards left over. How many were in the box he bought?

7. In the Chinese New Year parade, Mei Mei saw 6 dragons with 5 people inside each. She also saw 3 dragons with 4 people inside each. How many people in all did she see in dragon costumes?

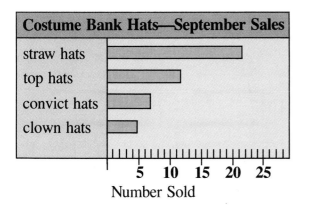

**Costume Bank Hats—September Sales**

straw hats
top hats
convict hats
clown hats

5   10   15   20   25
Number Sold

# 0 and 1 as Factors
## Using Mental Math

**LEARN ABOUT IT**

**EXPLORE  Study the Situation**

You can think about hamsters and cages to help you understand multiplying by 0 and 1. Jan said she would use 3 cages with 1 hamster in each cage to show $3 \times 1 = 3$.

**TALK ABOUT IT**

1. Do you agree with Jan? Use objects to show how you would multiply a number by 1.

2. Suppose you used 3 cages, with 0 hamsters in each cage. Which fact would you be showing?

3. How would you explain to someone the difference between adding 0 to 3 and multiplying 0 times 3?

You may have discovered the following.

- When 1 is a factor, the product is the same as the other factor. $4 \times 1 = 4$

- When 0 is a factor, the product is 0. $4 \times 0 = 0$

- $4 \times 1 = 1 \times 4$ and $4 \times 0 = 0 \times 4$

**TRY IT OUT**

Find the products using mental math.

**1.** $5 \times 0$     **2.** $9 \times 1$     **3.** $1 \times 7$     **4.** $0 \times 9$     **5.** $6 \times 0$

**6.** $1 \times 8$     **7.** $0 \times 6$     **8.** $1 \times 1$     **9.** $7 \times 0$     **10.** $0 \times 0$

**11.** $8 \times 0$     **12.** $9 \times 1$     **13.** $0 \times 7$     **14.** $1 \times 6$     **15.** $9 \times 0$

Multiply.

| | | | | | |
|---|---|---|---|---|---|
| **1.** 0<br>×8 | **2.** 1<br>×3 | **3.** 5<br>×0 | **4.** 8<br>×5 | **5.** 7<br>×1 | **6.** 0<br>×3 |
| **7.** 7<br>×2 | **8.** 4<br>×0 | **9.** 1<br>×4 | **10.** 7<br>×0 | **11.** 9<br>×1 | **12.** 0<br>×2 |

**13.** $8 \times 1$          **14.** $0 \times 9$          **15.** $5 \times 7$          **16.** $2 \times 1$

**17.** Find the product of 9 and 4.          **18.** Multiply 0 by 6.

**APPLY**

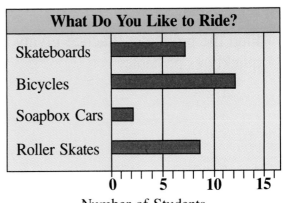

MATH REASONING  Write $\times$ or $+$ in the ▥ .

**19.** 5 ▥ 1 = 5          **20.** 5 ▥ 0 = 5          **21.** 5 ▥ 0 = 0

**PROBLEM SOLVING**

**22.** Jamie is taller than Dale. Nina is shorter than Dale. Jamie is shorter than Brad. Who is tallest?

**23.** Dustin had 5 rows of plants. Each row had 7 plants. There was 1 tomato plant in each row. How many tomato plants were in the garden?

**MIXED REVIEW**

A third grade class voted for wheel toys that they liked to ride. Use the graph to answer the questions.

**24.** Name 2 toys that got about the same number of votes.

**25.** Which toy got the most votes?

**26.** What does the graph tell you?

**What Do You Like to Ride?**

Skateboards

Bicycles

Soapbox Cars

Roller Skates

0      5      10      15

Number of Students

*More Practice, page 505, set D*

# Mastering the Facts

Sometimes there are special clues to help you remember the facts.

**EXPLORE** **Understand the Process**
As you have learned, you can use addition doubles when 2 is a factor. There are helpful patterns when 9 is a factor. Can you find a pattern to help you remember the product when 5 is multiplied by an even number?

**TALK ABOUT IT**

1. What pattern did you find in the examples above?

2. Does the pattern work for $10 \times 5$? Explain.

$2 \times 5 = 10$
$4 \times 5 = 20$
$6 \times 5 = 30$
$8 \times 5 = 40$

Here are some clues for remembering the product when 5 is multiplied by an odd number.

| 5 more than $2 \times 5$ | 5 more than $4 \times 5$ | 5 more than $6 \times 5$ | 5 more than $8 \times 5$ |
|---|---|---|---|
| $3 \times 5$ | $5 \times 5$ | $7 \times 5$ | $9 \times 5$ |

**TRY IT OUT**

Find the products. Practice giving the products as quickly as possible.

| | | | | |
|---|---|---|---|---|
| **1.** $4 \times 5$ | **2.** $9 \times 3$ | **3.** $2 \times 8$ | **4.** $5 \times 5$ | **5.** $6 \times 9$ |
| **6.** $9 \times 9$ | **7.** $7 \times 5$ | **8.** $9 \times 8$ | **9.** $7 \times 2$ | **10.** $2 \times 9$ |
| **11.** $2 \times 5$ | **12.** $5 \times 9$ | **13.** $5 \times 8$ | **14.** $4 \times 2$ | **15.** $9 \times 5$ |
| **16.** $5 \times 6$ | **17.** $9 \times 4$ | **18.** $6 \times 2$ | **19.** $3 \times 5$ | **20.** $7 \times 9$ |

Find the products. Practice giving the products as quickly as possible.

**1.** $4 \times 2$     **2.** $9 \times 4$     **3.** $5 \times 7$     **4.** $2 \times 8$     **5.** $4 \times 5$

**6.** $2 \times 9$     **7.** $5 \times 0$     **8.** $2 \times 2$     **9.** $9 \times 8$     **10.** $2 \times 5$

**11.** $\begin{array}{r} 5 \\ \times 3 \\ \hline \end{array}$  **12.** $\begin{array}{r} 7 \\ \times 2 \\ \hline \end{array}$  **13.** $\begin{array}{r} 9 \\ \times 7 \\ \hline \end{array}$  **14.** $\begin{array}{r} 8 \\ \times 5 \\ \hline \end{array}$  **15.** $\begin{array}{r} 1 \\ \times 4 \\ \hline \end{array}$  **16.** $\begin{array}{r} 2 \\ \times 9 \\ \hline \end{array}$

**17.** $\begin{array}{r} 2 \\ \times 8 \\ \hline \end{array}$  **18.** $\begin{array}{r} 9 \\ \times 0 \\ \hline \end{array}$  **19.** $\begin{array}{r} 9 \\ \times 9 \\ \hline \end{array}$  **20.** $\begin{array}{r} 6 \\ \times 5 \\ \hline \end{array}$  **21.** $\begin{array}{r} 6 \\ \times 2 \\ \hline \end{array}$  **22.** $\begin{array}{r} 9 \\ \times 6 \\ \hline \end{array}$

**23.** Find the product of 7 and 2.

**24.** Multiply 3 by 8.

**25.** What is 6 multiplied by 1?

**26.** What is the product of 9 and 4?

### MATH REASONING

**27.** Find the product. Think about addition doubles. $2 \times 379$

### PROBLEM SOLVING

**28.** Kyle figured that he eats 3 boxes of whole wheat cereal every two weeks. How many weeks will it take Kyle to eat 18 boxes of cereal?

**29.** Shauna eats fruit for breakfast and snacks. If she eats 2 pieces of fruit every day for 7 days, how much fruit does she eat in a week?

 **USING CRITICAL THINKING**

**30.** Sort these numbers into 2 equal sets. The numbers in each set should be alike in some way. Be prepared to explain how you did the sorting.

*More Practice, page 505, set E*

# Data Collection and Analysis
## Group Decision Making

UNDERSTAND
FIND DATA
PLAN
ESTIMATE
SOLVE
CHECK

**Doing a Survey**

**Group Skill:**
Listen to Others

How old do you think most children are when they lose their first tooth? You can find out by making a survey of the students in your school.

| How old were you when you lost your first tooth? | |
|---|---|
| Name | Age |
| Jennifer | 5 |
| Phillip | 6 |
| Joanie | don't know |
| Tatum | 4 |
| ____ | ____ |
| ____ | ____ |

**Collecting Data**

1. Have each person in your group write down the question that you will ask in the survey.
   How old were you when you lost your first tooth?

2. Ask the survey question to at least 20 people. Keep a record of their answers.

Age

6

5

4

**3.** Count how many children lost a tooth at each age. Start with the youngest age at which someone lost a tooth. Make a table.

**4.** Make a pictograph using the data in your table. Decide how many children will be represented by the picture of one tooth on your graph.

**5.** At what age do you think most children lose their first tooth?

**6.** Write at least three true sentences about the information in your graph.

Children

231

# WRAP UP

## Multiplication Riddles

Find a multiplication term from the chapter that answers each riddle.

1. You always find at least two of me in a multiplication equation. What am I?

2. When you multiply one factor by another factor, you get me. What am I?

3. When you multiply by me, you get me as an answer. What number am I?

4. When you multiply by me, one factor equals the product. What number am I?

## Sometimes, Always, Never

Decide which word should go in the blank, *sometimes*, *always*, or *never*. Explain your choices.

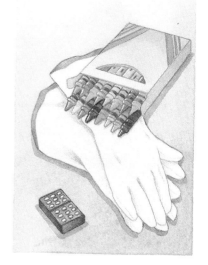

5. The product of two numbers _____ changes when the order of the numbers is changed.

6. A 5 _____ is in the ones place of a product when you multiply by a factor of 5.

7. When there are many pieces of data in a problem, you _____ work backward to find the solution.

8. Adding 0 to any number except 0 _____ gives you the same answer as multiplying the same number by 0.

## Project

Copy and complete this multiplication table. Shade in all the products that have factors of 2. What patterns do you see?

| × | 1 | 2 | 3 | 4 | 5 | 6 | 7 | 8 | 9 |
|---|---|---|---|---|---|---|---|---|---|
| 0 |   |   |   |   |   |   |   |   |   |
| 1 |   |   |   |   |   |   |   |   |   |
| 2 |   |   |   |   |   |   |   |   |   |
| 5 |   |   |   |   |   |   |   |   |   |
| 9 |   |   |   |   |   |   |   |   |   |

# CHAPTER REVIEW/TEST

## Part 1   Understanding

Can you multiply to find how many? Explain.

**1.**      **2.**

Write a multiplication equation for each.

**3.** $3 + 3 = 6$   **4.** $5 + 5 + 5 + 5 = 20$   **5.** $9 + 9 + 9 + 9 + 9 = 45$

**6.** What do these equations show you about 1 as a factor? $1 \times 4 = 4$ and $7 \times 1 = 7$

**7.** What do these equations show you about 0 as a factor? $2 \times 0 = 0$ and $0 \times 5 = 0$

## Part 2   Skills

Find the products.

**8.** $2 \times 4$   **9.** $7 \times 1$   **10.** $8 \times 0$   **11.** $2 \times 5$   **12.** $9 \times 6$

**13.** $1 \times 1$   **14.** $9 \times 4$   **15.** $6 \times 5$   **16.** $5 \times 9$   **17.** $0 \times 2$

## Part 3   Applications

**18.** Cynthia wrote a poem. She erased 5 lines she did not like. After a while, she added 8 new lines. Her poem now has 16 lines. How many lines did the poem have at first?

**19.** **Challenge** Alvin went to the Dino Fun Shop. He bought 2 rubber dinosaurs, 3 dinosaur stickers, and 2 dinosaur masks. Then he put back 1 mask and picked up 3 more stickers. He paid for the items and got $2.30 in change. How much money did he give the clerk?

| Dino Fun Shop Price List | |
|---|---|
| Rubber Dinosaur | $2.35 |
| Dino Stickers | 3 for $0.75 |
| Dinosaur Mask | $1.50 |

# ENRICHMENT
## What is the Rule?

Look at the columns of red and blue numbers. Think about the first three number pairs. If you look closely, you will see that they all are related in the same way.

1. How do you find the blue number if you know the red number? Can you give a rule?

2. Fill in the blue numbers that are missing.

3. Continue the chart so that you have the numbers 1 through 10 in the red column.

4. Suppose the chart looked like this. Can you give a rule for the new chart? You can test your rule by filling in the chart.

Work with a partner.

5. Begin a chart that uses a rule to give a number pair. Fill in four lines of your chart. Exchange charts with your partner. Complete your partner's chart for digits 1 through 10. Write the rule under the chart.

6. Begin a chart of large number pairs. Use a calculator to find your pairs. Remember to use the same rule for every pair of numbers in your chart. Exchange with your partner and complete the charts.

| 3 | 7 |
|---|---|
| 5 | 9 |
| 2 | 6 |
| 4 | ||||| |
| 7 | ||||| |
| 6 | ||||| |
| ||||| | ||||| |
| ||||| | ||||| |

| 3 | 15 |
|---|---|
| 5 | 25 |
| 2 | 10 |
| 4 | ||||| |
| ||||| | ||||| |

234

# CUMULATIVE REVIEW

**1.** $704 - 368$

   **A.** 336    **B.** 446

   **C.** 346    **D.** 306

**2.** $7,284 - 2,975$

   **A.** 5,300    **B.** 4,319

   **C.** 4,309    **D.** 5,309

**3.** Which method is probably best for finding $5,493 - 3,789$?

   **A.** compatible numbers

   **B.** compensation

   **C.** mental math

   **D.** calculator

**4.** Follow the pattern. How many puzzles did Jill do in week 4?

| Week | 1 | 2 | 3 | 4 |
|---|---|---|---|---|
| Puzzles done | 2 | 4 | 6 | ||||| |

   **A.** 7    **B.** 8

   **C.** 10    **D.** 12

**5.** Which game is most fair?

| Players | 1 | 2 | 3 |
|---|---|---|---|
| Game 1 | ‖‖ | ‖‖ | // |
| Game 2 | ‖‖ | ‖‖ | ‖‖ |
| Game 3 | ‖‖ | //// | / |

   **A.** Game 1    **B.** Game 2

   **C.** Game 3    **D.** All are fair.

**6.** What is the perimeter of a figure if the sides measure 23 yards, 23 yards, 14 yards, and 14 yards?

   **A.** 37 yards    **B.** 322 yards

   **C.** 60 yards    **D.** 74 yards

**7.** You need curtains for a window. What kind of measurement do you need?

   **A.** perimeter    **B.** thickness

   **C.** exact    **D.** estimate

**8.** Which is the same as 1 gallon?

   **A.** 8 pints    **B.** 12 cups

   **C.** 8 quarts    **D.** 6 quarts

**9.** Ms. Sandoval's age is between 31 and 41. It is less than 34. It is an even number. What is her age?

   **A.** 30    **B.** 32

   **C.** 33    **D.** 36

**10.** Who is older? How much older?

| Our Dogs | Rover | Punch |
|---|---|---|
| Born | 1982 | 1985 |
| Weight | 45 lb | 27 lb |
| Color | Black | Brown |

   **A.** Punch, 3 yr    **B.** Punch, 4 yr

   **C.** Rover, 3 yr    **D.** Rover, 4 yr

# 9

## MORE MULTIPLICATION FACTS

**M**ATH AND LANGUAGE ARTS

DATA BANK

Use the Language Arts Data Bank on page 479 to answer the questions.

**1** If each of the noble-men nibbled 2 nuts, how many nuts did they nibble in all? If each of them nibbled 3 nuts, how many did they nibble in all?

**2** If Fanny also fried 5 fish for Francis's mother and 5 fish for herself, how many fish did she fry in all?

**3** If 8 more gray geese gazed into Greece, how many geese would be gazing into Greece?

**4** **Use Critical Thinking** If you found 3 thumbs thumping, 5 fairies falling, and 7 sailors surfing, how many nannies would you find napping? Explain.

# 3 as a Factor
## Using Mental Math

**EXPLORE** **Use Graph Paper**

Work in Groups. You can see in the picture that $3 \times 4$ is the same as $2 \times 4$ and 4 more. Cut rectangles that show $3 \times 5$, $3 \times 6$, $3 \times 7$, $3 \times 8$, and $3 \times 9$. Cut off one row of squares. With your group, decide on a sentence that describes each fact.

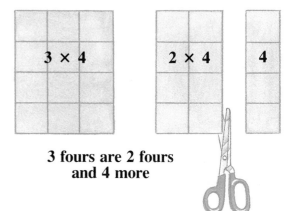

**3 fours are 2 fours and 4 more**

**TALK ABOUT IT**

1. How would you complete this sentence?
   $3 \times 8$ is $2 \times 8$ and ||||| more.
   Use your graph paper rectangles to explain.

2. Mary said "To multiply by 3, I break it apart." What do you think Mary meant by this?

Here is a method that you can use to find the product when 3 is a factor.

- Double the factor that is not a 3.

- Add that factor to the double.

   To find $3 \times 6$ or $6 \times 3$, double 6 and add 6.

   $3 \times 6 = 2 \times 6$ and 6 more
   $12 + 6 = 18$

These sums may help you multiply by 3. Solve.

( 14 + 7 )     ( 10 + 5 )     ( 18 + 9 )     ( 12 + 6 )     ( 16 + 8 )

**1.** $3 \times 7$     **2.** $3 \times 5$     **3.** $3 \times 9$     **4.** $3 \times 6$     **5.** $3 \times 8$

Multiply.

$(18 + 9)$

**1.** $\begin{array}{r} 9 \\ \times 3 \\ \hline \end{array}$ 　**2.** $\begin{array}{r} 3 \\ \times 0 \\ \hline \end{array}$ 　**3.** $\begin{array}{r} 3 \\ \times 8 \\ \hline \end{array}$ 　**4.** $\begin{array}{r} 2 \\ \times 6 \\ \hline \end{array}$ 　**5.** $\begin{array}{r} 9 \\ \times 1 \\ \hline \end{array}$ 　**6.** $\begin{array}{r} 3 \\ \times 2 \\ \hline \end{array}$

**7.** $\begin{array}{r} 8 \\ \times 3 \\ \hline \end{array}$ 　**8.** $\begin{array}{r} 3 \\ \times 7 \\ \hline \end{array}$ 　**9.** $\begin{array}{r} 6 \\ \times 3 \\ \hline \end{array}$ 　**10.** $\begin{array}{r} 7 \\ \times 5 \\ \hline \end{array}$ 　**11.** $\begin{array}{r} 3 \\ \times 3 \\ \hline \end{array}$ 　**12.** $\begin{array}{r} 9 \\ \times 4 \\ \hline \end{array}$

**13.** $3 \times 2$ 　　**14.** $2 \times 3$ 　　**15.** $3 \times 7$ 　　**16.** $9 \times 3$ 　　**17.** $3 \times 0$

**18.** $3 \times 8$ 　　**19.** $8 \times 5$ 　　**20.** $6 \times 2$ 　　**21.** $3 \times 3$ 　　**22.** $2 \times 9$

**23.** $4 \times 3$ 　　**24.** $9 \times 4$ 　　**25.** $3 \times 1$ 　　**26.** $5 \times 3$ 　　**27.** $3 \times 6$

**28.** Find the product of 3 and 8.　　**29.** Multiply 6 by 3.

APPLY

### MATH REASONING

**30.** Another name for 18 is $3 \times 6$.
Write two other names for 18.

### PROBLEM SOLVING

**31.** Solve this riddle. If you multiply me by 3, the product is the same as $16 + 8$. Who am I?

**32.** Julie and her friends went on a picnic. There were 2 station wagons with 7 students in each one. How many students went on the picnic?

### USING CRITICAL THINKING　Use Careful Reasoning

**33.** Can you find the secret number? Here are some clues.

- It is a 3 fact.
- The sum of its two digits is 9.
- It is even.

*More Practice, page 506, set A*

# Mental Math
## Finding Larger Doubles

LEARN ABOUT IT

The sums you find in this lesson will help you multiply by 4 in the next lesson.

**EXPLORE** **Think About the Process**

Jodi said, "I can find $12 + 12$ in my head. I just add $2 + 2$ to $10 + 10$." Can you use Jodi's method to find $14 + 14$?

**TALK ABOUT IT**

1. Why is $12 + 12$ the same as $20 + 4$?

2. What two numbers did you add to find $14 + 14$?

3. Why is a double like $16 + 16$ harder to find mentally than one like $13 + 13$?

$20 + 4$

$12 + 12$

Here is a method that you might use to find larger doubles in your head.

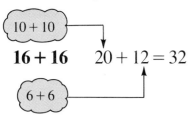

$16 + 16 \quad 20 + 12 = 32$

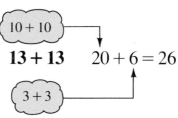

$13 + 13 \quad 20 + 6 = 26$

You may want to make up methods of your own for finding these larger doubles using mental math.

TRY IT OUT

Find the sums using mental math. Write answers only.

1. $11 + 11$    2. $17 + 17$    3. $15 + 15$    4. $12 + 12$

Find the sums using mental math. Write answers only.

**1.** $9 + 9$  **2.** $13 + 13$  **3.** $16 + 16$  **4.** $8 + 8$

**5.** $17 + 17$  **6.** $7 + 7$  **7.** $10 + 10$  **8.** $11 + 11$

Choose the letter of the sum that will help you find the double.

**9.** $18 + 18$  **10.** $15 + 15$

**11.** $11 + 11$  **12.** $13 + 13$

**13.** $16 + 16$  **14.** $14 + 14$

| | |
|---|---|
| **a.** $20 + 2$ | **b.** $20 + 16$ |
| **c.** $20 + 12$ | **d.** $20 + 18$ |
| **e.** $20 + 6$ | **f.** $20 + 8$ |
| **g.** $20 + 14$ | **h.** $20 + 10$ |

Find the products.

**15.** $\begin{array}{r} 6 \\ \times 2 \\ \hline \end{array}$  **16.** $\begin{array}{r} 5 \\ \times 6 \\ \hline \end{array}$  **17.** $\begin{array}{r} 9 \\ \times 7 \\ \hline \end{array}$  **18.** $\begin{array}{r} 2 \\ \times 7 \\ \hline \end{array}$  **19.** $\begin{array}{r} 8 \\ \times 5 \\ \hline \end{array}$  **20.** $\begin{array}{r} 9 \\ \times 9 \\ \hline \end{array}$

**21.** $\begin{array}{r} 8 \\ \times 9 \\ \hline \end{array}$  **22.** $\begin{array}{r} 2 \\ \times 8 \\ \hline \end{array}$  **23.** $\begin{array}{r} 5 \\ \times 7 \\ \hline \end{array}$  **24.** $\begin{array}{r} 8 \\ \times 0 \\ \hline \end{array}$  **25.** $\begin{array}{r} 5 \\ \times 2 \\ \hline \end{array}$  **26.** $\begin{array}{r} 9 \\ \times 6 \\ \hline \end{array}$

**MATH REASONING** Write $<$, $>$, or $=$ for each ▥.

**27.** $18 + 18$ ▥ $9 \times 4$  **28.** $3 \times 7$ ▥ $11 + 11$  **29.** $14 + 14$ ▥ $3 \times 9$

**PROBLEM SOLVING**

**30.** A triceratops ate 17 leaves. A brontosaurus ate 2 times as many. How many leaves did the brontosaurus eat?

▶ **USING CRITICAL THINKING** **Drawing a Conclusion**

Find the sum of the large double.

**31.** Since $17 + 17 = 34$, we know that $217 + 217 =$ ▥

**32.** Since $139 + 139 = 278$, we know that $439 + 439 =$ ▥

*More Practice, page 506, set B*

# 4 as a Factor
## Using Mental Math

**LEARN ABOUT IT**

**EXPLORE** **Think About the Process**
*Pure food for four mules*
Jenny said this tongue twister
6 times. How many mules is
4 times 6?

Jenny thought about egg cartons
to help her find $4 \times 6$. She said, "I
can find $4 \times 6$ by doubling $2 \times 6$."

**TALK ABOUT IT**

1. Using Jenny's method, what addition do you
   need to do to find $4 \times 6$? What is the product
   of $4 \times 6$?

2. Mary said, "To find $4 \times 8$, I think about my
   crayons!" What do you think she meant by this?

3. How would you find $4 \times 7$?

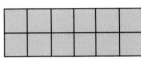

$2 \times 6 = 12$

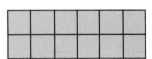

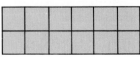

$4 \times 6 = ?$

To find the product when 4 is a factor, you can
think of doubling a double.

$4 \times 8 \rightarrow \boxed{2 \times 8} + \boxed{2 \times 8} \rightarrow 16 + 16 = 32$

**TRY IT OUT**

Practice finding these sums in your head. They
may help you multiply by 4. Solve.

$(14 + 14)$     $(10 + 10)$     $(18 + 18)$     $(12 + 12)$     $(16 + 16)$

**1.** $4 \times 7$     **2.** $4 \times 5$     **3.** $4 \times 9$     **4.** $4 \times 6$     **5.** $4 \times 8$

Multiply.

| | | | | | | |
|---|---|---|---|---|---|---|
| **1.** 4<br>×7 | **2.** 3<br>×4 | **3.** 4<br>×5 | **4.** 4<br>×9 | **5.** 4<br>×4 | **6.** 2<br>×6 | **7.** 8<br>×1 |
| **8.** 8<br>×4 | **9.** 4<br>×6 | **10.** 5<br>×4 | **11.** 5<br>×6 | **12.** 1<br>×4 | **13.** 5<br>×8 | **14.** 3<br>×7 |

**15.** $4 \times 7$    **16.** $9 \times 5$    **17.** $4 \times 5$    **18.** $6 \times 4$    **19.** $6 \times 9$

**20.** $3 \times 9$    **21.** $3 \times 4$    **22.** $4 \times 8$    **23.** $7 \times 3$    **24.** $4 \times 4$

**25.** Find the product of 9 and 4.    **26.** Multiply 7 by 4.

**MATH REASONING** Which product is greater? Use mental math.

**27.** $3 \times 6$ or $4 \times 4$    **28.** $3 \times 7$ or $4 \times 5$    **29.** $3 \times 9$ or $4 \times 7$

**PROBLEM SOLVING**

**30.** *Fanny Finch fried five floundering fish for Francis Fowler's father.* If Fanny fried 4 times as many fish, how many fish would she have fried?

**31.** **Language Arts Data Bank** If 4 times as many sand castles were sitting in the sand, how many sand castles were there? See page 479.

Choose the word that gives the better estimate.

**32.** A soup bowl holds about 2 _____ of soup.    **A.** cups    **B.** quarts
**33.** An aquarium holds about 7 _____ of water.    **A.** pints    **B.** gallons
**34.** A large book weighs about 3 _____.    **A.** ounces    **B.** pounds
**35.** A doorknob is about 1 _____ off the floor.    **A.** foot    **B.** yard
**36.** A strawberry weighs about 1 _____.    **A.** ounce    **B.** pound

# Basic Facts That Are Squares

**EXPLORE Model with Graph Paper**
Work in groups. Ben cut a square from graph paper
to show a fact where both factors are the same. He
wrote the fact on the front and the product on the
back. With your group, make squares like this for
all the square facts from $5 \times 5$ to $9 \times 9$.

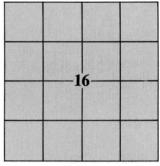

**Front**

**Back**

**TALK ABOUT IT**

1. Which of the square facts are new facts?

2. The number 25 is a square fact in the
   twenties. Is there a square fact in the thirties?
   forties? fifties? sixties? seventies? eighties?

Here are rhymes that may help you remember the
larger square facts.

- $6 \times 6$ tried some tricks, and changed its name
  to 36.

- $7 \times 7$ did just fine, so they called it 49.

- $8 \times 8$ went to the store, to get its name tag 64.

- $9 \times 9$ had some fun, when it used its
  nickname 81.

**TRY IT OUT**

Find the products.

| 1. $3 \times 3$ | 2. $7 \times 7$ | 3. $8 \times 8$ | 4. $5 \times 5$ | 5. $9 \times 9$ |

| 6. $\begin{array}{r} 1 \\ \times 1 \end{array}$ | 7. $\begin{array}{r} 2 \\ \times 2 \end{array}$ | 8. $\begin{array}{r} 4 \\ \times 4 \end{array}$ | 9. $\begin{array}{r} 7 \\ \times 7 \end{array}$ | 10. $\begin{array}{r} 6 \\ \times 6 \end{array}$ |

Find the products.

**1.** 7    **2.** 7    **3.** 8    **4.** 4    **5.** 4    **6.** 5
    $\times 7$     $\times 3$     $\times 8$     $\times 8$     $\times 4$     $\times 5$

**7.** 5    **8.** 3    **9.** 2    **10.** 9    **11.** 3    **12.** 6
    $\times 7$     $\times 3$     $\times 9$     $\times 9$     $\times 8$     $\times 6$

**13.** $2 \times 2$     **14.** $6 \times 6$     **15.** $9 \times 3$     **16.** $8 \times 8$     **17.** $4 \times 8$

**18.** $5 \times 5$     **19.** $5 \times 7$     **20.** $9 \times 9$     **21.** $6 \times 3$     **22.** $7 \times 7$

**23.** What is the product of 8 and 8?     **24.** Multiply 7 by 9.

**MATH REASONING**

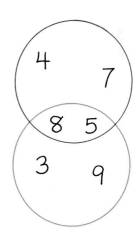

**25.** What is the product of the numbers in the red ring but not the blue ring?

**26.** What is the product of the numbers not in the red ring?

**27.** What is the product of the numbers in both the red and the blue rings?

**PROBLEM SOLVING**

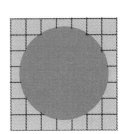

**28.** The blue cloth covers part of the table top. How many tiles are on the table?

▶ **ESTIMATION**

Each tank holds 10 gallons. Look at the gauges.
Estimate how many gallons of fuel oil are in each tank.

**29.** 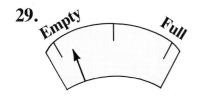    **30.** 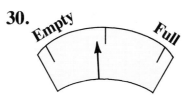    **31.**

*More Practice, page 506, set D*

# Problem Solving
## Multiple-Step Problems

UNDERSTAND
FIND DATA
PLAN
ESTIMATE
SOLVE
CHECK

### LEARN ABOUT IT

You may need to use more than one operation in order to solve some problems. These problems are called multiple-step problems.

> Jimmy bought some fish to stock his aquarium. He bought 5 zebra danios at $1 each and 4 velvet swordtails at $2 each. How much did he spend?

First I'll find out how much the zebra danios cost.

$$5 \times \$1 = \$5$$

Next I'll find out how much the velvet swordtails cost.

$$4 \times \$2 = \$8$$

Now I'll add the prices together.

$$\$5 + \$8 = \$13$$

Jimmy spent $13.

### TRY IT OUT

Solve.

1. Gail bought 6 pom pom orandas for $6 each and 3 red fantails for $4 each. What was the total cost of all the fish?

2. Stones for the bottom of aquariums are $3 for a bag of black stones and $2 for a bag of white stones. Joe's dad needs 5 bags of each. How much will it cost?

3. The price of admission at the Berkhart Aquarium is $3 for children. Dan brought along $12.85 to buy 7 children's tickets. How much more money did he need?

4. Paco's mother was given 47 fish. The next day, she bought 4 bags with 4 fish in each bag. How many fish did she own in all?

Solve. Use any problem solving strategy.

**1.** Jeff bought 3 bags of stones for his aquarium. Each bag weighed 5 pounds. How many pounds of stones did he have in all?

**2.** Mai came to the store with $40. She bought a $3 \frac{1}{2}$ -gallon aquarium and 2 castles. How much money did she have left?

**3.** How much more does a 10-gallon aquarium cost than a $3 \frac{1}{2}$ -gallon aquarium?

**4.** Kissing gourami are 3 inches long. Angelfish are 4 inches long. Samantha decided to buy 4 gourami at $4 each because of the size of her aquarium. How much did she pay in all?

**5.** José had 5 fish when he was 8 years old, 7 fish when he was 9, and 9 fish when he was 10. At this rate, how many fish will he have when he is 13 years old?

| Aquarium Supplies | |
| --- | --- |
| 55-gallon aquarium | $129.98 |
| 10-gallon aquarium | $34.50 |
| $3 \frac{1}{2}$ -gallon aquarium | $24.99 |
| pump | $21.79 |
| filter | $19.00 |
| fish food | $5.00 |
| castles | $3.00 |
| plastic plants | $2.50 |
| net | $1.39 |
| seaweed | $1.25 |

**6.** At the school carnival, prizes at the Go Fish booth are goldfish. There are 9 big bowls with 4 fish in each. There are 8 small bowls with 3 fish in each. How many goldfish are there all together?

**7.** How much would it cost for Latania to buy the $3 \frac{1}{2}$ -gallon aquarium, the pump, filter, and net?

**8. Understanding the Operations**
Tell what operations you would use. Use objects to solve the problem.

John had 4 bowls of guppies with the same number in each bowl. He gave away 1 bowlful. Now he has 9 guppies. How many guppies did he have to start?

*More Practice, page 521, set A*

# The Last Three Facts
$6 \times 7, 6 \times 8, 7 \times 8$

**LEARN ABOUT IT**

You have now studied the facts for 0, 1, 2, 3, 4, 5, and 9, and the square facts. There are only 3 more facts to learn, $6 \times 7$, $6 \times 8$, and $7 \times 8$.

**EXPLORE  Think About the Process**

*Silly Sally swiftly shooed seven silly sheep.* Tom said this tongue twister 6 times. How many sheep is $6 \times 7$?

Tom used graph paper to help him figure out $6 \times 7$.

> Break apart $6 \times 7$ into $3 \times 7$ and $3 \times 7$.

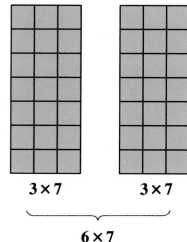

$3 \times 7$      $3 \times 7$

$6 \times 7$

**TALK ABOUT IT**

1. What is the product of $6 \times 7$? How do you know?

2. How could you use Tom's idea to figure out $6 \times 8$?

$6 \times 7$ is the same as $3 \times 7$ and $3 \times 7$.

$6 \times 8$ is the same as $3 \times 8$ and $3 \times 8$.

$7 \times 8$ is the same as 6 eights and 8 more.

Sometimes it helps to reverse the factors to find a fact. Changing the order of factors does not change the product.

$7 \times 6 = 6 \times 7$      $8 \times 6 = 6 \times 8$      $8 \times 7 = 7 \times 8$

**TRY IT OUT**

Find the products.

**1.** $6 \times 7$    **2.** $7 \times 6$    **3.** $6 \times 8$    **4.** $8 \times 6$    **5.** $7 \times 8$    **6.** $8 \times 7$

Find the products.

| | | | | | |
|---|---|---|---|---|---|
| **1.** $\begin{array}{r} 6 \\ \times 7 \\ \hline \end{array}$ | **2.** $\begin{array}{r} 9 \\ \times 8 \\ \hline \end{array}$ | **3.** $\begin{array}{r} 7 \\ \times 8 \\ \hline \end{array}$ | **4.** $\begin{array}{r} 8 \\ \times 4 \\ \hline \end{array}$ | **5.** $\begin{array}{r} 8 \\ \times 6 \\ \hline \end{array}$ | **6.** $\begin{array}{r} 6 \\ \times 9 \\ \hline \end{array}$ |
| **7.** $\begin{array}{r} 7 \\ \times 5 \\ \hline \end{array}$ | **8.** $\begin{array}{r} 8 \\ \times 7 \\ \hline \end{array}$ | **9.** $\begin{array}{r} 7 \\ \times 7 \\ \hline \end{array}$ | **10.** $\begin{array}{r} 7 \\ \times 6 \\ \hline \end{array}$ | **11.** $\begin{array}{r} 5 \\ \times 7 \\ \hline \end{array}$ | **12.** $\begin{array}{r} 6 \\ \times 8 \\ \hline \end{array}$ |

**13.** $9 \times 5$  **14.** $6 \times 7$  **15.** $8 \times 6$  **16.** $5 \times 5$  **17.** $7 \times 8$

**18.** $6 \times 8$  **19.** $7 \times 9$  **20.** $8 \times 7$  **21.** $7 \times 6$  **22.** $2 \times 9$

**23.** Find the product of 6 and 9.   **24.** Multiply 5 by 6.

**APPLY**

**MATH REASONING**

**25.** 24 is a 3 fact ($3 \times 8$). 24 is also a 4 fact ($4 \times 6$). What other number is both a 3 and a 4 fact?

**PROBLEM SOLVING**

**26. Language Arts Data Bank** Find a tongue twister with the number 6, 7, or 8. Write and answer a question about it. See page 479.

**MIXED REVIEW**

The pictograph shows how long Kim practices piano each day.

**27.** How many minutes does Kim practice on Monday?

**28.** Which day does Kim practice only one-half hour?

**29.** Write something you can tell from the graph.

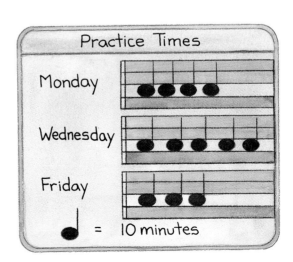

*More Practice, page 506, set E*

# Exploring Algebra
## Missing Factors

You can use multiplication facts to help you find a missing factor.

Statement A can be read as, "2 times what number is 12?"

**A.** $2 \times \boxed{\phantom{6}} = 12$

Statement B shows that the missing factor is 6.

**B.** $2 \times \boxed{6} = 12$

Fill in the missing factor in statement C.

**C.** $4 \times \boxed{\phantom{6}} = 20$

**TALK ABOUT IT**

1. Which fact did you use to complete statement C?

2. Is the missing factor in $4 \times \boxed{\phantom{6}} = 12$ the same as the missing factor in $\boxed{\phantom{6}} \times 4 = 12$? How do you know?

3. Give two missing factor problems you can solve if you know $2 \times 7 = 14$.

**TRY IT OUT**

Find the missing factor.

**1.** $3 \times \boxed{\phantom{6}} = 15$    **2.** $\boxed{\phantom{6}} \times 2 = 14$    **3.** $4 \times \boxed{\phantom{6}} = 32$    **4.** $\boxed{\phantom{6}} \times 8 = 40$

**5.** $5 \times \boxed{\phantom{6}} = 20$    **6.** $\boxed{\phantom{6}} \times 7 = 28$    **7.** $2 \times \boxed{\phantom{6}} = 18$    **8.** $\boxed{\phantom{6}} \times 4 = 24$

**9.** $3 \times \boxed{\phantom{6}} = 21$    **10.** $\boxed{\phantom{6}} \times 6 = 36$    **11.** $5 \times \boxed{\phantom{6}} = 30$    **12.** $\boxed{\phantom{6}} \times 9 = 27$

**13.** $7 \times \boxed{\phantom{6}} = 35$    **14.** $\boxed{\phantom{6}} \times 3 = 24$    **15.** $5 \times \boxed{\phantom{6}} = 25$    **16.** $\boxed{\phantom{6}} \times 3 = 12$

# MIDCHAPTER REVIEW/QUIZ

**1.** $3 \times 5$     **2.** $4 \times 8$     **3.** $8 \times 8$     **4.** $6 \times 6$     **5.** $6 \times 7$

**6.** $9 \times 4$     **7.** $7 \times 3$     **8.** $4 \times 4$     **9.** $3 \times 0$     **10.** $8 \times 7$

**11.** $\begin{array}{r} 5 \\ \times 4 \\ \hline \end{array}$
**12.** $\begin{array}{r} 9 \\ \times 9 \\ \hline \end{array}$
**13.** $\begin{array}{r} 8 \\ \times 3 \\ \hline \end{array}$
**14.** $\begin{array}{r} 1 \\ \times 4 \\ \hline \end{array}$
**15.** $\begin{array}{r} 7 \\ \times 6 \\ \hline \end{array}$
**16.** $\begin{array}{r} 4 \\ \times 7 \\ \hline \end{array}$

**17.** $\begin{array}{r} 6 \\ \times 8 \\ \hline \end{array}$
**18.** $\begin{array}{r} 7 \\ \times 8 \\ \hline \end{array}$
**19.** $\begin{array}{r} 6 \\ \times 4 \\ \hline \end{array}$
**20.** $\begin{array}{r} 7 \\ \times 7 \\ \hline \end{array}$
**21.** $\begin{array}{r} 9 \\ \times 3 \\ \hline \end{array}$
**22.** $\begin{array}{r} 8 \\ \times 6 \\ \hline \end{array}$

Use mental math to find these doubles.

**23.** $12 + 12$     **24.** $18 + 18$     **25.** $19 + 19$     **26.** $16 + 16$     **27.** $14 + 14$

Fill in the blanks. Show that your answer works.

**28.** To find 3 sevens, find 2 sevens and add ‖‖‖.

**29.** To find $3 \times 8$, double ‖‖‖ and add 8.

**30.** To find $4 \times 6$, think: if 2 sixes are 12, then 4 sixes are ‖‖‖.

**31.** To find $4 \times 9$, think: if 2 nines are ‖‖‖, then 4 nines are ‖‖‖.

**32.** To find $6 \times 7$, think: if 3 sevens are 21, then 6 sevens are ‖‖‖.

## PROBLEM SOLVING

**33.** Glenna had 7 nickels. Suzi had 6 dimes. Together could they buy a greeting card for $1?

**34.** Barry buys six 3¢-stamps and eight 4¢-stamps. How much change should he get from 50¢?

**35.** John brought seven 6-packs of canned juice for the class party. If there are 19 students, can they each have 2 cans of juice?

# Mastering the Facts

It is useful to work toward remembering the basic multiplication facts.

**EXPLORE**  **Remember the Facts**

Work with a partner. Each of you choose a different box of facts. Stare at your fact box for 30 seconds. Have your partner time you. Look away and try to say the facts to your partner. Take turns. Then trade boxes and try again.

**TALK ABOUT IT**

1. Which of the boxes was easiest to remember? Why?
2. When you looked away, were you able to see your products in your mind?
3. Have you found any easy methods for remembering facts? Explain.

■ Sometimes you can remember a fact by saying it over and over to yourself.

■ Sometimes you can remember a fact by picturing it in your mind's eye.

■ Sometimes you can make up your own trick for remembering a fact.

$6 \times 7 = 42$
$7 \times 8 = 56$

**Box A**

$6 \times 8 = 48$
$7 \times 9 = 63$

**Box B**

**TRY IT OUT**

Find the products. Practice giving the products as quickly as possible.

**1.** $5 \times 8$  **2.** $8 \times 4$  **3.** $3 \times 6$  **4.** $9 \times 6$  **5.** $8 \times 7$

**6.** $6 \times 4$  **7.** $8 \times 9$  **8.** $7 \times 7$  **9.** $5 \times 8$  **10.** $7 \times 5$

Find the products. Practice giving the products quickly.

**1.** $8 \times 6$     **2.** $7 \times 9$     **3.** $6 \times 3$     **4.** $3 \times 0$     **5.** $3 \times 7$

**6.** $6 \times 6$     **7.** $4 \times 3$     **8.** $4 \times 8$     **9.** $9 \times 5$     **10.** $7 \times 5$

**11.** $\begin{array}{r} 4 \\ \times 3 \\ \hline \end{array}$   **12.** $\begin{array}{r} 5 \\ \times 0 \\ \hline \end{array}$   **13.** $\begin{array}{r} 8 \\ \times 9 \\ \hline \end{array}$   **14.** $\begin{array}{r} 4 \\ \times 7 \\ \hline \end{array}$   **15.** $\begin{array}{r} 6 \\ \times 7 \\ \hline \end{array}$   **16.** $\begin{array}{r} 8 \\ \times 8 \\ \hline \end{array}$

**17.** Find the product of 3 and 9.     **18.** Multiply 6 by 0.

**APPLY**

**MATH REASONING**

**19.** Give the missing number.
Since $85 \times 6 = 510$, we know that
$86 \times 6 = $ ‖‖‖.

**PROBLEM SOLVING**

**20.** In *Alice in Wonderland,* Alice asked the Mock Turtle how many hours a day he did lessons. "Ten hours the first day," said the Mock Turtle: "nine the next, and so on."

"What a curious plan!" exclaimed Alice.

How many hours did the Mock Turtle do lessons in the first three days?

**CRITICAL THINKING  Discover a Relationship**

Study the factors in each box and find their products.

| $5 \times 5$ | $6 \times 6$ | $7 \times 7$ | $8 \times 8$ |
| $4 \times 6$ | $5 \times 7$ | $6 \times 8$ | $7 \times 9$ |

**21.** What do you notice about the two pairs of factors in each box?

**22.** What do you notice about the two products?

# Mental Math
## Multiplying 3 Factors

LEARN ABOUT IT

When you use mental math to multiply 3 factors, you must decide which 2 factors to multiply first.

### EXPLORE Model with Blocks
There are 3 tennis balls in each can and 2 cans in each box. Jon has enough money to buy 4 boxes. How many tennis balls will he have in all?

### TALK ABOUT IT

1. How many tennis balls are in each box?

2. How many balls are in all 4 boxes?

3. Can you use blocks to show another way to use multiplication to find the total number of balls?

Grouping symbols ( ) tell which multiplication to do first. When none are shown, you can pick any two factors to multiply. Notice that you can multiply any two factors first, and the product will be the same.

Find the product of the factors 3, 2, and 4.

| Multiply these. | Multiply these. | Pick any two. Try these. |
|---|---|---|
| $(3 \times 2) \times 4$ | $3 \times (2 \times 4)$ | $3 \times 2 \times 4$ |
| $6 \quad \times 4 = 24$ | $3 \times \quad 8 = 24$ | $12 \times 2 = 24$ |

### TRY IT OUT

Use mental math to find the products.

**1.** $(3 \times 2) \times 5$　　**2.** $3 \times (5 \times 4)$　　**3.** $(2 \times 5) \times 3$　　**4.** $4 \times (2 \times 4)$

254

Use mental math to find the products. Use the groupings shown.

**1.** $(2 \times 4) \times 1$
**2.** $2 \times (4 \times 1)$
**3.** $(2 \times 1) \times 4$
**4.** $2 \times (1 \times 4)$

Multiply. Use any groupings.

**5.** $4 \times 1 \times 5$
**6.** $4 \times 2 \times 4$
**7.** $3 \times 2 \times 5$
**8.** $4 \times 1 \times 4$

**9.** $4 \times 2 \times 3$
**10.** $6 \times 1 \times 3$
**11.** $3 \times 2 \times 2$
**12.** $2 \times 4 \times 2$

**13.** $8 \times 9 \times 0$
**14.** $4 \times 1 \times 5$
**15.** $6 \times 0 \times 7$
**16.** $3 \times 3 \times 2$

**17.** $2 \times 4 \times 4$
**18.** $9 \times 1 \times 4$
**19.** $3 \times 1 \times 3$
**20.** $2 \times 2 \times 2$

**21.** Use counters to show $(3 \times 2) \times 4$. Next, show $3 \times (2 \times 4)$. Compare the products.

**APPLY**

**MATH REASONING**

**22.** Since $3 \times 4 \times 5 = 60$, we know that $5 \times 3 \times 4 =$ ▏▏▏▏.

**23.** Since $8 \times 7 \times 6 = 336$, we know that $7 \times 8 \times 6 =$ ▏▏▏▏.

**PROBLEM SOLVING**

**24.** Erica put 2 shoes in each box. She put 4 boxes of shoes on each shelf. She filled 3 shelves. How many shoes were on the shelves?

**25.** The team scored 4 touchdowns. Each touchdown was worth 6 points. They also made 3 extra points. How many points did the team score in all?

▶ **ALGEBRA**

In the following problems, each shape stands for one number. Find the number for each shape.

**26.** $\square \times \square \times \square = 1$
$\square + \triangle = 5$

**27.** $\bigcirc \times \bigcirc \times \bigcirc = 8$
$\bigcirc + \triangle + \triangle = 10$

*More Practice, page 507, set A*

# Multiples

## LEARN ABOUT IT

**EXPLORE** **Use a Calculator**

Jane used a calculator and these key codes to complete these tables. Use a calculator to check her work.

| ON/AC | 0 | × | 4 | = | 1 | = | 2 | = | 3 | = |

| × | 0 | 1 | 2 | 3 | 4 | 5 | 6 | 7 | 8 | 9 |
|---|---|---|---|---|---|---|---|---|---|---|
| 4 | 0 | 4 | 8 | 12 | | | | | | |

Multiples of 4

| ON/AC | 0 | × | 5 | = | 1 | = | 2 | = | 3 | = |

| × | 0 | 1 | 2 | 3 | 4 | 5 | 6 | 7 | 8 | 9 |
|---|---|---|---|---|---|---|---|---|---|---|
| 5 | 0 | 5 | 10 | 15 | | | | | | |

Multiples of 5

## TALK ABOUT IT

1. What are the missing numbers in the table for 4? for 5?
2. What key code would you use to multiply by 6?

■ A **multiple** of a number is the product of that number and another factor.

The products in the first table are multiples of 4. The products in the second table are multiples of 5.

**Example** Some multiples of 3: 0, 3, 6, 9, 12, 15, . . .

## TRY IT OUT

1. These numbers are multiples of what number?
   0, 7, 14, 21, 28, . . .

2. Find the first 10 multiples of 8 and of 9.

Copy the pyramid and its title. Give the missing numbers.

**1.**

```
        0
      2   4
    6   8  ▨
  ▨  ▨  16  18
```

Multiples of 2

**2.**

```
        0
      5   10
   15   ▨   25
  30  ▨  ▨  ▨
```

Multiples of _____

**3.**

```
        ▨
      ▨   ▨
   24  32  40
  48  ▨  ▨  ▨
```

Multiples of _____

Find the products.

**4.**  4
     ×6

**5.**  9
     ×6

**6.**  4
     ×7

**7.**  9
     ×9

**8.**  5
     ×9

**9.**  8
     ×4

**10.**  5
      ×8

**11.**  7
      ×3

**12.**  9
      ×7

**13.**  8
      ×7

**14.**  5
      ×7

**15.**  6
      ×5

**MATH REASONING**

**16.** Give the missing number. The multiples of 6 are also multiples of 2 and ▨.

**PROBLEM SOLVING**

**17.** Suppose a grasshopper and a cricket start jumping along a number line. They start at 0. The grasshopper jumps 3 units and the cricket jumps 2 units. Both insects landed on the number 6. What other numbers less than 40 can they both land on?

▶ **COMMUNICATION**

**18.** Copy and complete this sentence. The number 0 is a multiple of every number because _____.

# Practicing the Facts

Sometimes when you forget a fact, you may be able to fall back on one of the strategies.

**EXPLORE** **Think About the Process**
Pretend you have forgotten each of the products in the box. Can you figure out the products using the strategies you have learned so far?

3X9  4X8
5X8  9X4
7X2  5X3

**TALK ABOUT IT**

1. Were you able to think of a strategy for each of the facts?

2. Is there more than one strategy that could be used for some of the facts? Explain.

Here are some of the strategies that may be helpful when you forget a fact.

| Using Doubles | Skip Counting | Multiplying by 9 | Multiplying in Parts |
|:---:|:---:|:---:|:---:|
| $7 + 7$ | $5, 10, 15, 20, 25, 30$ | $80 - 8$ | $14 + 14$ |
| $2 \times 7$ | $6 \times 5$ | $9 \times 8$ | $4 \times 7$ |

**TRY IT OUT**

Find the products. Practice giving the products as quickly as possible.

1. $7 \times 8$
2. $9 \times 7$
3. $5 \times 9$
4. $5 \times 8$
5. $3 \times 6$

6. $8 \times 8$
7. $6 \times 2$
8. $6 \times 9$
9. $6 \times 7$
10. $3 \times 5$

11. $7 \times 5$
12. $9 \times 9$
13. $3 \times 8$
14. $6 \times 4$
15. $4 \times 4$

16. $5 \times 5$
17. $8 \times 4$
18. $6 \times 6$
19. $7 \times 7$
20. $8 \times 9$

Find the products. Practice giving the products as quickly as possible.

**1.** $9 \times 7$   **2.** $8 \times 6$   **3.** $5 \times 0$   **4.** $7 \times 4$   **5.** $9 \times 4$

**6.** $7 \times 6$   **7.** $8 \times 8$   **8.** $9 \times 5$   **9.** $1 \times 9$   **10.** $8 \times 3$

**11.** $5 \times 6$   **12.** $0 \times 7$   **13.** $6 \times 9$   **14.** $6 \times 2$   **15.** $3 \times 1$

**16.**  3
$\times 5$
**17.**  1
$\times 8$
**18.**  2
$\times 2$
**19.**  9
$\times 9$
**20.**  7
$\times 8$
**21.**  7
$\times 7$

**22.**  4
$\times 4$
**23.**  8
$\times 6$
**24.**  8
$\times 9$
**25.**  5
$\times 2$
**26.**  5
$\times 7$
**27.**  5
$\times 8$

**28.**  4
$\times 1$
**29.**  7
$\times 4$
**30.**  2
$\times 8$
**31.**  6
$\times 0$
**32.**  5
$\times 5$
**33.**  9
$\times 5$

**34.** Find the product of 6 and 3.   **35.** Multiply 6 by 9.

**MATH REASONING**

**36.** Study the pattern. Give the missing product.

| 0 | → | 57 | → | 114 | → | 171 | → | 228 | → | ____ |
|---|---|----|----|-----|---|-----|---|-----|---|------|
| $0 \times 57$ | | $1 \times 57$ | | $2 \times 57$ | | $3 \times 57$ | | $4 \times 57$ | | $5 \times 57$ |

**PROBLEM SOLVING**

**37.** Trevor wrote a poem. His poem had 6 stanzas. Each stanza was 4 lines. How many lines did Trevor's poem have?

▶ **MENTAL MATH**

Break apart the numbers to find the products. Use mental math.

**38.** $3 \times 14$   **39.** $4 \times 16$   **40.** $5 \times 12$   **41.** $5 \times 16$   **42.** $3 \times 16$

259

# Problem Solving
## Problems with More Than One Answer

UNDERSTAND
FIND DATA
PLAN
ESTIMATE
SOLVE
CHECK

### LEARN ABOUT IT

Some problems have more than one answer. After you find one answer, ask yourself if there may be other answers.

**4 rows**

> In 1779, John Paul Jones' ship flew a flag with 12 stars. If there were the same number of stars in each row, how many rows could there have been?

**3 rows**

Look at the pictures to see some ways the stars could have been arranged.

This problem has more than one answer. There could have been 4 rows, 3 rows, 6 rows, or 2 rows of stars.

**6 rows**

### TRY IT OUT

Use objects to solve.

**2 rows**

1. At the football game, 10 students performed with flags. The same number of students lined up in each row. How many rows could there have been?

2. At the United Nations' Day assembly, 21 students held flags of different countries. If they lined up on the stage in equal rows, how many rows could there have been?

3. After Alaska became a state in 1959, the United States flag had 49 stars. Each row had the same number of stars. How many rows were there?

4. The Confederate flag has 13 stars forming an X. One star is in the middle. Each arm of the X has the same number of stars. How many stars are there in each arm?

Choose a strategy from the list or use other strategies that you know.

1. The flag that the *Star-Spangled Banner* was written about had 15 stars. How many rows of stars could there have been if there were the same number of stars in each row?

2. The first United States Navy Jack flag had a rattlesnake and 13 stripes. It was adopted in 1775. How many years ago was that?

**1775-Navy Jack**

3. On the Fourth of July, Johann bought some tiny flags. He gave away 5 to friends. He got 3 more at the school assembly. He ended with 9. How many flags did he buy in the first place?

4. Ana made a flag out of cloth. It had 14 stars. She glued the same number of felt stars in each row. How many rows could there have been?

5. How many more stars did the flag have in 1960 than in 1818?

6. If each star stands for a state, how many states were added to the United States since 1861?

7. For sharing, Mona brought 35 small flags from her father's factory. In her classroom, there are 7 rows of desks with 4 students in each row. If she gave one flag to each student, how many flags did she have left over?

| Flags in U.S. History | | |
|---|---|---|
| Year Adopted | Number of Stars | Number of Stripes |
| 1777 | 13 | 13 |
| 1795 | 15 | 15 |
| 1818 | 20 | 13 |
| 1861 | 34 | 13 |
| 1912 | 48 | 13 |
| 1959 | 49 | 13 |
| 1960 | 50 | 13 |

8. How many years has it been since the United States has added a new state star to the flag?

# Applied Problem Solving
## Group Decision Making

UNDERSTAND
FIND DATA
PLAN
ESTIMATE
SOLVE
CHECK

**Group Skill:**

Encourage and Respect Others

Judy's mom just got a raise and wants to spend $60 on a day of fun. Judy has invited you to come along. Three people in Judy's family want to go to the baseball game. Three want to go to the water slides. You must break the tie. To help you decide, compare what the family could do at the game and at the slides with $60.

## Facts to Consider

| Baseball game | |
|---|---|
| **Tickets** | |
| Box seats | $8.00 |
| Regular seats | $4.00 |
| **Food** | |
| Hotdog | $2.00 |
| Caramel Corn | $1.50 |
| Popcorn | $1.25 |
| Drink | $1.00 |

| Water Slides | |
|---|---|
| **Tickets** | |
| All day (9 a.m.-9 p.m.) | $9.00 |
| Half day (3 p.m.-9 p.m.) | $5.00 |
| **Food** | |
| Hamburger | $2.50 |
| Hotdog | $2.00 |
| Nachos | $1.50 |
| Fries | $1.25 |
| Drink | $1.00 |

1. At the ballgame, would you buy box seats for everyone? Why or why not?

2. How much would regular seats for everyone at the game cost?

3. How much would it cost to buy regular seats and a hot dog for everyone? How much money would be left over?

4. Which tickets would you buy at the slides? How much money would be left over for food?

Where did you decide to go for the day of fun? Use a list or table to show Judy's family what they can do at the baseball game and at the slides for $60.

NO PEPPER GAMES

# WRAP UP

## Multiple Match Game

Match each phrase on the left with a number on the right. Use each letter only once.

1. the least multiple of two and one      **a.** 48
2. a multiple of zero and one      **b.** 4
3. a multiple of three and five      **c.** 2
4. the least multiple of two and four      **d.** 81
5. a multiple of six and nine      **e.** 7
6. a product of three doubled and four doubled      **f.** 15
7. a multiple of one and seven      **g.** 0
8. a product of two equal factors      **h.** 18

## Sometimes, Always, Never

Decide which word should go in the blank, *sometimes, always,* or *never.* Explain your choices.

9. Five times a nonzero number is _____ 1 more of that number than 4 times the number.

10. You _____ need grouping symbols to tell which operation to do first.

11. Knowing multiplication facts _____ is helpful when you want to find the missing number in a problem such as $3 \times \square = 18$.

12. A problem _____ has more than one correct answer.

## Project

Use graph paper. Draw a picture for each of the square facts from $1 \times 1$ through $9 \times 9$. Write each fact with its product under its picture. What patterns can you find in the pictures and the products?

# CHAPTER REVIEW/TEST

## Part 1  Understanding

Tell what mental math method you use to solve
each problem. Solve.

**1.** $3 \times 7$  **2.** $15 \times 2$  **3.** $4 \times 5$  **4.** $6 \times 7$  **5.** $8 \times 8$

**6.** Write two missing factor problems for $3 \times 2 = 6$.

**7.** Show two ways to group to find the product of $2 \times 3 \times 4$.

**8.** These numbers are all multiples of what number? Give the largest number that you can find. 0, 9, 18, 27, 36 . . .

**9.** Wanda bought 4 tulips for $2 each and 3 rose bushes for $5 each. What operations help you find how much she spent?

## Part 2  Skills

Multiply.

**10.**  $\begin{array}{r} 8 \\ \times 3 \\ \hline \end{array}$  **11.**  $\begin{array}{r} 17 \\ \times 2 \\ \hline \end{array}$  **12.**  $\begin{array}{r} 7 \\ \times 4 \\ \hline \end{array}$  **13.**  $\begin{array}{r} 3 \\ \times 3 \\ \hline \end{array}$  **14.**  $\begin{array}{r} 14 \\ \times 2 \\ \hline \end{array}$

**15.** $6 \times 8$  **16.** $8 \times 7$  **17.** $4 \times 2 \times 3$  **18.** $3 \times 2 \times 5$

Find the missing factors.

**19.** $8 \times \square = 32$  **20.** $\square \times 6 = 42$  **21.** $3 \times \square = 21$

## Part 3  Applications

**22.** Elsie had 18 marbles. She bought 3 bags with 5 marbles in each bag. How many marbles does she have now?

**23.** In June Tim had to water the lawn every four days, starting on June 4. On what dates did Tim water the lawn?

**24.** **Challenge** Jim is planting a total of 12 tulip and daffodil bulbs in his garden. He wants an equal number of daffodils and tulips. He wants to plant tulips and daffodils in separate rows. What arrangements can he make?

# ENRICHMENT
## Make a 9s Finger Factor Machine

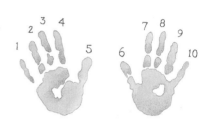

- Place your two hands, palms down, on your desk or table.

  Start at the left. Give each finger a factor from 1 to 10.

  Here's how to use your Finger Factor Machine to find the product of any 9s fact.

- Turn under the finger with the factor that you want to multiply by 9.

  $9 \times 4 = ?$    Turn under finger **4**.

- The finger turned under will separate the tens place from the ones place.

- The number of fingers to the **left** of the finger turned under tells you the number of tens, 3.

  The number of fingers to the **right** of the finger turned under tells you the number of ones, 6.

 tens        ones

**3 tens**          **6 ones**

$9 \times 4 = 36$

Now use your Finger Factor Machine to find these products.

**1.** $9 \times 7$

**2.** $9 \times 8$

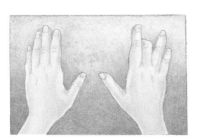

**3.** $9 \times 2$

**4.** $9 \times 6$

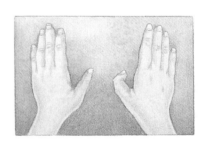

# CUMULATIVE REVIEW

1. What does a bar graph help you do with two amounts?

   A. estimate    B. round

   C. compare    D. add

2. Which shows 5 on a tally chart?

   A. ///    B. ####

   C. ###    D. #### /

3. On a picture graph, 1 picture stands for 2 students. How many pictures do you draw to show 6 students?

   A. 3    B. 12

   C. 2    D. 1

4. Measure to find the perimeter.

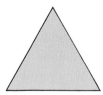

   A. 9 inches    B. 3 inches

   C. 1 inch    D. 4 inches

5. Which weighs over 1 pound?

   A. raindrop    B. ruler

   C. sock    D. book

6. Which is more than 1 gallon?

   A. 4 pints    B. 15 cups

   C. 5 quarts    D. 7 pints

7. Which temperature suggests weather for a swim in a lake?

   A. 17°F    B. 32°F

   C. 55°F    D. 78°F

8. Which suggests multiplication?

   A. 4 groups of 2

   B. 4 and 7

   C. 1 less than 9

   D. 3 more than 8

9. Which fact shows seven nines?

   A. $7 + 9$    B. $7 + 7 + 9$

   C. $7 \times 9$    D. $63 - 7$

10. $8 \times 2$

   A. 8    B. 10

   C. 14    D. 16

11. $9 \times 9$

   A. 18    B. 72

   C. 81    D. 99

12. Steven collected travel posters. He gave away 8 to friends. He lost 2. There were 15 left. How many did Steven start with?

   A. 15    B. 17

   C. 23    D. 25

267

# 10

## GEOMETRY

**M**ATH AND
FINE ARTS

### DATA BANK

Use the Fine Arts
Data Bank on
page 470 to answer
the questions.

1 Which painter used
straight lines to plan
the painting?

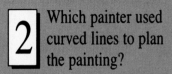 **2** Which painter used curved lines to plan the painting?

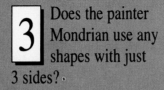 **3** Does the painter Mondrian use any shapes with just 3 sides?

**4** **Use Critical Thinking** Painters often use patterns. What patterns can you find in the paintings by Cassatt and Mondrian?

# Space Figures

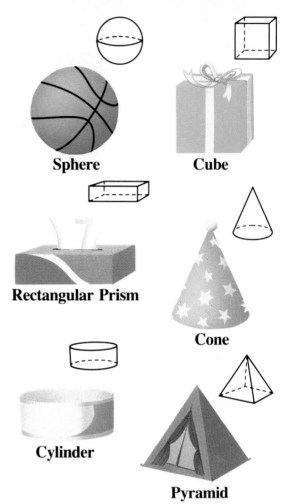

**Sphere**

**Cube**

**Rectangular Prism**

**Cone**

**Cylinder**

**Pyramid**

## LEARN ABOUT IT

**EXPLORE Make a Decision**

These familiar objects are all **space figures.** They have special names. Imagine how you would sort them into two boxes.

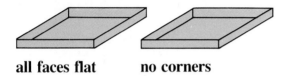

**all faces flat**          **no corners**

## TALK ABOUT IT

**1.** Which space figures did you put into each box? Explain.

**2.** Think of another way to sort the figures. What label would you put on the boxes?

You can talk about space figures in different ways.

- How many and what kind of faces does the figure have?
  Example: A rectangular prism has all flat faces.

- How many corners does the figure have?
  Example: A cylinder has no corners.

- What kind of edges does the figure have? Will it roll?
  Example: A sphere has no edges. It will roll.

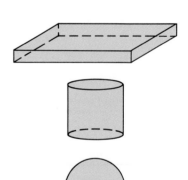

## TRY IT OUT

What examples of each figure are in your classroom?

**1.** sphere          **2.** cube          **3.** cylinder          **4.** rectangular prism

270

Write sphere, cube, cylinder, rectangular prism,
cone, or pyramid for each object.

**1.**

**2.**

**3.**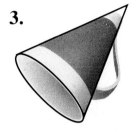

**4.**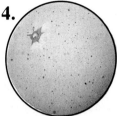

**5.**

**6.**

**7.**

**8.**

**APPLY**

<u>MATH REASONING</u> Which space figures do the
following describe?

**9.** It has only flat faces. They are all the same.

**10.** It has only 1 flat face. It can roll. It has no
straight edges.

<u>PROBLEM SOLVING</u>

**11.** Which space figure would you find most often
on a playground? in a kitchen?

▶ <u>COMMUNICATION</u> **Write to Learn**

**12.** Write about a cube. Use the words **corners**
and **flat faces.**

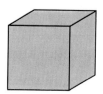

*More Practice, page 521, set B*

# Plane Figures

**EXPLORE** **Discover a Relationship**

To make the flat shapes in the mobile, Celia drew around the edges of space figures. What shapes can you make by drawing around the edges of 2 different space figures?

**TALK ABOUT IT**

1. What shapes can you draw with a rectangular prism?
2. What space figure can you use to draw a 3-sided figure?
3. Compare and describe the other figures that you can draw.

The figures you draw are **plane figures.** Plane figures lie on a flat surface.

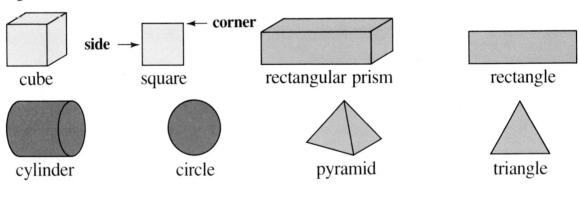

cube    side →    corner    square    rectangular prism    rectangle

cylinder    circle    pyramid    triangle

Draw each plane figure.

1. A figure with 4 equal sides and square corners

2. A figure with 3 sides and 3 corners

3. A figure with 4 sides, 2 longer than the others

272

What plane figure does each picture suggest?

**1.**

**2.**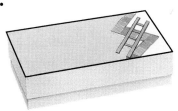

**3.**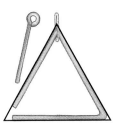

Name each figure. Tell how many straight sides and how many corners it has.

**4.**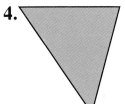

**5.**

**6.**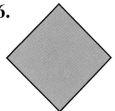

**7.**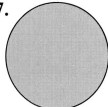

**APPLY**

## MATH REASONING

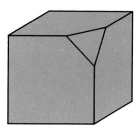

**8.** Suppose you cut off the corner of a cube like this. What plane figure could you draw using the piece you cut off?

## PROBLEM SOLVING

**9.** One side of a square rabbit pen is 9 feet long. How many feet of fence is needed for the pen?

**10. Fine Arts Data Bank** What plane figures do you see in the painting by Mondrian? See page 470.

▶ **USING CRITICAL THINKING  Take a Look**

**11.** How many triangles can you see? There are more than you may see at first.

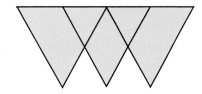

# Polygons and Segments

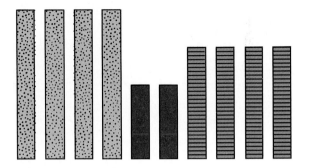

**EXPLORE  Make Models**
Work in groups. Use strips like these. Think of them as the sides of plane figures. Join the ends. What different figures can you make?

**TALK ABOUT IT**

1. Can you make a triangle with no equal sides? two equal sides? three equal sides?

2. Can you make a square? A rectangle?

3. Describe another figure that you can make.

These shapes are **polygons.** The sides are **segments.** The corners are **points.** Polygons are closed plane figures with segments that join at points.

**Some Polygons**

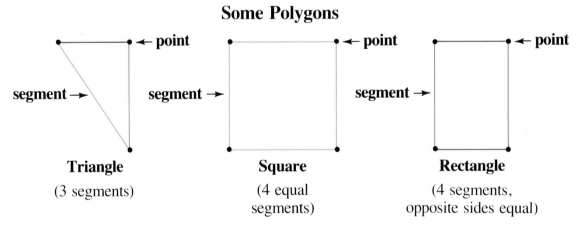

| **Triangle** | **Square** | **Rectangle** |
|:---:|:---:|:---:|
| (3 segments) | (4 equal segments) | (4 segments, opposite sides equal) |

**TRY IT OUT**

1. Use your strips to make a triangle. Draw it.

2. How many segments and points does your triangle have?

**Katie's strips**

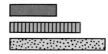
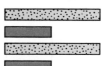

**Jeff's strips**

Tell who made each shape.

1. A rectangle

2. A square

3. A triangle with 3 equal sides

4. A triangle with 2 equal sides

5. A triangle with 0 equal sides

**Lila's strips**

**Phil's strips**

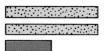

**Jen's strips**

## APPLY

**MATH REASONING** Write **polygon** or **not polygon** for each figure. Tell how you decided.

6.

7.

8.

## PROBLEM SOLVING

9. A, B, C, and D are names of segments. A is shorter than C. D is longer than C and shorter than B. Which is longest? Which is shortest?

 ## MIXED REVIEW

Use mental math to add, subtract, or multiply. Write only the answers.

10. $8 + 2$     11. $7 \times 1$     12. $6 \times 0$     13. $5 + 3$     14. $11 - 3$

15. $1 \times 5$     16. $1 - 0$     17. $10 - 2$     18. $1 \times 0$     19. $1 + 9$

20. $1 + 8$     21. $8 + 0$     22. $8 \times 1$     23. $0 \times 8$     24. $8 - 1$

25. $7 - 1$     26. $4 \times 1$     27. $0 \times 0$     28. $0 + 9$     29. $10 - 0$

*More Practice, page 521, set D*

# Polygons and Angles

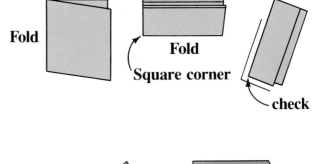

Fold

Fold

Square corner

check

## EXPLORE  Fold Paper

Make a square corner by folding your paper this way. Use your square corner to find the square corners of these polygons.

## TALK ABOUT IT

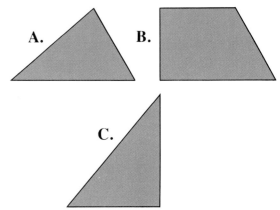

A.

B.

C.

**1.** Which polygon has no square corners? 1 square corner? 2 square corners?

**2.** Can you name a polygon that has 4 square corners?

The corners of polygons are **angles.** You can see angles in many things. Square corners are **right angles.**

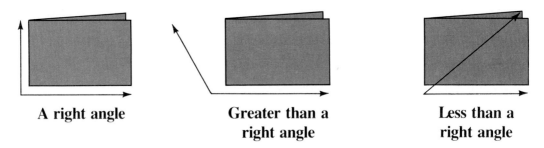

**A right angle**        **Greater than a**        **Less than a**
                         **right angle**           **right angle**

Decide if the angle is a right angle, greater than a right angle, or less than a right angle.

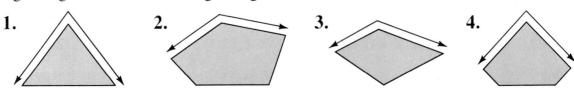

**1.**            **2.**            **3.**            **4.**

Decide if the angle is a right angle, greater than a
right angle, or less than a right angle.

**1.**    **2.**    **3.**

How many right angles does each figure have?

**4.** 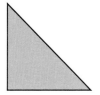   **5.**    **6.**    **7.**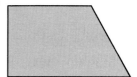

**APPLY**

**MATH REASONING**  Draw each angle.

**8.** less than a right angle       **9.** more than a right angle

**PROBLEM SOLVING**

**10.** Write two different times when the hands of
a clock form a right angle.

▶ **ESTIMATION**
Estimate whether each angle is a right angle.
Check using the square corner you made.

**11.** 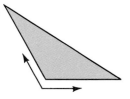       **12.** 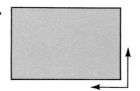       **13.**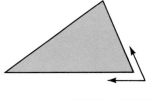

*More Practice, page 522, set A*

# Problem Solving
## Determining Reasonable Answers

UNDERSTAND
FIND DATA
PLAN
ESTIMATE
SOLVE
CHECK

**LEARN ABOUT IT**

When you solve a problem, the last step is to check your work. You can ask yourself questions.

- Is my arithmetic correct?
- Did I use a strategy correctly?
- Is my answer reasonable?

Maria and her class made space figures. Her teacher, Mrs. Rose, bought a package of straws and a ball of string. How much did she spend? Maria's calculator showed this answer.

| 1.85 |

| **Price List** | |
|---|---|
| String | $0.95 |
| Glue | $0.75 |
| Tape | $0.39 |
| Clay | $0.98 |
| Pipecleaners | $0.49 |
| Package of straws | $0.45 |
| Single straws | $0.05 |
| Box of toothpicks | $0.30 |

Maria thought: "I can round $0.95 to $1. The string and the package of straws together cost about $1.45." $1.85 is not a reasonable answer. It is too high.

**TRY IT OUT**

Do not solve the problems. Decide if each answer is reasonable. If it is not reasonable, tell why.

1. The class used 1 box of toothpicks and a piece of clay to make pyramids. How much did the supplies cost?

   Answer: | 1.25 |

2. Mrs. Rose bought tape to make cardboard cylinders. She gave the clerk a dollar bill. How much change did she get back?

   Answer: | 0.61 |

Solve. Use any problem solving strategy.

1. Virginia's class made a large pyramid. Their teacher took $5 to buy 2 rolls of paper. How much money was left over if each roll of paper cost $1.79?

2. At 2:52 p.m., Mr. Harmon went to get more art supplies. He said, "I will be back when the hands of the clock next form a right angle." What time did he return?

| Some Strategies |
| :---: |
| Act It Out |
| Choose an Operation |
| Make an Organized List |
| Look for a Pattern |
| Use Logical Reasoning |
| Use Objects |
| Draw a Picture |
| Guess and Check |
| Make a Table |
| Work Backward |

3. Joe used rolled up newspaper to make a triangle. Each roll was 12 inches long. What was the perimeter of his polygon in feet?

4. There were 15 students who made pyramids and 17 who made cubes. How many more students made cubes?

5. Tom had 2 kits for making space figures. Each kit cost $9.39. About how much did they cost together?

6. Ann, Jose, Lula, and Cindy made large pyramids. Jose's was larger than Cindy's. Ann's was smaller than Cindy's. Jose's was smaller than Lula's. Whose pyramid was largest?

7. Mrs. Rose bought dowels to make mobiles. The dowels were 9¢ each. If she bought 9 of them, how much did she pay?

8. **Suppose** To make a wooden cube, Angelo bought 6 balsa wood squares for $0.39 each, a box of nails for $1.49, and paint for $1.60. What was his total bill? Decide which one of the following facts would change your answer to the problem.

   A. Angelo paid with a $10 bill.
   B. Angelo bought 2 boxes of nails.

# Using Critical Thinking

LEARN ABOUT IT

Patrick made a Polygon People Store for his class project. Here are some of the polygon people that he had for sale.

Price List

| | |
|---|---|
| 3 sides | 5¢ |
| 4 sides | 10¢ |
| 5 sides | 15¢ |
| Over 5 sides | 20¢ |
| Extras Each right angle | 8¢ |
| All sides equal | 12¢ |
| Some parallel sides | 9¢ |

Nancy looked at the price list. She said "I'll take the one that costs the most."

## TALK ABOUT IT

1. Which one did Nancy buy? Explain.

2. What did it cost to buy a triangle?

3. Which shape do you think cost the least? Why?

TRY IT OUT

1. Jason wanted to buy all the polygon people that have at least 1 right angle. Which ones did he want to buy?

2. Which polygon people are worth the extra 12¢?

280

# MIDCHAPTER REVIEW/QUIZ

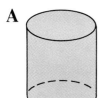  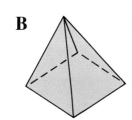  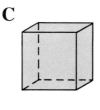  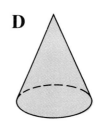  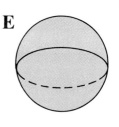

A  B  C  D  E

1. What is the name of each space figure?

2. Choose two figures. How are they alike? How are they different?

3. Which two figures could you trace to make a circle?

4. Which figure could you trace to make both a triangle and a square?

Find the correct polygon.

5. It has exactly three angles.

6. It has more than four segments.

7. It has a right angle.

8. It has an angle that is greater than a right angle.

9. It has an angle that is less than a right angle.

  W

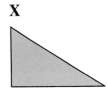  X

  Y

  Z

## PROBLEM SOLVING

10. How many right angles are in this polygon?

11. Do not solve. Decide if the answer is reasonable or not.

Antonio bought 2 packages of erasers. Each cost $0.42 with tax. How much did he spend?

Answer: | 1.04 |

281

# Symmetry

**EXPLORE  Fold Paper**

Fold a square so that one part fits exactly on the other part. How many ways can you fold your square so that one part exactly fits on another? Try the same thing with a triangle and a rectangle.

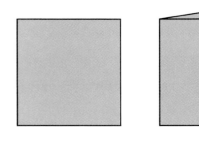

**TALK ABOUT IT**

1. How many ways did you fold the rectangle to have matching parts? What about the triangle?

2. What other shape can you think of that you can fold so that 2 parts match exactly?

A figure has a **line of symmetry** if it can be folded so that its two parts match exactly. Some figures have more than one line of symmetry. A square is one of these figures.

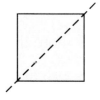

**Line of symmetry**

Some pictures of real objects appear to have lines of symmetry, too.

Tell whether each dashed line appears to be a line of symmetry.

1.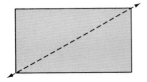

2.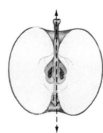

282

Tell whether the dashed line appears to be a line
of symmetry.

**1.**   **2.**   **3.**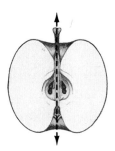

Trace the figures and draw a line of symmetry on
each. Check by folding.

**4.**   **5.**   **6.**

**APPLY**

**MATH REASONING**

**7.** Visualize a circle. Predict how many lines of
symmetry it will have. Fold a circle. Is there
more than one line of symmetry?

**PROBLEM SOLVING**

**8.** Draw a triangle that has exactly one line of
symmetry. Draw a natural object that has one
line of symmetry.

▶ **USING CRITICAL THINKING  Make a Prediction**

**9.** Predict what each figure will look like when
cut out and unfolded. Then fold a paper in
half. Draw a figure like the picture. Next, cut
out the figure to test your prediction.

**10.** Some triangles have no lines of symmetry,
some have exactly 1, and some have 3 lines
of symmetry. Can you draw one of each?

*More Practice, page 522, set C*

# Congruence

**EXPLORE** **Use Tangram Pieces**

Use tangram puzzle pieces labeled like this. Find two pieces that are exactly alike. Are there two others exactly alike? Find two pieces that will fit together to be the same size and shape as G.

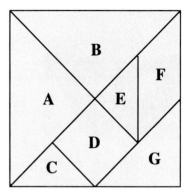

**Tangram Puzzle**

## TALK ABOUT IT

1. How can you show that two pieces are exactly alike?
2. How can you show that two pieces fit together to be the same size and shape as another piece?

Figures that are the same size and shape are **congruent** to each other.

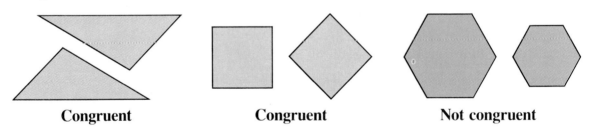

Congruent          Congruent          Not congruent

**TRY IT OUT**

Which figure is congruent to the first?

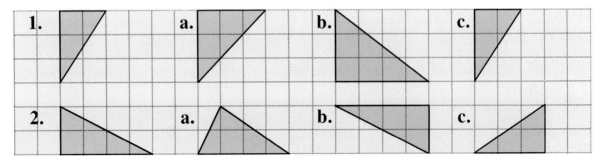

Which figure is congruent to the first?

**1.**

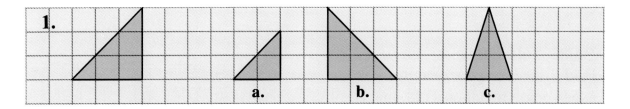

a.   b.   c.

Are the two figures congruent? Prove it by tracing.

**2.**

**3.**

### APPLY

#### MATH REASONING

**4.** If you cut a rectangle this way, would you have two congruent triangles? Convince a classmate.

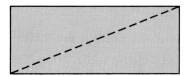

#### PROBLEM SOLVING

**5. Fine Arts Data Bank** In Mondrian's painting, are any 2 red rectangles congruent? See page 470.

▶ USING CRITICAL THINKING **Make a Prediction**

Use dot paper to show what this figure will look like if you make each turn.

**6.** a quarter turn

**7.** a half turn

**8.** a three-quarter turn

**Start**

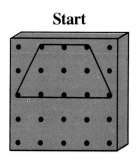

*More Practice, page 522, set D*

**285**

# Parallel Lines

## EXPLORE  Fold Paper

Fold a sheet of paper carefully so the
edges come together. Fold again
the same way. Can you predict how
the folds in your paper will look
before you unfold it? Draw lines
along the folds.

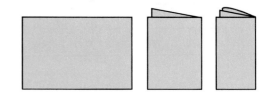

## TALK ABOUT IT

1. Was your prediction correct?

2. Look at your lines along the
   folds. Would you say that any
   two of the lines are the same
   distance apart at all places?

You have drawn **parallel lines.**
Parallel lines are lines that are the
same distance apart at all points.
They have no common point.
**Intersecting lines** are lines that
cross. They have a common point.

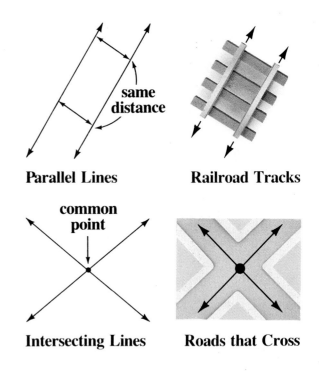

**Parallel Lines**          **Railroad Tracks**

**Intersecting Lines**      **Roads that Cross**

## TRY IT OUT

1. Write the names of two things in your
   classroom that suggest parallel lines.

2. Write the names of two things that suggest
   intersecting lines.

Does each picture suggest parallel or intersecting lines? Tell why.

1.

2.

3.

4.

5.

6.

7. Use your ruler to draw a pair of parallel lines like this. Draw a line that intersects both of the parallel lines you drew.

## APPLY

### MATH REASONING

8. How many angles can you find in the figure that you drew for Exercise 7?

### PROBLEM SOLVING

9. This polygon is a rectangle. Give the colors for the pairs of parallel lines.

### ▶ ESTIMATION

10. Estimate how many small squares are in the large polygon.

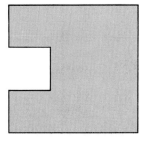

# Coordinate Geometry

**EXPLORE  Work with a Partner**

- Each of you secretly write the name of one of the objects shown on the graph.

- Take turns asking questions like these: Is it located 4 units to the right? Is it up 7 units?

- First one to guess the other person's object wins the game.

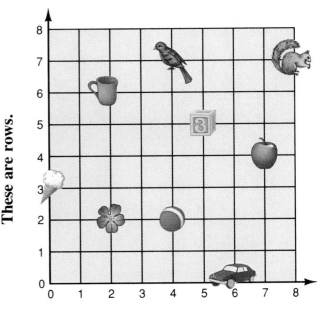

These are rows.

These are columns.

**TALK ABOUT IT**

1. What objects are located 4 units to the right? 2 units up?

2. Did you work out a strategy for playing the game? Can you explain it?

**Number pairs** give the location of a point on a graph. The number pair (4, 2) gives you the location of the ball on the graph above. The number 4 tells you to go over 4 units to the right. The number 2 tells you to go up 2.

Here are two ways you can think about number pairs on a graph.

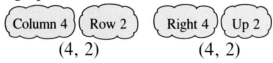

Column 4  Row 2   Right 4  Up 2
(4, 2)              (4, 2)

**TRY IT OUT**

Answer these questions for the graph above.

1. What object is at (5, 5)?

2. What object is at (6, 0)?

288

Give a number pair to locate each of these figures.

1. 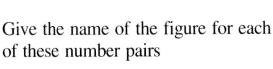   2. ✖   3. ➚

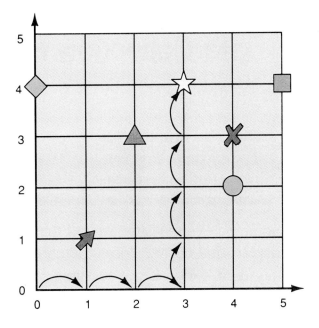

Give the name of the figure for each of these number pairs

4. (4,2)   5. (2,3)   6. (5,4)

**MATH REASONING**

7. Think about these number pairs on a graph.
(0, 0) (1, 1) (2, 2) (3, 3) (4, 4) (5, 5) . . .
What can you say about the pattern they form?

**PROBLEM SOLVING**

8. Pedro's father waved to him from his office window. Use a number pair to tell from which window he waved.

---

**MIXED REVIEW**

Suppose you drop a pebble several times on each game board.

A      B      C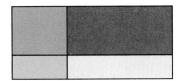

9. For which board is a pebble most likely to land on red?

10. For which board is a pebble not very likely to land on green?

11. Which board is a fair board?

*More Practice, page 523, set A*

**289**

# Data Collection and Analysis
## Group Decision Making

UNDERSTAND
FIND DATA
PLAN
ESTIMATE
SOLVE
CHECK

**Doing an Investigation**

**Group Skill:**
Check for Understanding

Some people measure around their fists to find out the correct length to buy their socks! Can you guess why? For this activity you will need some string, a pair of scissors, and a ruler.

### Data Collection

1. Work with a partner in your group. Take your shoe off. Have your partner help you cut a string to match the length of your foot. Cut another string to go around the biggest part of your fist. Mark the string that shows the length of your foot with a piece of tape.

Foot and Fist Measurements

Donna
17 cm Fist
16 cm Foot

Toby
18 cm Fist

June

2. Have each person tape their two strings on a large piece of paper. Record which is the fist measure and which is the foot measure.

3. Measure each string and write the measure on the chart.

4. What do you notice about your foot and fist measurements?

5. Why do you think you may be able to measure your fist to know the length to buy your socks?

6. Do you think someone with large hands will also have long feet? Why do you think so? How could you find out?

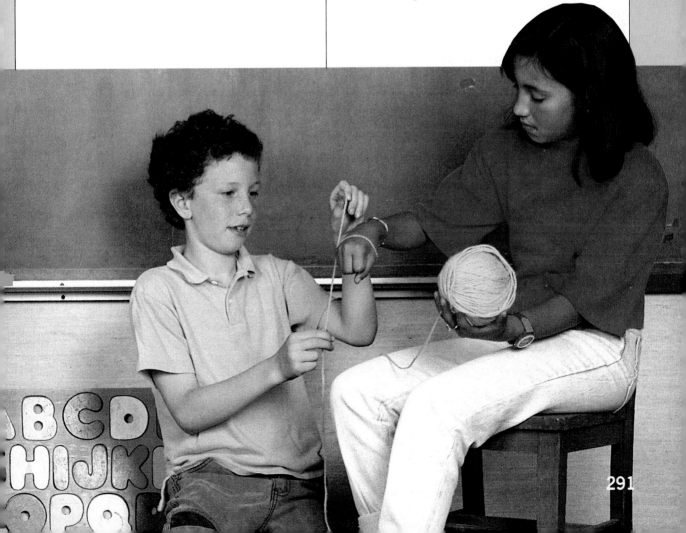

291

# WRAP UP

## What's My Name?

Which space figure am I?

**1.** I have no edges at all and roll easily.

**2.** I can have five flat faces and five corners.

Which plane figure am I?

**3.** I am formed by drawing around one face of a cube.

**4.** I am formed by drawing around one face of a cylinder.

What kind of lines are we?

**5.** We have a common point.

**6.** We are the same distance apart at all points.

## Sometimes, Always, Never

Decide which word should go in the blank, *sometimes*, *always*, or *never*. Explain your choices.

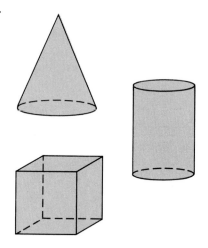

**7.** A circle is _____ a sphere.

**8.** A rectangle is _____ a polygon.

**9.** An angle in a triangle is _____ a right angle.

**10.** A square folded in half _____ makes two rectangles.

## Project

Find an example of each of these in your classroom. Tell how you use its special form.

■ rectangular prism

**Example:** A box is a rectangular prism. It sits flat without rolling. You can stack rectangles of paper neatly inside.

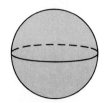

■ sphere        ■ cylinder

■ right angle    ■ symmetry

292

# CHAPTER REVIEW/TEST

## Part 1    Understanding

1. Name each space figure. How are they alike?

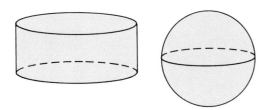

2. To what space figure can you relate a triangle? Explain your answer.

3. Is the red line or the blue line a line of symmetry? Why?

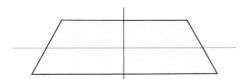

4. Are the blue lines parallel or intersecting? Why?

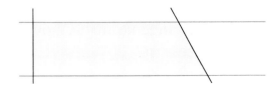

## Part 2    Skills

5. Which figures are polygons?

6. Which polygons have right angles?

7. Which polygons have angles greater than a right angle?

8. Which are congruent figures?

9. How many segments are in D?

10. How many angles are in E?

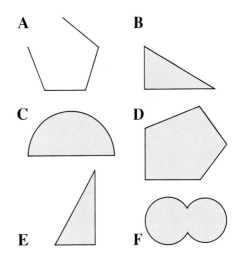

A    B

C    D

E    F

## Part 3    Applications

11. Sue bought crayons for 84¢ and paper for 49¢. Is $1.90 a reasonable answer for the amount she spent? Explain.

12. **Challenge** Find the points (2, 2), (4, 6), and (6, 2) on a grid. Draw lines from point to point. What shape do you see?

# ENRICHMENT
## Take a Good Look

Brent and Carrie are looking at the can of tomatoes from different directions. Do you think the can looks the same to both of them?

1. Which student do you think sees the can in the shape of a rectangle?

2. Which student sees the can in the shape of a circle?

3. What space figure would look like a circle no matter how you looked at it?

Look at the space figures. Tell what shape you would see from each view. Draw the shape.

4.

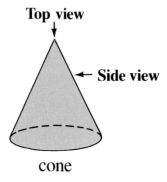

cone

5.

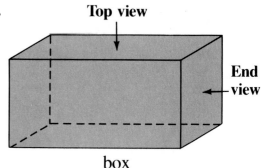

box

6.

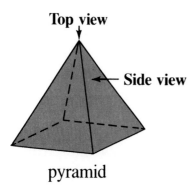

pyramid

7.

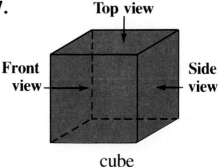

cube

# CUMULATIVE REVIEW

1. What is the length to the nearest inch?

   A. 1 in.   B. 2 in.

   C. 3 in.   D. 4 in.

2. Which is less than one mile?

   A. 2,000 yd   B. 6,000 ft

   C. 4,000 ft   D. 3,000 yd

3. Which unit would you use to weigh a table tennis ball?

   A. feet   B. inches

   C. pounds   D. ounces

4. Which fact relates to $4 + 4$?

   A. $1 \times 4$   B. $2 \times 4$

   C. $3 \times 4$   D. $4 \times 1$

5. What is the product of 5 and 8?

   A. 30   B. 35

   C. 40   D. 58

6. When 0 is the product, one factor is always this number.

   A. 0   B. 1

   C. 10   D. 100

7. Give the product of $9 \times 6$.

   A. 45   B. 54

   C. 63   D. 69

8. $3 \times 6$ is the same as 2 sixes and how much more?

   A. 3   B. 0

   C. 6   D. 1

9. $4 \times 7$ is double 2 times this number.

   A. 2   B. 14

   C. 4   D. 7

10. Find $8 \times 7$.

    A. 53   B. 56

    C. 58   D. 64

11. Find $2 \times (4 \times 5)$.

    A. 11   B. 18

    C. 22   D. 40

12. Ellen had $7.55 to buy 4 movie tickets at $2 each. How much more money did she need?

    A. $0.25   B. $0.45

    C. $0.55   D. $1.45

13. There are 16 glasses in a box. They are lined up in equal rows. How many rows can there be?

    A. 7 or 9
    B. $1 \times 6$
    C. 1, 2, 4, 8, or 16
    D. 2, 4, or 6

# 11

**M**ATH AND SCIENCE

DATA BANK

Use the Science Data Bank on page 473 to answer the questions.

**1** Animal footprints show both the heel and toe prints. How many toe prints would you see in 2 raccoon footprints? In 3 raccoon footprints?

# DIVISION
# CONCEPTS
# AND
# FACTS

**2** A forest ranger found 4 bobcat footprints. How many toe prints did she see? How many toe prints did she see later in 6 coyote footprints?

**3** The group of animal babies born together is called a litter. If 3 deer each had a litter, how many babies would there be in all?

**4** Use Critical Thinking Find the total number of toe prints in an armadillo's front footprint and hind footprint. Can you multiply to find the answer?

# Problem Solving
## Understanding Division

| UNDERSTAND |
| FIND DATA |
| PLAN |
| ESTIMATE |
| SOLVE |
| CHECK |

## LEARN ABOUT IT

The action in a problem tells you which operation to use to solve it.

Look at the action in the problem. Show the action with counters. Complete a number sentence about the action.

| **Problem** | **Action** | **Operation** |
|---|---|---|
| A chipmunk fit 12 small acorns in his 2 cheek pouches. He shared the acorns equally between the 2 pouches. How many acorns are in each pouch? | | Divide<br><br>$12 \div 2 = ?$ |

Sharing equally tells you to **divide.**

**We see:** 12 things shared equally by 2.

**We write:** $12 \div 2$.

**We say:** Twelve divided by two.

## TRY IT OUT

Use counters to show the action. Decide if the problem suggests multiplication or division. Explain why.

1. A chipmunk gathered 14 corn kernels. He shared them equally between his 2 cheek pouches. How many did he put in each pouch?

2. In the park, 3 chipmunks each had a litter of 5 babies. How many baby chipmunks were there in all?

Here is a list of the key actions you have learned so far. The action tells you what operation to use to solve the problem.

In each problem, show the action with objects. Tell which action you are showing. Decide which operation you would use to solve the problem.

| Action | Operation |
|---|---|
| Put together | Add |
| Take away Compare Find a missing part | Subtract |
| Put together same-size sets | Multiply |
| Share a set equally | Divide |

1. In a sandbank, 3 mink footprints all look alike. These footprints show 15 toeprints in all. How many toeprints does each footprint have?

2. At a wildlife park there were 18 baby wolves. The 3 mother wolves each had the same number of babies. How many babies did each mother have?

3. A woodchuck that weighed 8 pounds in the summer weighed 14 pounds in November. How much more did it weigh in November than in the summer?

4. A park ranger found 3 fox dens in the meadow and 2 dens near the woods. How many fox dens did the ranger find that day?

5. A squirrel buried 15 acorns in Connie's flower bed. By January it had dug up 9 of the acorns. How many were left?

6. Travis' cat put 6 footprints on his homework page. Each footprint had 4 toeprints. How many toeprints were on the homework page?

7. A chipmunk brought 34 beechnuts home in her cheek pouches. She took out 15 of them. How many were still in her pouches?

▶ **COMMUNICATION Write to Learn**

8. Write a sentence about something at your school that shows a set shared equally.

*More Practice, page 523, set B*

# More About Division

**12 counters**

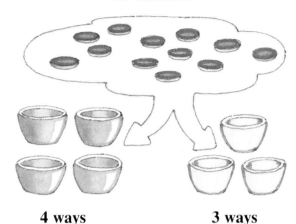

## LEARN ABOUT IT

**EXPLORE Use Counters and Cups**
Work in groups. The picture
suggests two ways you might share
12 counters equally. Find how many
counters are in each cup when you
share 12 equally.

**4 ways**          **3 ways**

## TALK ABOUT IT

1. How many are in each cup when you share
   the counters 4 ways? 3 ways?

2. Why could you not share them equally 5
   ways?

You can write about sharing using a division
equation. The answer to a division problem is
called the **quotient.**

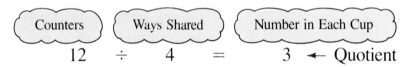

$$12 \div 4 = 3 \leftarrow \text{Quotient}$$

Counters    Ways Shared    Number in Each Cup

**Read:** Twelve divided by four equals three.

8 shared equally 2 ways          15 shared equally 3 ways

$8 \div 2 = 4$                    $15 \div 3 = 5$

## TRY IT OUT

Show using counters and cups. Give the quotient.

1. Share 6 acorns equally 3 ways.      2. Share 10 walnuts equally 2 ways.

   $6 \div 3 = $ |||||                    $10 \div 2 = $ |||||

**300**

Find the quotients. Use objects if you need help.

**1.**

Share 8 equally 4 ways.
$8 \div 4 = $ |||||

**2.**

Share 6 equally 2 ways.
$6 \div 2 = $ |||||

**3.**

$15 \div 5 = $ |||||

**4.**

$14 \div 2 = $ |||||

**APPLY**

**MATH REASONING**

**5.** Which of these can you share
equally 2 ways?

3 apples    4 oranges
  5 bananas   6 plums
7 pears    8 peaches

**PROBLEM SOLVING**

**6.** After sharing equally, Ann got
5 strawberries, Don got 5, and
Sue got 5. How many berries did
they share?

**7. Use Objects** Mario shared
18 grapes equally among 2
friends and himself. How many
grapes did each person get?

**MIXED REVIEW**

Use mental math to find the products.

| × 2 | |
|---|---|
| 3 | 6 |
| **8.** 4 | ||||| |
| **9.** 6 | ||||| |
| **10.** 5 | ||||| |

| × 3 | |
|---|---|
| 7 | 21 |
| **11.** 0 | ||||| |
| **12.** 6 | ||||| |
| **13.** 8 | ||||| |

| × 5 | |
|---|---|
| 4 | 20 |
| **14.** 2 | ||||| |
| **15.** 7 | ||||| |
| **16.** 8 | ||||| |

| × 9 | |
|---|---|
| 3 | 27 |
| **17.** 4 | ||||| |
| **18.** 7 | ||||| |
| **19.** 9 | ||||| |

*More Practice, page 507, set C*

# Dividing by 2 and 3

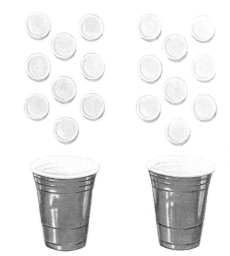

**EXPLORE** Use Cups and Counters

Work in groups. The figure shows 18 counters put in 2 equal sets. You can use this model to think about division or multiplication.

Can you show both a multiplication and a division equation using 18 counters and 3 cups?

**TALK ABOUT IT**

1. How many were in each cup when you put 18 counters into 3 equal sets?

2. Explain how you showed $3 \times 6$ and $18 \div 3$.

- When you divide, you know the whole and one factor. You want to find the other factor.

- You can think of finding a quotient as finding a missing factor.

**Examples**

| **2** | **×** | **3** | **=** | **6** | → | **6** | **÷** | **2** | **=** | ‖‖‖ |
|---|---|---|---|---|---|---|---|---|---|---|
| factor | | factor | | product | | product | | factor | | missing factor |

$3 \times$ what number $= 18$?  $\quad 18 \div 3$  $\qquad$ $3 \times 6 = 18$  $\quad 18 \div 3 = 6$

$2 \times$ what number $= 14$?  $\quad 14 \div 2$  $\qquad$ $2 \times 7 = 14$  $\quad 14 \div 2 = 7$

## TRY IT OUT

Find the quotients. You may use counters and cups.

$2 \times$ what number $= 10$? $\qquad$ $3 \times$ what number $= 12$? $\qquad$ $2 \times$ what number $= 18$?

**1.** $10 \div 2 = $ ‖‖‖ $\qquad$ **2.** $12 \div 3 = $ ‖‖‖ $\qquad$ **3.** $18 \div 2 = $ ‖‖‖

Divide.

| 3 × what number = 24? | 3 × what number = 27? | 2 × what number = 16? |

**1.** $24 \div 3 = $ |||||  **2.** $27 \div 3 = $ |||||  **3.** $16 \div 2 = $ |||||

**4.** $8 \div 2$  **5.** $18 \div 3$  **6.** $12 \div 2$  **7.** $27 \div 3$  **8.** $12 \div 3$

**9.** $3 \div 3$  **10.** $4 \div 2$  **11.** $24 \div 3$  **12.** $16 \div 2$  **13.** $10 \div 2$

**14.** $6 \div 2$  **15.** $6 \div 3$  **16.** $14 \div 2$  **17.** $2 \div 2$  **18.** $15 \div 3$

**19.** Divide 21 by 3.  **20.** Divide 14 by 2.  **21.** Divide 18 by 2.

Give the missing numbers.

| Divide by 2 | 12 | 8 | 10 | 16 | 18 |
|---|---|---|---|---|---|
| | 6 | 4 | ||||| | ||||| | ||||| |

**22. 23. 24.**

| Divide by 3 | 18 | 27 | 12 | 21 | 15 |
|---|---|---|---|---|---|
| | 6 | 9 | ||||| | ||||| | ||||| |

**25. 26. 27.**

**MATH REASONING**

**28.** Is the missing number 2 or 3?  $12 \div $ ||||| $ = 6$

**PROBLEM SOLVING**

**29.** Mallory, Kara, and Felicia shared a box of 18 plums equally. How many did each girl get?

**30. Unfinished Problem** Write a division question about a sack of apples. There were 18 apples in the sack.

▶ **USING CRITICAL THINKING  Support your conclusion**

**31.** Would you rather share a small bag of popcorn equally with 1 other person or with 2 other persons? Tell why.

**32.** Would you rather share a really hard job equally with 1 person or with 2 persons? Tell why.

*More Practice, page 507, set D*

# Dividing by 4

Knowing multiplication facts for 4 helps you divide by 4.

**EXPLORE  Use Counters**

Ramon watched 4 mother beavers and their babies by a dam. He estimated that there were 15 to 25 babies. If each mother had the same number of babies, how many babies might Ramon have seen? Use counters to show your answer.

**TALK ABOUT IT**

1. Could there have been 20 babies? How?

2. Could there have been 18 babies? Explain.

3. Check the Science Data Bank to find out how many babies Ramon most likely saw.

Thinking about related multiplication facts will help you find quotients.

$4 \times$ what number $= 20$?

$4 \times \mathbf{5} = 20$

$20 \div 4 = \text{|||||} \longrightarrow 20 \div 4 = \mathbf{5}$

**TRY IT OUT**

Find the quotients.

$4 \times$ what number $= 24$?

$4 \times$ what number $= 28$?

$4 \times$ what number $= 16$?

**1.** $24 \div 4 = \text{|||||}$

**2.** $28 \div 4 = \text{|||||}$

**3.** $16 \div 4 = \text{|||||}$

**4.** $12 \div 4$

**5.** $36 \div 4$

**6.** $32 \div 4$

Divide.

**1.** $18 \div 2$     **2.** $32 \div 4$     **3.** $21 \div 3$     **4.** $12 \div 4$     **5.** $14 \div 2$

**6.** $20 \div 4$     **7.** $10 \div 2$     **8.** $18 \div 3$     **9.** $4 \div 4$     **10.** $28 \div 4$

**11.** Divide 32 by 4.     **12.** Divide 21 by 3.     **13.** Divide 24 by 4.

Is the missing number 2, 3, or 4?

**14.** $18 \div \text{|||||} = 6$     **15.** $14 \div \text{|||||} = 7$     **16.** $16 \div \text{|||||} = 4$

## APPLY

**MATH REASONING** Answer without dividing.

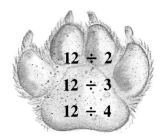

**17.** Which quotient is largest?

**18.** Which quotient is smallest?

**19.** Which quotient is between the others?

$12 \div 2$
$12 \div 3$
$12 \div 4$

## PROBLEM SOLVING

**20.** In the meadow, 4 mother rabbits each had the same number of babies. There were 11 male and 9 female babies. How many babies did each mother have?

**21. Science Data Bank** Ramon saw 3 red fox mothers and their pups playing on a hill. If each mother had a litter, how many pups were there? See page 473.

DATA BANK

## ▶ CALCULATOR

**22.** Estimate to decide the number of the first step that will be greater than 999. Then use a calculator to find out. You may be surprised.

$2 \boxed{\times} 2 \boxed{=} 4 \boxed{=} \boxed{=} \boxed{=} \dots$

| | Step 1 | Step 2 | Step 3 | Step 4 | . . . |
|---|---|---|---|---|---|
| 2 | 4 | 8 | 16 | 32 | . . . |

# Problem Solving
## Using Data from a Graph

UNDERSTAND
FIND DATA
PLAN
ESTIMATE
SOLVE
CHECK

### LEARN ABOUT IT

To solve some problems, you will need to use data from a graph.

> Firestation Number 1 fought the same number of fires in each of 4 weeks in May. About how many fires did they fight each week that month?

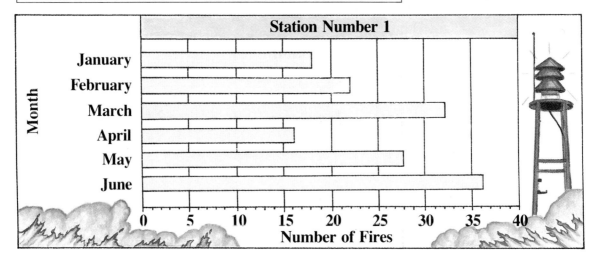

Station Number 1

Month / Number of Fires

January, February, March, April, May, June

0  5  10  15  20  25  30  35  40

I'll find the data I need in the graph.  There were a total of 28 fires in May.

I'll divide to find the answer.  $28 \div 4 = 7$

Firestation Number 1 fought about 7 fires each week.

### TRY IT OUT

1. About how many fires were there each week in March?

2. How many more fires were there in the busiest month than in the slowest month?

3. What month had about double the number of fires in January?

4. About how many fires were there each week in June?

Solve. Use data from the graph on page 306.

**1.** Marty was on duty for 29 of the fires in February and March. How many fires in all did she miss in February and March?

**2.** How many fires were there in the first six months of the year?

**3.** Engine 1, Engine 2, the rescue van, and the ladder truck went to a fire. Engine 1 was between Engine 2 and the ladder truck. Engine 2 was behind the rescue van. In what order did they go?

**4.** One fire engine has a ladder that is 32 feet long. It is divided into 4 equal sections. How long is each section?

**5.** Three stations sent all of their firefighters to a huge fire. There were 21 firefighters in all. Which stations sent firefighters? Use the picture graph.

**7. Talk About Your Solution**
Manny is a firefighter. This is his schedule. He begins by working 24 hours. Then he is off 24 hours. Again, he works 24 hours and then he is off 24 hours. Next, he works 24 hours, but then Manny is off for 4 days. After that, the schedule repeats.

**6.** How many more fires were there in the months of April, May, and June than in the months of January, February, and March? Use the graph on page 306.

**Number of Firefighters in Each Station**

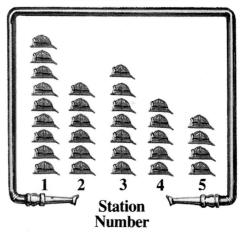

**Station Number**

Each  stands for a firefighter.

What will Manny be doing on the 12th day? HINT: One day equals 24 hours.

■ Explain your solution to a classmate.

■ Compare your solutions.

# Dividing by 5

**LEARN ABOUT IT**

**EXPLORE  Look for a Pattern**

Knowing the multiplication facts for 5 is your best help when you are dividing by 5. Look at the numbers in the box. Do you see a pattern?

**TALK ABOUT IT**

| Can you divide these by 5? |
|:---:|
| 40    30    5 |
| 35    25 |
| 15    20 |
| 45    10 |

1. What pattern did you find?

2. What can you say about the product when you multiply 5 by an even number? by an odd number?

3. Do you think $30 \div 5$ is odd or even? What about $35 \div 5$?

When you divide, it helps to think about related products.

$5 \times$ what number $= 20$?          $5 \times 4 = 20$

$20 \div 5 = ||||| \quad \longrightarrow \quad 20 \div 5 = 4$

$5 \times$ what number $= 35$?          $5 \times 7 = 35$

$35 \div 5 = ||||| \quad \longrightarrow \quad 35 \div 5 = 7$

**TRY IT OUT**

Find the quotients.

$5 \times$ what number $= 25$?          $5 \times$ what number $= 35$?          $5 \times$ what number $= 40$?

1. $25 \div 5 = |||||$          2. $35 \div 5 = |||||$          3. $40 \div 5 = |||||$

4. $15 \div 5$          5. $45 \div 5$          6. $30 \div 5$

Divide.

**1.** $25 \div 5$    **2.** $20 \div 4$    **3.** $40 \div 5$    **4.** $30 \div 5$    **5.** $24 \div 3$

**6.** $16 \div 2$    **7.** $15 \div 5$    **8.** $18 \div 3$    **9.** $20 \div 5$    **10.** $36 \div 4$

**11.** $35 \div 5$    **12.** $24 \div 4$    **13.** $45 \div 5$    **14.** $15 \div 3$    **15.** $14 \div 2$

**16.** $18 \div 2$    **17.** $30 \div 5$    **18.** $21 \div 3$    **19.** $32 \div 4$    **20.** $35 \div 5$

**21.** Divide 40 by 5.    **22.** Divide 32 by 4.    **23.** Divide 30 by 5.

**24.** Divide 21 by 3.    **25.** Divide 35 by 5.    **26.** Divide 12 by 2.

**APPLY**

**MATH REASONING**

**27.** Sort these numbers into three sets, the multiples of 3, of 4, and of 5.

27   16   28
9   25   35
21   8   10

**PROBLEM SOLVING**

**28. Extra Data** Solve. Tell what data was extra. Stephanie babysat 3 days for a total of 12 hours. She earned $30. She used the money to buy 5 play tickets for her family. How much was each ticket?

**29.** There were 387 people at the play the first night and 429 the second night. How many more people went to the play on the second night than on the first night?

▶ **ALGEBRA**

What number was divided?

**30.** $\boxed{\phantom{0}} \div 2 = 6$    **31.** $\boxed{\phantom{0}} \div 3 = 6$    **32.** $\boxed{\phantom{0}} \div 4 = 6$    **33.** $\boxed{\phantom{0}} \div 5 = 6$

**34.** Did you find a short cut for solving these problems? What is it?

*More Practice, page 507, set F*

# Problem Solving

UNDERSTAND
FIND DATA
PLAN
ESTIMATE
SOLVE
CHECK

**MIXED PRACTICE**

Choose a strategy from the strategies list or other strategies you know to solve these problems.

**Some Strategies**

Act It Out
Use Objects
Choose an Operation
Draw a Picture
Make an Organized List
Guess and Check
Make a Table
Look for a Pattern
Use Logical Reasoning
Work Backward

1. Morris spends 3 hours each week on his paper route. How much time does he spend in 8 weeks?

2. Jenny can choose which type of paper route she wants. There are large, medium, and small routes. There are morning and evening routes. How many different choices does she have?

3. Mrs. Rosenberg gave Bart a 10-dollar bill. She said, "Here is $8.95 for the paper this month. Keep $0.75 for the tip." How much change should Bart give back?

4. The newspapers must be delivered by 5:00 p.m. It takes Joe 16 minutes to deliver his papers. If he starts at 3:35 p.m., when will he be done?

5. Chia can fold 5 newspapers in a minute. It takes her 9 minutes to fold all her papers. How many papers does she fold each day?

6. There is a 3-story apartment building on Ted's route. There is the same number of customers on each floor. If there are 27 customers in all, how many live on each floor?

7. Betsy makes about $36 a month from her paper route. How much is this each week? HINT: A month is about 4 weeks.

8. How much more did Joe Gross pay than Jean Li?

| Times Tribune |
| --- |
| Li, Jean |
| $8.75 |
| 12/5/91 |
| 1 month |

| Times Tribune |
| --- |
| Gross, Joe |
| $26.25 |
| 12/5/91 |
| 3 months |

*More Practice, page 523, set D*

# MIDCHAPTER REVIEW/QUIZ

A class of 24 students is forming teams. Draw a picture and tell which operation is needed in each problem.

**1.** If there must be 8 teams, what size will the teams be?

**2.** If each student scores 2 points, how many points will that be?

**3.** The Reds scored 8 points and the Blues scored 7 points. How many points were scored?

**4.** If one team scores 6 points and another scores 10 points, how many points are scored?

Copy the picture. Show the sharing. Solve.

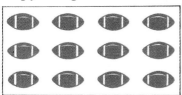

**5.** $12 \div 3 =$ ||||

**6.** $12 \div 4 =$ ||||

**7.** $12 \div 2 =$ ||||

**8.** $12 \div 6 =$ ||||

Think about missing factors to find the quotients.

**9.** $15 \div 3$   **10.** $16 \div 2$   **11.** $15 \div 5$   **12.** $16 \div 4$

**13.** $20 \div 5$   **14.** $36 \div 4$   **15.** $2 \div 2$   **16.** $9 \div 3$

**17.** $10 \div 5$   **18.** $24 \div 3$   **19.** $24 \div 4$   **20.** $27 \div 3$

**21.** $28 \div 4$   **22.** $10 \div 2$   **23.** $21 \div 3$   **24.** $20 \div 4$

**25.** $40 \div 5$   **26.** $35 \div 5$   **27.** $14 \div 2$   **28.** $32 \div 4$

## PROBLEM SOLVING

**29.** How many students live between one-half mile and 1 mile away?

**30.** How many students live closer to school than 1 mile?

**31.** If students who live 1 mile or more away ride in 2 cars in equal groups, how many are in each car?

The picture graph shows how far some third graders live from school.

| Less than one-half mile | 👣 👣 👣 👣 |
| Between one-half and 1 mile | 👣 👣 👣 |
| 1 mile or more | 👣 👣 👣 👣 |

👣 = 2 students

311

# Fact Families

There are fact families for multiplication and division, just as there are for addition and subtraction.

factor    factor
2         3
product
6

**EXPLORE  Discover a Relationship**
The ring shows 2 factors and their product. How many different multiplication and division equations can you write using just these three numbers?

**TALK ABOUT IT**

1. Why are you able to write two multiplication equations?

2. Is there a related division equation for each multiplication equation?

Using 2 factors and their products, you can write a fact family.

**Fact Family Numbers**

factor    factor
3         4
product
12

| Fact Family |
| --- |
| $3 \times 4 = 12$ |
| $4 \times 3 = 12$ |
| $12 \div 4 = 3$ |
| $12 \div 3 = 4$ |

**TRY IT OUT**

Write a fact family for each set of factors and products.

1. factor  factor
   2      7
   product
   14

2. factor  factor
   3      8
   product
   24

3. factor  factor
   3      3
   product
   9

Write a fact family for each set of factors and products.

**1.** $4 \times 5 = 20$                    **2.** $6 \times 6 = 36$

Find the missing fact family number. Then write the complete fact family.

**3.**
    **3**        **5**
   factor   factor

    ‖‖‖
   product

**4.**
    **2**        ‖‖‖
   factor   factor

    **16**
   product

**5.**
    ‖‖‖        **5**
   factor   factor

    **30**
   product

**APPLY**

**MATH REASONING**

**6.** Don't try to divide. Just tell which quotient is greater.

    **A.** $72 \div 3 = $ ‖‖‖
    **B.** $72 \div 4 = $ ‖‖‖

**PROBLEM SOLVING**

**7. More Than One Answer** There are 15 students to play a game. How can they form teams so that there are the same number of players on each team?

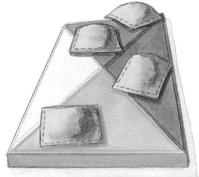

**MIXED REVIEW**

Find the perimeter of each polygon.

**8.**

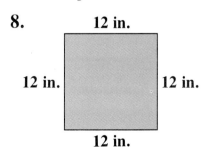

12 in. / 12 in. / 12 in. / 12 in.

**9.**

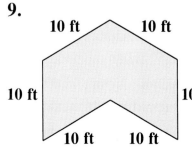

10 ft / 10 ft / 10 ft / 10 ft / 10 ft / 10 ft

**10.**

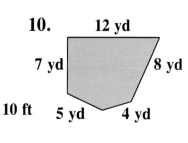

12 yd / 7 yd / 8 yd / 5 yd / 4 yd

*More Practice, page 523, set E*

**313**

# 0 and 1 in Division

**EXPLORE** **Discover a Relationship**

Can you find numbers that the paw prints are hiding that would make the equation correct?

Remember, when 0 is a factor, the product is 0.

**A.**  $\times 0 = 0$

**B.** $\times 5 = 0$

**C.** $\times 0 = 5$

**TALK ABOUT IT**

1. Which equation has just one number that makes it true? Why?

2. Which equation is true no matter what number the paw print is hiding? Why?

3. Which equation is always false no matter what number the paw print is hiding? Why?

Here are some division rules.

- Never divide by 0.

- 0 divided by any number (except 0) is 0.

- Any number (except zero) divided by itself is 1.

$1 \times 5 = 5$

$5 \div 5 = 1$

- Any number divided by 1 is that number.

$5 \times 1 = 5$

$5 \div 1 = 5$

Here is a new sign for division. The number you divide by is the **divisor.**

$32 \div 4 = 8$  quotient $\rightarrow 8$ $4\overline{)32}$ divisor

**TRY IT OUT**

Find the quotients. Circle the divisors.

1. $4 \div 4$  2. $0 \div 4$  3. $1\overline{)2}$  4. $8\overline{)8}$

314

Find the quotients.

**1.** $3\overline{)24}$    **2.** $6\overline{)0}$    **3.** $4\overline{)32}$    **4.** $5\overline{)35}$    **5.** $7\overline{)7}$    **6.** $1\overline{)8}$

**7.** $2\overline{)10}$    **8.** $5\overline{)30}$    **9.** $2\overline{)14}$    **10.** $9\overline{)0}$    **11.** $3\overline{)18}$    **12.** $2\overline{)16}$

**13.** $8 \div 8$    **14.** $36 \div 4$    **15.** $0 \div 3$    **16.** $40 \div 5$    **17.** $6 \div 1$    **18.** $0 \div 15$

**19.** Divide 16 by 4.    **20.** Divide 0 by 8.    **21.** Divide 7 by 7.

## APPLY

**MATH REASONING** Write $+$, $-$, $\times$, or $\div$ for each ▊.

**22.** $6$ ▊ $2 = 4$    **23.** $6$ ▊ $2 = 12$    **24.** $6$ ▊ $2 = 3$    **25.** $6$ ▊ $2 = 8$

### PROBLEM SOLVING

**26.** Which equation would solve the problem? Jean, Jim, and Jodi decided to share the 18 shells they found at the beach. When they opened the sack to share them, they found all of the shells had been lost through a hole in the sack. How many shells did each person get?

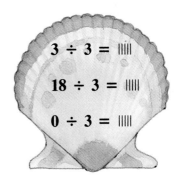

$3 \div 3 = $ ▊

$18 \div 3 = $ ▊

$0 \div 3 = $ ▊

**27.** **Science Data Bank** Hikers saw 3 mother armadillos and their babies playing in the dirt. How many baby armadillos do you think they saw? See page 473.

## ▶ ESTIMATION

**28.** About how many pieces of chalk are in your classroom? 1, 10, or 100?

**29.** About how many textbooks are in your classroom? 2, 20, or 200?

**30.** About how many math exercises do you solve in a day? 3, 30, or 300?

*More Practice, page 507, set G*

# Problem Solving
## Extending the Division Concept

UNDERSTAND
FIND DATA
PLAN
ESTIMATE
SOLVE
CHECK

### LEARN ABOUT IT

You have used division to find how many are in each of a given number of equal sets. Now you will learn to use division to find how many equal sets you can make from a given set.

Look at the action in the problem. Show the action with counters. Complete a number sentence about the action.

| **Problem** | **Action** | **Operation** |
|---|---|---|

There are 12 Morks who need rides to Ugl. A spaceship holds 2 Morks. How many spaceships are needed to take the Morks to Ugl?

Divide

$12 \div 2 = ?$

Taking away same-size sets repeatedly tells you to divide.

**We see:** 12 separated into groups of 2 each.

**We write:** $12 \div 2$.

take away 2 . . .

take away 2 . . .

### TRY IT OUT

Use objects to tell which division action goes with each problem. Do not solve the problem but tell whether you are sharing or repeatedly taking away.

**1.** Morks love bananas. 5 Morks ate 15 bananas. Each ate the same number. How many bananas did each Mork eat?

**2.** Mandy Mork picked 18 pears. She gave 2 to each of her friends until they were all gone. How many friends got pears?

Here is a list of the key actions that tell you what operation to use to solve a problem.

Tell which action and operation you would use to solve the problem. You may want to use counters to decide. Then solve the problem.

| Action | Operation |
| --- | --- |
| Put together | Add |
| Take away<br>Compare<br>Find a missing part | Subtract |
| Put together<br>  same-size sets | Multiply |
| Share a set equally<br>Take away same-size<br>  sets repeatedly | Divide |

1. Keith found out that only 7 students could be in the class play. There were 25 students in Keith's class. How many could not be in the play?

2. Samuel has 12 eggs. He needs 3 for each omelet he makes. How many omelets can he make?

3. Darren has a box of 30 strawberries that he plans to share equally among some friends. If he gives himself and each friend 5 strawberries, how many shares does that make?

4. Bonnie must learn 8 more lines of music before her piano lesson in 4 days. How many lines should Bonnie learn each day?

5. April collects stamps. She has just 4 empty pages left in her stamp book. If she has 24 stamps, how many should she plan to put on each page?

6. Charles took 36 pictures at the picnic. He took 24 at the fair. How many did he take at both events?

7. Chad has 45 cents in his pocket. All of his coins are nickels. How many nickels does he have?

8. Lauren said, ''I have 16 shoes in my closet. Can you tell me how many pairs that is?''

▶ **CALCULATOR**

9. You can use the subtraction constant on a calculator to find how many 3s are in 42. Subtract and count until you reach 0.

$\boxed{C}$ 42 $\boxed{-}$ 3 $\boxed{=}$ $\boxed{=}$ $\boxed{=}$ . . .

*More Practice, page 508, set A*

# Applied Problem Solving
## Group Decision Making

UNDERSTAND
FIND DATA
PLAN
ESTIMATE
SOLVE
CHECK

**Group Skill:**

Disagree in an Agreeable Way

In the crafts workshop your group will make Big Heads. These are big stuffed heads with short arms and feet but no body. You pay for the ornaments you use to decorate the Big Head your way.

Here is the price list.

**Sequins**
40¢ for a package of 20 assorted colors

**Large Buttons**
5 for 40¢

**Small Buttons**
8 for 40¢

**Beads**
6 for 36¢

**Yarn**
1 yard for 8¢

**Felt fabric**
6 inch by 6 inch squares at 25¢ each (red, blue, pink, black, green, and brown)

**Facts to Consider**

■ You really like sequins and plan to use some on your Big Head.

■ You only have $4.00 to spend decorating the Big Head.

■ You need about 10 to 12 yards of yarn to give your Big Head a long hair style. You need only about 6 to 8 yards of yarn for a short hair style.

■ You can cut out pieces of a felt square, use buttons, or use beads to make lips, nose, freckles, and other face parts.

**Some Questions to Answer**

**1.** How much does it cost to buy 1 large button for a nose?

**2.** How much does it cost to buy 4 small buttons for earrings?

**3.** Estimate the cost of a long hair style.

**4.** What does it cost to buy 3 beads for a mouth?

**What is Your Plan?**

Draw a picture of your Big Head. Make a list to show the number and cost of each item you used. Show the total cost.

# WRAP UP

## Divide or Multiply?

Tell whether the problem suggests multiplication or division. Explain.

1. Erica had 20 toy cars. She wanted to share them equally with 4 friends. How many did each friend get?

2. Carlos had 5 toy cars. Each car had 4 tires. How many tires were there in all?

3. Rita has 16 tires. She needs 4 tires for each toy car. On how many cars can she put tires?

## Sometimes, Always, Never

Decide which word should go in the blank, *sometimes, always,* or *never*. Explain your choices.

4. You _____ can use subtraction to find a quotient such as 24 ÷ 2.

5. The quotient for a number divided by 1 is _____ the number.

6. An odd number divided by 5 _____ has an odd quotient.

7. A fact family _____ has just two equations.

8. Zero divided by any number except 0 is _____ 0.

## Project

Start with 45 counters. Show two different ways to use the counters to find the quotient 45 ÷ 3.

# CHAPTER REVIEW/TEST

## Part 1   Understanding

Decide which operation you would use to solve
the problem. Draw a picture to explain.

1. Ron found 6 different kinds of
   wildflowers. He took pictures of
   4 of each kind. How many
   pictures did Ron take?

2. Pam bought 20 packs of trail
   food to last over 4 days of
   hiking. How many packs could
   she use each day?

3. By which number can you divide
   and get a quotient that is the
   same as the dividend?

4. Which number when divided by
   any number except 0 gives a
   quotient of 0?

5. Tell how you can use a missing
   factor to find the quotient.
   Use $15 \div 5$ to explain.

## Part 2   Skills

Divide.

6. $24 \div 3$      7. $16 \div 4$      8. $28 \div 4$      9. $35 \div 5$      10. $14 \div 2$

11. $5\overline{)45}$      12. $6\overline{)0}$      13. $1\overline{)9}$      14. $2\overline{)18}$      15. $3\overline{)18}$

16. Write the fact family for the factors 2 and 8.

## Part 3   Applications

17. What is the total rainfall for
    January and February? Which
    2 months had equal rainfall?

    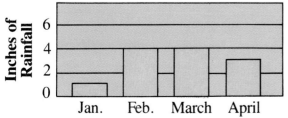

    **Rainfall in Dusky Woods Park**

18. Ranger Dan checks 27 nests
    every month to count baby
    birds. Each day he checks 3
    nests. How many days does he
    need to check all of the nests?

19. **Challenge** Five fat figs fit in
    a box. Phil has fifty figs. How
    many boxes can Phil fill?
    Draw a picture of your answer.

# ENRICHMENT
## A Division Time Saver

Sara knows several numbers that she can divide evenly by 3. she says these numbers are **divisible** by 3. Find the missing numbers.

**1.** $3 \div 3 = $ |||||

**2.** ||||| $\div 3 = 2$

**3.** $9 \div$ ||||| $= 3$

**4.** ||||| $\div 3 = 4$

**5.** $15 \div$ ||||| $= 5$

**6.** $18 \div 3 = $ |||||

**7.** $21 \div$ ||||| $= 7$

**8.** ||||| $\div 3 = 8$

**9.** $27 \div 3 = $ |||||

Bret showed Sara how to find out if large numbers are divisible by 3. Here is what he showed her.

> sum of the digits

**36**      $3 + 6 = $ **9**      9 is divisible by 3, so 36 is divisible by 3.

**141**      $1 + 4 + 1 = $ **6**      6 is divisible by 3, so 141 is divisible by 3.

**186**      $1 + 8 + 6 = $ **15**      15 is divisible by 3, so 186 is divisible by 3.
　　　　or, $1 + 5 = $ **6**      6 is divisible by 3, so 186 is divisible by 3.

Find out if each number is divisible by 3.

**10.** 423

**11.** 1,502

**12.** 259

**13.** 1,584

**14.** 9,102

**15.** 1,114

**16.** 127

**17.** 3,609

**18.** Bret found a box with 258 old baseball cards in his attic. Can he share them evenly with Sara and another friend?

**19.** Describe how you would explain to a friend when a number is divisible by 3.

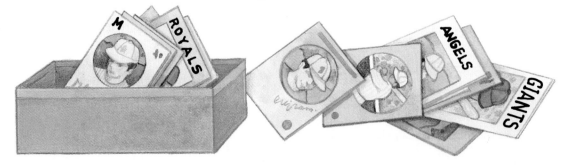

# CUMULATIVE REVIEW

**1.** Find $9 \times 8$.

   **A.** 72     **B.** 78

   **C.** 81     **D.** 89

**2.** Find $0 \times 5$.

   **A.** 5     **B.** 0

   **C.** 15     **D.** 1

**3.** Find $17 \times 2$.

   **A.** 20     **B.** 24

   **C.** 32     **D.** 34

**4.** 0, 8, 16, 24, and 32 are multiples of which number?

   **A.** 0     **B.** 3

   **C.** 6     **D.** 8

**5.** Find the multiples of 9.

   **A.** 3, 6, 9     **B.** 9, 18, 27

   **C.** 9, 19, 90     **D.** 19, 29, 39

**6.** Name this space figure.

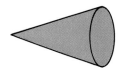

   **A.** pyramid     **B.** prism

   **C.** cone     **D.** sphere

**7.** The sides of polygons are _____.

   **A.** corners     **B.** spaces

   **C.** points     **D.** segments

**8.** What plane figure do you get by tracing one face of a cylinder?

   **A.** circle     **B.** triangle

   **C.** rectangle     **D.** square

**9.** Figures that are the same size and shape are _____.

   **A.** angles     **B.** polygons

   **C.** symmetrical     **D.** congruent

**10.** Kevin needed 3 dollar bills, 2 quarters, 3 dimes, and some nickels and pennies to buy a book that cost between \$3 and \$4. Which is a possible price?

   **A.** \$3.98     **B.** \$3.69

   **C.** \$4.95     **D.** \$4.98

**11.** Elaine picked the only round flower in the garden. Which number pair locates the round flower?

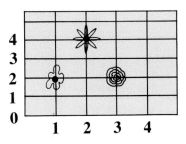

   **A.** (1, 2)     **B.** (2, 2)

   **C.** (2, 3)     **D.** (3, 2)

# 12

## MORE DIVISION FACTS

**M**ATH AND
SOCIAL STUDIES

DATA BANK

Use the Social Stud-
ies Data Bank on
page 476 to answer
the questions.

1 A bus company
planned to buy all
new tires for the
wheels of 9 of their buses.
How many tires should the
company buy?

**2** If workers greased all of the wheels of 7 coach cars, what is the number of wheels that they greased?

**3** Garage workers are replacing all of the wheels on 6 taxicabs. How many new wheels should they order?

**4** Use Critical Thinking Compare a bicycle, a taxicab, and a bus. How are these forms of transportation alike? How are they different?

# Dividing by 6

LEARN ABOUT IT

Knowing the multiplication facts for 6 can help you divide by 6.

| Name | Number of Stickers | Number of Cards |
|------|-------------------|-----------------|
| Stacy | 42 | 7 |
| Aaron | 18 | \|\|\|\|\| |
| Inez | 24 | \|\|\|\|\| |
| Bob | 54 | \|\|\|\|\| |
| Amber | 36 | \|\|\|\|\| |

**EXPLORE  Study the Table**

A pack of monster stickers has 6 stickers on each card. Look at the table. Can you find how many cards of stickers each student bought?

**TALK ABOUT IT**

1. How can you find how many cards Aaron must buy to have 18 stickers?

2. What multiplication fact helps you find the number of cards Aaron bought?

3. Give the multiplication facts that will help you find how many cards the other students bought.

Think of the missing factor to find the quotient.

$6 \times ? = 30$

$30 \div 6 = $ \|\|\|\|\|

$6 \times 5 = 30$

$30 \div 6 = 5$

$6 \times ? = 48$

\|\|\|\|\|
$6)\overline{48}$

$6 \times 8 = 48$

$8$
$6)\overline{48}$

**TRY IT OUT**

Find the quotients.

$6 \times ? = 24$

1. $24 \div 6 = $ \|\|\|\|\|

$6 \times ? = 36$

2. $36 \div 6 = $ \|\|\|\|\|

$6 \times ? = 42$

3. $42 \div 6 = $ \|\|\|\|\|

Find the quotients.

**1.** $42 \div 6$     **2.** $20 \div 4$     **3.** $18 \div 6$     **4.** $12 \div 6$     **5.** $40 \div 5$

**6.** $24 \div 3$     **7.** $24 \div 6$     **8.** $24 \div 4$     **9.** $54 \div 6$     **10.** $18 \div 3$

**11.** $48 \div 6$     **12.** $30 \div 5$     **13.** $14 \div 2$     **14.** $36 \div 6$     **15.** $32 \div 4$

**16.** $5\overline{)35}$     **17.** $6\overline{)48}$     **18.** $4\overline{)28}$     **19.** $6\overline{)36}$     **20.** $6\overline{)42}$

**21.** $3\overline{)27}$     **22.** $6\overline{)54}$     **23.** $6\overline{)24}$     **24.** $5\overline{)45}$     **25.** $2\overline{)16}$

**26.** Divide 48 by 6.     **27.** Divide 40 by 5.     **28.** Divide 30 by 6.

**29.** How many 6s are in 54?     **30.** How many 5s are in 40?

**31.** Continue the pattern.
6, 12, 18, |||||, |||||, |||||, |||||, |||||, |||||

**APPLY**

**MATH REASONING** Do not try to find the quotient. Just decide which is greater.

**32.** $168 \div 4$ or $168 \div 6$     **33.** $216 \div 6$ or $216 \div 3$

**PROBLEM SOLVING**
Use the table on page 326.

**34.** Aaron dropped 12 stickers in a puddle. How many cards got wet?

**35. Write Your Own Problem**
Write and answer a question about the data in the table.

▶ **CALCULATOR**

**36.** Use the subtraction constant on a calculator. Count how many 6s you can subtract from 108. Then write the equation. $108 \div 6 = $ |||||
Check your answer by multiplying.

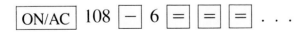

| ON/AC | 108 | − | 6 | = | = | = | . . . |

# Dividing by 7

**Trains on Time**

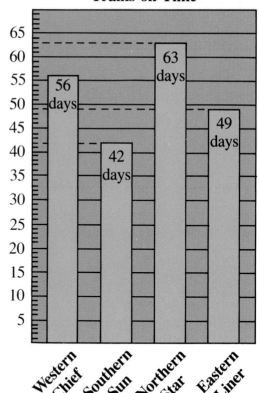

## LEARN ABOUT IT

**EXPLORE** **Study the graph**

The station master's records show when the trains are on time. For how many weeks was the Western Chief on time every day? Remember, there are 7 days in a week.

**TALK ABOUT IT**

1. How can you find the number of weeks the Western Chief was on time?

2. Can you use multiplication to help you solve this problem? Explain.

3. What multiplication fact will help you find $63 \div 7$?

Think of missing factors to find quotients.

$7 \times ? = 56$
$56 \div 7 = \text{||||}$
$7 \times ? = 35$
$\text{||||}$
$7 \overline{)35}$

$7 \times 8 = 56$
$56 \div 7 = 8$
$7 \times 5 = 35$
$5$
$7 \overline{)35}$

## TRY IT OUT

Divide.

$7 \times ? = 49$
**1.** $49 \div 7 = \text{||||}$

$7 \times ? = 28$
**2.** $28 \div 7 = \text{||||}$

$7 \times ? = 63$
**3.** $63 \div 7 = \text{||||}$

Divide.

**1.** $42 \div 7$    **2.** $30 \div 6$    **3.** $21 \div 7$    **4.** $56 \div 7$    **5.** $40 \div 5$

**6.** $48 \div 6$    **7.** $28 \div 7$    **8.** $36 \div 6$    **9.** $63 \div 7$    **10.** $35 \div 7$

**11.** $5\overline{)30}$    **12.** $7\overline{)49}$    **13.** $7\overline{)56}$    **14.** $6\overline{)54}$    **15.** $2\overline{)16}$

**16.** How many 6s are in 42?    **17.** How many 7s are in 42?

Use mental math to choose which is greater.

**18.** $14 - 7$ or $42 \div 7$   **19.** $2 + 7$ or $56 \div 7$   **20.** $36 \div 6$ or $11 - 6$

**MATH REASONING**

**21.** Find one example that proves Jack is wrong. Jack said, "When you divide by 7, the quotient is always odd."

**PROBLEM SOLVING**

**22.** Coretta rode the Eastern Liner to see her aunt. How many weeks in a row was this train on time? Use the graph on page 328.

**23. Social Studies Data Bank**

At a factory, 7 taxicabs are ready to have the wheels put on. How many wheels will the factory use? See page 476.

### MIXED REVIEW

Multiply.

**24.** $3 \times 3$    **25.** $5 \times 4$    **26.** $6 \times 4$    **27.** $5 \times 5$    **28.** $8 \times 4$

**29.** $8 \times 8$    **30.** $7 \times 8$    **31.** $6 \times 7$    **32.** $6 \times 8$    **33.** $7 \times 7$

**34.** $4 \times 9$    **35.** $4 \times 4$    **36.** $8 \times 7$    **37.** $9 \times 9$    **38.** $4 \times 0$

**39.** $6 \times 6$    **40.** $8 \times 6$    **41.** $4 \times 6$    **42.** $7 \times 6$    **43.** $4 \times 7$

# Problem Solving
## Finding Related Problems

| UNDERSTAND |
|---|
| FIND DATA |
| PLAN |
| ESTIMATE |
| SOLVE |
| CHECK |

**LEARN ABOUT IT**

Many problems are related. That means you can solve them using the same strategy or strategies. When you are solving a problem, it helps to think about a related problem.

Here are two related problems. They are both solved using logical reasoning.

| Su-Lin plays marbles at recess. In her bag she keeps more than 20 marbles, but fewer than 30 marbles. She has more than $3 \times 9$. She has an odd number of marbles. How many marbles does Su-Lin have? | Jake found some coins on the playground. He found between 55¢ and 65¢. He found less than 58¢. It wasn't an odd number of cents. He didn't find 57¢. How much money did Jake find? |
|---|---|

20 21 22 23 24 25 26 27 28 29 30      55 56 57 58 59 60 61 62 63 64 65

20 21 22 23 24 25 26 27 28 (29) 30      55 (56) 57 58 59 60 61 62 63 64 65

**TRY IT OUT**

Solve.

1. Rae, Darla, Adam, and Nico lined up to play tetherball. Rae was behind Adam. Nico was in front of Darla. Nico was between Rae and Darla. Who were first and last in line?

3. Are the problems related? Tell why or why not.

2. On the first day of school, the secretary said, "Faraz, your 4th-grade classroom is between the 3rd- and 6th-grade rooms. The 5th-grade room is right before the 3rd-grade room." What order are the rooms in?

Solve.

1. In the lunch line, Jeff stood between Betsy and Ferman. Ferman was in front of Jeff. Caryl was after Betsy. Who were first and last in line?

2. On Play Day, students formed 8 teams of 3 runners each for a relay race. How many students ran the race?

3. Isabel can play either outside or inside at recess. She likes to play with either Denise or Linnette. How many different choices does Isabel have?

4. Each day 4 students help at recess. Since there are 5 days in a school week, how many students help each week?

5. Greg figured out that he is at school 33 hours each week. He spends 8 of those hours on the playground. It takes him 15 minutes to walk home. How many hours does Greg spend in the classroom each week?

6. In a 3-Legged Race, 2 partners tie their inside legs together to run the race. Each team of 2 partners runs on 3 legs. Complete the chart to find how many legs 10 people run on.

| Partners | 2 | | | | |
|----------|---|---|---|---|---|
| Number of legs | 3 | | | | |

7. The 4th, 5th, and 6tn graders at Fairview School have recess at the same time. About how many students are on the playground then? Use the graph.

8. About how many more 2nd graders than 1st graders attend Fairview School? Use the graph.

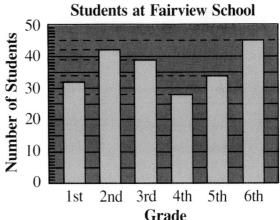

**Students at Fairview School**

9. **Understanding the Operations** Tell what operation you would use. Use objects to solve.

   After lunch, 29 students met for a game. How many more players must join them to make 7 even teams?

# Exploring Algebra

△ + ■ = 10

You can replace the △ and ■ with pairs of numbers to make a true equation.

- List all of the ways that you can replace the △ and ■. Copy the table. Write the numbers under the △ and ■ in the table.

- Write each of the number pairs in the table as an ordered pair. Use this order (△, ■).

- Make a graph. Number the sides of the graph from 1 to 10. Make a point for each ordered pair that you listed.

- Connect the points in the graph.

| △ + ■ = 10 | |
|---|---|
| 1 | 9 |
|  |  |
|  |  |
|  |  |
|  |  |
|  |  |
|  |  |
|  |  |
|  |  |

## TALK ABOUT IT

1. How many ways could you find to make the sum of 10?

2. What do you notice about the graph?

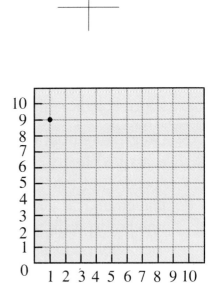

Repeat the activity with this equation. △ − ■ = 2
What was the result when you connected the points in the graph?

332

# MIDCHAPTER REVIEW/QUIZ

Find the quotients.

**1.** $18 \div 6$     **2.** $35 \div 7$     **3.** $35 \div 5$     **4.** $36 \div 6$

**5.** $21 \div 7$     **6.** $7 \div 7$     **7.** $54 \div 6$     **8.** $49 \div 7$

**9.** $12 \div 6$     **10.** $30 \div 6$     **11.** $63 \div 7$     **12.** $7 \div 1$

**13.** $42 \div 6$     **14.** $42 \div 7$     **15.** $40 \div 5$     **16.** $32 \div 4$

**17.** $28 \div 4$     **18.** $30 \div 5$     **19.** $14 \div 7$     **20.** $12 \div 2$

**21.** $6 \div 1$     **22.** $14 \div 2$     **23.** $24 \div 4$     **24.** $0 \div 6$

**25.** How many 6s are in 24?

**26.** What is 48 divided by 6?

**27.** What is 28 divided by 7?

**28.** How many 7s are in 56?

**29.** How many 3s are in 21?

**30.** How many 3s are in 18?

## PROBLEM SOLVING

Solve.

**31.** Read problems 32–35. Which two problems are related?

**32.** A store is shipping 51 books. They must be shipped in cartons that hold no more than 7 books each. How many cartons will the store use?

**33.** At 8 a.m. some buses were in the garage for repairs. 7 were fixed. 3 more buses came in the afternoon. At 5 p.m. 8 buses were in the garage. How many buses were in the garage in the morning?

**34.** Leon had some coins in his pocket. He spent 25¢ for bus fare. Tim paid him back the 15¢ he had borrowed. When Leon got home, he had 37¢. How much money did he start with?

**35.** All sides of this plane figure are of equal length. The perimeter measures 12 centimeters. What is the length of each side?

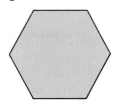

# Dividing by 8

**EXPLORE** **Think About the Situation**

Count the sides of the stop sign. Its sides are all of equal length. If the perimeter of this sign is 32 units, how long is each side?

**TALK ABOUT IT**

1. How would you explain the meaning of perimeter to someone else?

2. What action in the perimeter problem suggests that you should divide?

3. What multiplication fact will help you solve this problem?

Knowing multiplication facts for 8 will help you divide by 8.

$8 \times ? = 32$

$32 \div 8 = |||||$

$8 \times 4 = 32$

$32 \div 8 = 4$

$8 \times ? = 48$

$|||||$
$8)\overline{48}$

$8 \times 6 = 48$

$6$
$8)\overline{48}$

**TRY IT OUT**

Divide.

$8 \times ? = 40$

1. $40 \div 8 = |||||$

$8 \times ? = 72$

2. $72 \div 8 = |||||$

$8 \times ? = 24$

3. $24 \div 8 = |||||$

4. $8)\overline{56}$

5. $8)\overline{16}$

6. $8)\overline{32}$

Divide.

**1.** $32 \div 8$   **2.** $56 \div 7$   **3.** $16 \div 8$   **4.** $40 \div 5$   **5.** $56 \div 8$

**6.** $48 \div 6$   **7.** $48 \div 8$   **8.** $63 \div 7$   **9.** $40 \div 8$   **10.** $14 \div 2$

**11.** $64 \div 8$   **12.** $32 \div 4$   **13.** $24 \div 8$   **14.** $25 \div 5$   **15.** $36 \div 4$

**16.** $3\overline{)24}$   **17.** $8\overline{)16}$   **18.** $4\overline{)28}$   **19.** $8\overline{)72}$   **20.** $7\overline{)42}$

**21.** $6\overline{)54}$   **22.** $7\overline{)49}$   **23.** $8\overline{)64}$   **24.** $5\overline{)35}$   **25.** $8\overline{)32}$

**26.** Divide 72 by 8.   **27.** Divide 35 by 7.   **28.** Divide 40 by 8.

**29.** How many 8s are in 56?   **30.** How many 6s are in 42?

Do not find the quotient. Just tell which is greater.

**31.** $48 \div 8$ or $48 \div 6$   **32.** $56 \div 7$ or $56 \div 8$

**33.** Continue the pattern.

8, 16, 24, 32, |||||, |||||, |||||, |||||, |||||

**MATH REASONING** Write $+$, $-$, $\times$, or $\div$ for the ||||.

**34.** $48 \;||||\; 8 = 40$   **35.** $48 \;||||\; 8 = 6$   **36.** $48 \;||||\; 8 = 56$

**PROBLEM SOLVING**

**37.** A bus made 8 trips between Cedar Falls and Jackson. The bus went 72 miles in all. How many miles was each trip?

**38. Social Studies Data Bank**

Find the number of wheels on 3 cars of a train. How many buses would have that same number of wheels? See page 476.

▶ **ALGEBRA**

Replace ■, △, and |||| in the expressions. Solve.

■ = 6        △ = 8        |||| = 5

**39.** $42 \div \blacksquare$   **40.** $42 - ||||$   **41.** $42 + \triangle$   **42.** $\blacksquare \times \triangle$

*More Practice, page 508, set D*

# Dividing by 9

**EXPLORE** **Solve to Understand**

All of the numbers in the truck are products of 9. Divide each one by 9.

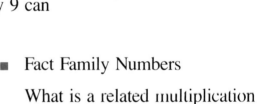

72   81   45
  54   63
9   36  27
    18

**TALK ABOUT IT**

1. What multiplication fact would help you find 45 ÷ 9?

2. What pattern could you use to find the product of 9 × 7?

3. How can that pattern help you find the quotient for 63 ÷ 9?

The patterns you used for multiplying by 9 can help you divide by 9.

Choose your own method for finding this quotient.

$$72 \div 9$$

- ■ Fact Family Numbers

  What is a related multiplication fact?

- ■ Missing Factors

  9 times what number gives 72?

- ■ Memory
- ■ Patterns

**TRY IT OUT**

Divide.

| | | | |
|---|---|---|---|
| **1.** 36 ÷ 9 | **2.** 63 ÷ 9 | **3.** 64 ÷ 8 | **4.** 18 ÷ 9 |
| **5.** 56 ÷ 7 | **6.** 45 ÷ 9 | **7.** 27 ÷ 9 | **8.** 81 ÷ 9 |
| **9.** 9)45 | **10.** 9)72 | **11.** 7)56 | **12.** 9)54 |

336

Divide.

**1.** $45 \div 9$   **2.** $28 \div 7$   **3.** $72 \div 9$   **4.** $64 \div 8$   **5.** $36 \div 9$

**6.** $48 \div 6$   **7.** $18 \div 9$   **8.** $40 \div 8$   **9.** $54 \div 9$   **10.** $56 \div 7$

**11.** $5\overline{)40}$   **12.** $9\overline{)81}$   **13.** $8\overline{)72}$   **14.** $9\overline{)36}$   **15.** $9\overline{)63}$

**16.** $9\overline{)45}$   **17.** $4\overline{)32}$   **18.** $6\overline{)36}$   **19.** $9\overline{)54}$   **20.** $7\overline{)21}$

**21.** Divide 42 by 7.   **22.** Divide 56 by 8.   **23.** Divide 35 by 7.

**24.** How many 6s are in 54?   **25.** How many 9s are in 81?

**APPLY**

**MATH REASONING**

Sort these numbers into three sets.

**26.** Multiples of 9

**27.** Multiples of 5

**28.** Square numbers

**PROBLEM SOLVING**

**29.** Tennis balls were selling at 3 balls for 4 dollars. Rafael wanted to buy 24 balls. How much would the balls cost?

**30.** The tennis coach bought her team some tennis shirts for $75.60. That price included $3.60 tax. If each shirt was $9, how many did she buy?

**MIXED REVIEW**

How many right angles does each figure have?
Use a square corner as a right angle checker.

**31.**   **32.**   **33.**   **34.**

   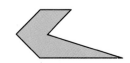

# Know Your Facts

Your best strategy for division facts is to think about related multiplication facts. But there are some patterns that will help you find quotients when dividing by certain numbers.

**EXPLORE  Discover a Relationship**

Study these quotients. Can you give a rule for finding the basic fact quotient when dividing by 5?

$$10 \div 5 = 2$$
$$15 \div 5 = 3$$
$$20 \div 5 = 4$$
$$25 \div 5 = 5$$
$$30 \div 5 = 6$$
$$35 \div 5 = |||||$$
$$40 \div 5 = |||||$$
$$45 \div 5 = |||||$$

**TALK ABOUT IT**

1. What rule did you find when you were dividing a number ending in 0 by 5?

2. What rule did you find when you were dividing a number ending in 5 by 5?

Here is a pattern for finding quotients when dividing by 9.

$$18 \div 9 = 2 \qquad 27 \div 9 = 3 \qquad 36 \div 9 = ||||| \qquad 45 \div 9 = ||||| \ldots$$
(+1 above each)

**TRY IT OUT**

Find the quotients. Practice giving the quotients as quickly as possible.

| | | | | |
|---|---|---|---|---|
| **1.** $35 \div 5$ | **2.** $42 \div 7$ | **3.** $72 \div 9$ | **4.** $64 \div 8$ | **5.** $45 \div 5$ |
| **6.** $24 \div 6$ | **7.** $54 \div 9$ | **8.** $40 \div 5$ | **9.** $21 \div 3$ | **10.** $32 \div 4$ |
| **11.** $10 \div 2$ | **12.** $27 \div 3$ | **13.** $81 \div 9$ | **14.** $56 \div 8$ | **15.** $30 \div 6$ |
| **16.** $49 \div 7$ | **17.** $28 \div 4$ | **18.** $30 \div 5$ | **19.** $45 \div 9$ | **20.** $16 \div 2$ |

338

Find the quotients. Practice giving the quotients as quickly as possible.

**1.** $56 \div 8$    **2.** $48 \div 6$    **3.** $63 \div 9$    **4.** $35 \div 7$    **5.** $40 \div 5$

**6.** $24 \div 4$    **7.** $49 \div 7$    **8.** $64 \div 8$    **9.** $45 \div 9$    **10.** $0 \div 6$

**11.** $48 \div 8$    **12.** $81 \div 9$    **13.** $30 \div 5$    **14.** $32 \div 8$    **15.** $56 \div 7$

**16.** $6\overline{)36}$   **17.** $9\overline{)36}$   **18.** $7\overline{)42}$   **19.** $8\overline{)8}$   **20.** $4\overline{)32}$   **21.** $5\overline{)35}$

**22.** $3\overline{)27}$   **23.** $8\overline{)72}$   **24.** $6\overline{)30}$   **25.** $9\overline{)54}$   **26.** $7\overline{)28}$   **27.** $8\overline{)40}$

**28.** Divide 42 by 7.              **29.** What is 0 divided by 8?

**30.** What is 72 divided by 9?      **31.** Divide 6 by 6.

**APPLY**

**MATH REASONING**

**32.** The two covered numbers are the same.    **A.** $36 \div$ ■    **B.** $48 \div$ ■
Which quotient is greater, A or B?

**PROBLEM SOLVING**

**33. Two-step Problem** Rico has 27 red marbles and 15 yellow marbles. He divides all of the marbles into 7 equal piles. How many are in each pile?

**34.** Jared and his family planned to bike 24 miles. They began to ride at 8:00 a.m. They stopped for lunch at 11:45 a.m. How long did Jared ride before lunch?

▶ **CRITICAL THINKING  Analyze the Situation**

Try the number trick. Use your calculator if the numbers get too large.

**35.** What do you notice about the answer each time?

**36.** Will it work for any number?

**37.** Use counters to show how the trick works.

| Number Trick |
| --- |
| Start with a number. |
| Multiply by 3. |
| Add 12. |
| Divide by 3. |
| Subtract 4. |

*More Practice, page 508, set F*

# Data Collection and Analysis
## Group Decision Making

UNDERSTAND
FIND DATA
PLAN
ESTIMATE
SOLVE
CHECK

**Doing a Survey**

**Group Skill:**
Listen to Others

Eating a variety of healthy food helps us grow strong bones and muscles. Every day we should eat foods from each of the four food groups: the protein group, the milk group, the grain group, and the fruit and vegetable group. Do the students in your school know which foods belong to each food group? Conduct a survey to find out.

**Collecting Data**

1. Work with your group to make a list of five or more different foods and their food groups. Try to choose some foods which may be hard to place in a food group. Don't include junk food, such as candy, on the list.

| Food | Food Group |
| --- | --- |
| Peanuts | Protein |
| Eggs | _____ |
| Bread | _____ |
| _____ | _____ |
| _____ | ___ |

340

**2.** Survey at least 15 students in your school. Show them the four food groups. Then read the name of each food on your list and ask them to which food group they think it belongs. Make a table to keep track of which foods they get right.

**Four Food Groups:**
Protein
Fruit and Vegetable
Milk
Grain

| Food | Number of Correct Answers |
|------|---------------------------|
| Peanuts | ## ## |
| Eggs | ## // |
| Bread | /// |

**3.** Count how many students place each food in the correct food group.

**4.** Make a bar graph using the data in your table.

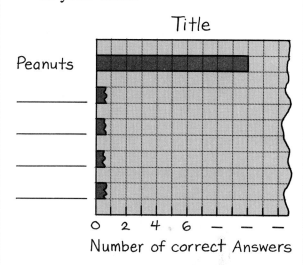

**Presenting Your Analysis**

**5.** Which food was named correctly by the most people?

**6.** Which food was named correctly by the fewest people?

**7.** Write at least three true sentences about the information in your graph.

# WRAP UP

## Put It Back

Some of the terms and numbers in the box dropped out of these sentences. Put them back where they belong. You will have some left over.

<table>
<tr><td>quotient</td></tr>
<tr><td>product</td></tr>
<tr><td>strategy</td></tr>
<tr><td>3 and 2</td></tr>
<tr><td>4 and 5</td></tr>
<tr><td>7 and 0</td></tr>
<tr><td>7    8    9</td></tr>
</table>

1. A missing factor in a multiplication fact is a _____ in the related division fact.

2. Knowing multiplication facts for 8 helps you divide by _____.

3. If two problems are related, you can often use the same _____ to solve both.

4. To make ▲ + ☐ = 7 a true equation, one pair of whole numbers you can use is _____.

## Sometimes, Always, Never

Decide which word should go in the blank, *sometimes*, *always*, or *never*. Explain your choices.

5. A multiplication fact _____ has only one related division fact.

6. Knowing that $6 \times 6 = 36$ _____ can help you solve $36 \div 6$.

7. You _____ can find how many equal sets are in a group by dividing.

8. A problem can _____ be solved by making a list.

## Project

Suppose your class is making books about Birds, Cats, Dogs, and Elephants. Plan a 60-page book that has 6 to 9 chapters. Name the chapters and tell about how many pages you will plan for each chapter.

# CHAPTER REVIEW/TEST

## Part 1    Understanding

1. How many 7s can you subtract from 28? Write a division fact to show this.

2. How many 8s are in 56? Write a division fact to show this.

3. Write two division facts related to $8 \times 9 = 72$.

4. Write the rest of the fact family for $6 \times 9 = 54$.

5. Write 3 number pairs that you can write in the ▢ and ▲ to make this a true equation.
$$▢ + ▲ = 5$$

6. On the graph, does point (3, 1) lie on a straight line with (4, 3) and (3, 2)?

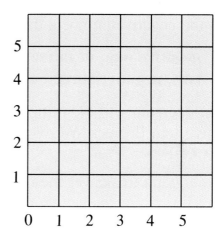

## Part 2    Skills

Divide.

| | | | | |
|---|---|---|---|---|
| **7.** $6\overline{)36}$ | **8.** $24 \div 6$ | **9.** $48 \div 6$ | **10.** $7\overline{)42}$ | **11.** $63 \div 7$ |
| **12.** $7\overline{)35}$ | **13.** $8\overline{)32}$ | **14.** $64 \div 8$ | **15.** $8\overline{)40}$ | **16.** $9\overline{)45}$ |
| **17.** $9\overline{)81}$ | **18.** $36 \div 9$ | **19.** $9\overline{)27}$ | **20.** $72 \div 8$ | **21.** $6\overline{)54}$ |

## Part 3    Applications

22. Carla gave 5 marbles to a friend. She lost 3. Then she had 17. How many marbles did Carla start with?

23. Nancy had between 40 and 49 marbles. She had an even number, more than 46. How many marbles did Nancy have?

24. Calvin gave 45¢ of his savings to his sister. After he spent 35¢ on a new pen, he had 87¢ left. How much had Calvin saved?

25. **Challenge** Of problems 22, 23, and 24, which two problems are related? How?

# ENRICHMENT
## Negative Numbers

Students in the Weather Club made a graph to show the temperatures from 12:00 midnight to 7:00 a.m. They used **negative numbers** to show temperatures below zero.

Read the negative numbers this way.

> **negative one, negative two, negative three . . .**

The temperature at 5:00 a.m. was negative four degrees. We write ⁻4°.

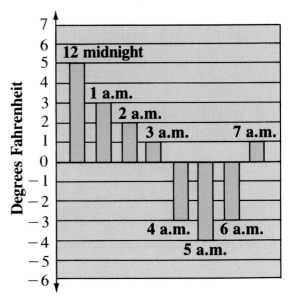

Use the graph to answer the questions.

1. What was the temperature at 12:00 midnight?

2. Had the temperature gone up or down by 1:00 a.m.?

3. At what times was the temperature 1°F?

4. What was the temperature at 4:00 a.m.?

5. What was the difference in temperature between 1:00 a.m. and 3:00 a.m.?

6. What was the difference in temperature between 5:00 a.m. and 6:00 a.m.?

7. At which two times was the temperature ⁻3°F?

8. If the temperature fell 3 degrees between 7:00 a.m. and 8:00 a.m., what would the temperature be at 8:00 a.m.?

344

# CUMULATIVE REVIEW

**1.** Find $7 \times 6$.

   **A.** 38    **B.** 42

   **C.** 46    **D.** 48

**2.** A polygon with 4 square corners and 4 equal sides is a _____.

   **A.** cube        **B.** rectangle

   **C.** right angles    **D.** square

**3.** A figure has a line of symmetry if it can be folded so that its two parts are _____.

   **A.** different    **B.** matching

   **C.** tangrams    **D.** right angles

**4.** Figures with the same size and shape are called _____.

   **A.** intersecting    **B.** squares

   **C.** parallel    **D.** congruent

**5.** Lines that are the same distance from each other at every point but never touch are _____.

   **A.** symmetrical    **B.** congruent

   **C.** parallel    **D.** intersecting

**6.** Which operation shows how objects are shared equally?

   **A.** addition    **B.** multiplication

   **C.** subtraction    **D.** division

**7.** Give the quotient. $18 \div 3$

   **A.** 6    **B.** 7

   **C.** 8    **D.** 9

**8.** $4\overline{)28}$

   **A.** 5    **B.** 6

   **C.** 8    **D.** 7

**9.** Find $30 \div 5$.

   **A.** 5    **B.** 6

   **C.** 7    **D.** 8

**10.** By which number can you never divide?

   **A.** 2    **B.** 10

   **C.** 1    **D.** 0

**11.** Look at the graph. How many students went on trips?

| Student Trips | |
| --- | --- |
| Circus | √√√√ |
| Rodeo | √√ |
| Dog Show | √√√ |

√ = 1 student

   **A.** 9    **B.** 4

   **C.** 10    **D.** 5

# 13

## MULTIPLICATION

**L**ANGUAGE ARTS

DATA BANK

Use the Language Arts Data Bank on page 480 to answer the questions.

1 One rainy day, all the girls in Ramona Quimby's class who had white boots put their boots in a pile. How many boots were in the pile?

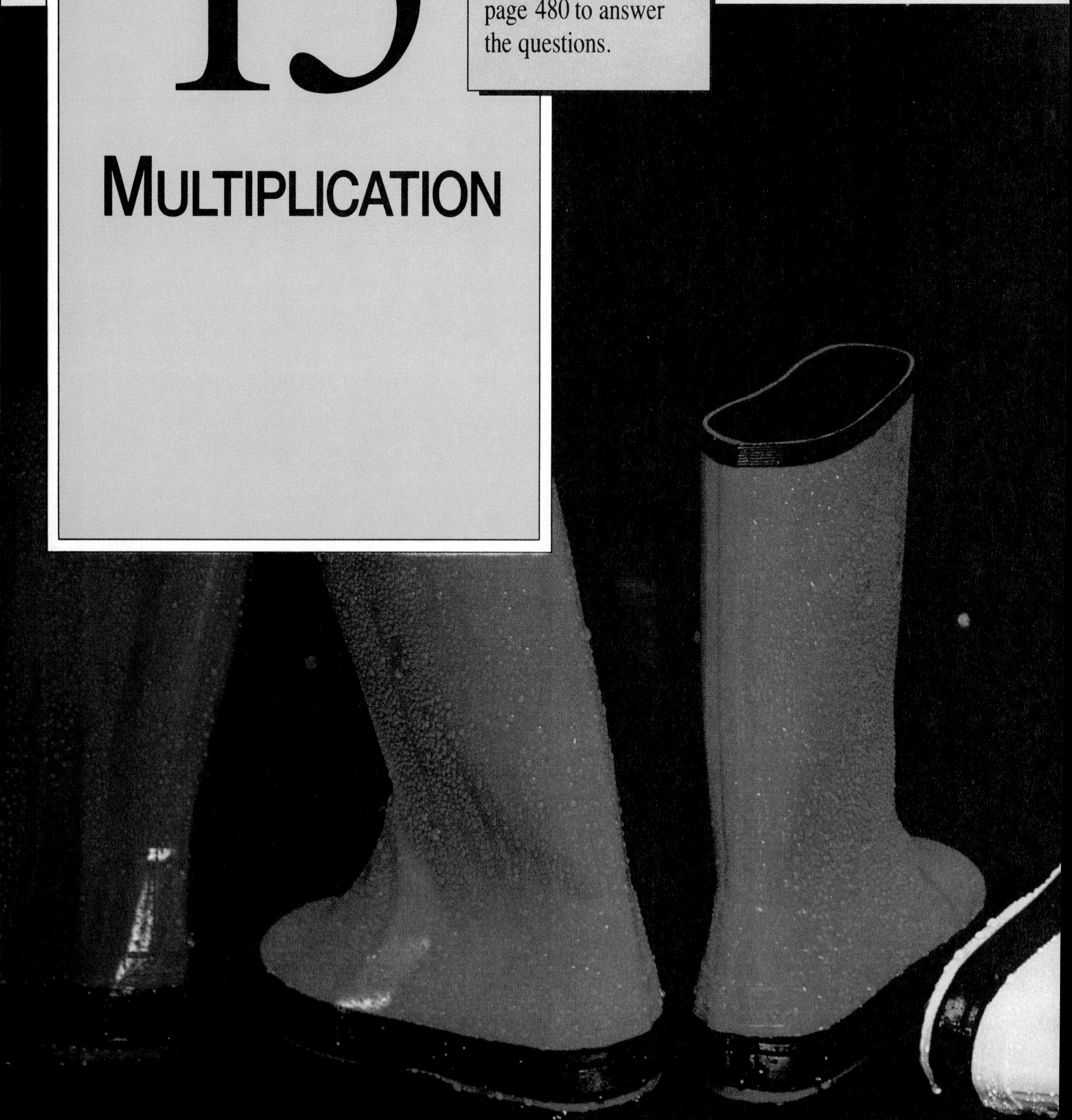

**2** How many pairs of boys' boots are in the classroom on rainy days? How many single brown boots are in the kindergarten?

**3** Ramona's mother bought her shiny new red boots. If all the girls wore their boots on the same day, how many red boots would there be in all?

**4** **Use Critical Thinking** 3 girls outgrew white boots and bought red ones, and 2 boys bought new red boots. Are there now more red or brown boots?

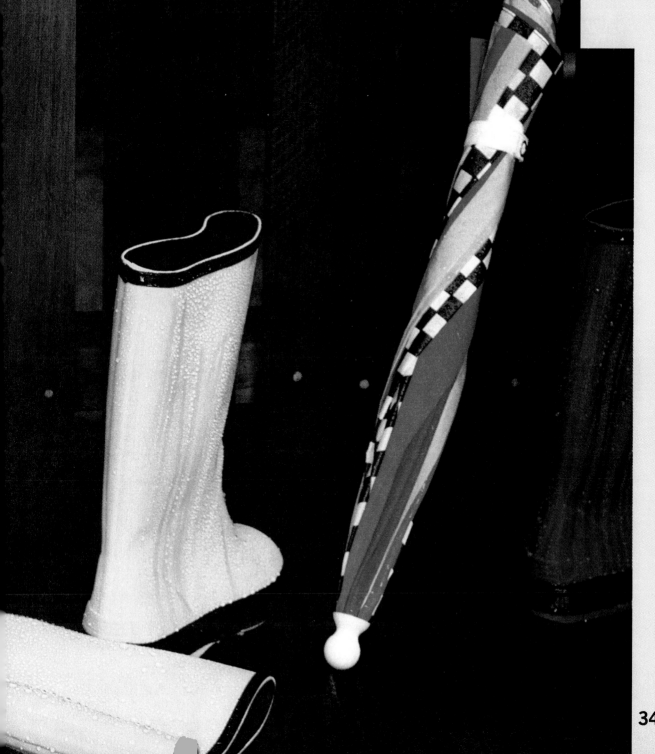

# Mental Math
## Multiply and Then Add

Nathan's Cards

**LEARN ABOUT IT**

**EXPLORE** **Think About the Process**

Work in groups. Use two sets of red cards with numbers 1 to 9 and one set of blue cards with numbers 1 to 8. Draw two red cards and one blue card. Multiply the numbers on the red cards. Add the product to the number on the blue card. Who in your group has the highest total after three turns?

Look at the cards that Cindy and Nathan drew. Whose score was higher?

**TALK ABOUT IT**

1. What is the highest score you can get in the game? the lowest?

2. Ted drew two red cards and a blue card with the number 1. His score was 25. What numbers do you think he might have drawn on the red cards?

Here is how Tina and Carlos found their scores.

**Tina** ( Think $6 \times 3 = 18$ / Then count 19, 20 )   **Carlos** ( Think $7 \times 4 = 28$ / $28 + 6$ is 34 )

**TRY IT OUT**

Multiply the first two numbers. Then add the product to the third number. Write answers only.

**1.** 2,7,1     **2.** 3,9,2     **3.** 5,4,3     **4.** 6,7,4     **5.** 5,9,5

Multiply the first two numbers. Then add the product to the third number. Write answers only.

**1.** 4,7,1      **2.** 5,7,7      **3.** 6,7,5      **4.** 7,7,4      **5.** 6,7,7

Multiply and then add 2. Write answers only.

| **6.** | 9 | **7.** | 2 | **8.** | 6 | **9.** | 4 | **10.** | 7 |
|---|---|---|---|---|---|---|---|---|---|
| | $\times 0$ | | $\times 9$ | | $\times 5$ | | $\times 6$ | | $\times 3$ |

Find the product and then add 4. Write answers only.

**11.** $3 \times 5$      **12.** $4 \times 8$      **13.** $8 \times 5$      **14.** $2 \times 3$      **15.** $7 \times 0$

**APPLY**

**MATH REASONING** What is the greatest number you can get if you multiply any two of these numbers and add the third?

**16.** 2,3,4            **17.** 1,2,3            **18.** 3,4,5            **19.** 4,5,6

**PROBLEM SOLVING**

**20. Find All Solutions** Kyle added 4 to the product of two numbers. His score was 22. What two numbers did he multiply?

**21.** Erin's score was 21. Her blue card was 6 and one of her red cards was 5. What was Erin's other red card? Use the rules on page 348.

▶ **USING CRITICAL THINKING  Make a Conjecture**

**22.** Which of these two games do you think is more likely to give higher scores?

**Game A**
Choose three numbers.
Multiply two of the numbers.
Add the third number.

**Game B**
Choose three numbers.
Find the sum of two of the numbers.
Multiply by the third number.

*More Practice, page 508, set G*

# Mental Math
## Special Products

**EXPLORE** **Discover a Relationship**

In the book *The Saturdays* by Elizabeth Enright, Mona, Rush, and Randy put their allowances together to pay for special Saturday outings. If each person put in 50¢, how much money was in the Saturday fund?

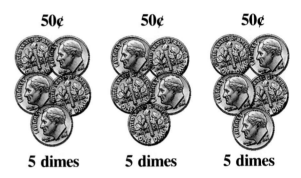

50¢          50¢          50¢

5 dimes      5 dimes      5 dimes

**3 × 5 dimes are 15 dimes**
**3 × 50 = ?**

**TALK ABOUT IT**

1. How many dimes did each person put in the fund? How many cents is each dime worth?

2. How many dimes in all are in the fund? How many cents are all the dimes worth? Explain.

3. Suppose Mona, Rush, and Randy each put in 60¢. How much would be in the fund then? Explain.

COINS

To find $7 \times 30$, break apart 30 to find the number of tens. Then write the product as a whole number.

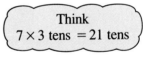

Think
$7 \times 3$ tens $= 21$ tens

$7 \times 30 = 210$

### Other Examples

$5 \times 60 = 300$ (30 tens $= 300$)          $4 \times 80 = 320$ (32 tens $= 320$)

Multiply by breaking apart numbers into tens.

**1.** $8 \times 10$          **2.** $4 \times 30$          **3.** $40 \times 6$          **4.** $5 \times 80$          **5.** $30 \times 2$

Multiply

**1.** $12 \times 10$      **2.** $24 \times 10$      **3.** $36 \times 10$      **4.** $40 \times 10$      **5.** $18 \times 10$

Find the products by breaking apart numbers into tens.

**6.** $6 \times 30$      **7.** $8 \times 30$      **8.** $8 \times 20$      **9.** $2 \times 30$      **10.** $3 \times 30$

**11.** $\begin{array}{r} 90 \\ \times\ 3 \\ \hline \end{array}$      **12.** $\begin{array}{r} 30 \\ \times\ 5 \\ \hline \end{array}$      **13.** $\begin{array}{r} 40 \\ \times\ 4 \\ \hline \end{array}$      **14.** $\begin{array}{r} 80 \\ \times\ 6 \\ \hline \end{array}$      **15.** $\begin{array}{r} 90 \\ \times\ 2 \\ \hline \end{array}$

**16.** Find the product of 7 and 80.      **17.** Multiply 40 by 9.

**APPLY**

**MATH REASONING**  You can use mental math to find
  these products if you are careful which two numbers
  you multiply first.

**Example:**   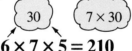      **18.** $4 \times 7 \times 5$      **19.** $8 \times 3 \times 5$

$$6 \times 7 \times 5 = 210$$

**PROBLEM SOLVING**

**20.** In *Henry Huggins* by Beverly
Cleary, Henry gets 25¢
allowance each week. If Henry
saves 20¢ of his allowance
every week for 7 weeks, how
much money will he save in all?

**21. Language Arts Data Bank**
Oliver is the youngest child in
*The Saturdays*. How much
money would he save if he
saved all his allowance for 6
weeks? See page 480.

▶ **ESTIMATION**

**22.** Sophie's Cafe needs to buy
about 20 dozen eggs each week
for breakfasts. About how many
dozen eggs does Sophie buy in
4 weeks?

**23.** A third grade class needed to
sell 175 snow cones at the
spring fair to pay for a field trip
to the zoo. They sold 2 snow
cones each to 80 people.
Estimate to decide if the class
sold enough snow cones.

# Estimating Products Using Rounding

**EXPLORE  Compare the Data**

Sue is 9 years old today. Her friend Willie says, "I am 400 weeks old!" Is Sue younger or older than Willie?

**TALK ABOUT IT**

1. How can you find the number of weeks in 9 years?

2. What is a good estimate of Sue's age in weeks? Is she older or younger than Willie?

When you want to compare a product to another number, you can often estimate the product. The number to which you compare the product is called the **reference point.**

## Another Example

Are there more or less than 300 days in 38 weeks?

40 nearest ten

$7 \times 38$ $\qquad$ $7 \times 40 = 280$

We rounded up, so the product is less than 280. There are less than 300 days in 38 weeks.

| Calendar Data | |
|---|---|
| 1 week | = 7 days |
| 1 year | = 365 days |
| | = 52 weeks |
| | = 12 months |
| 1 leap year | = 366 days |

Estimate the product. Is it more or less than the reference point in parentheses?

**1.** $4 \times 7(120)$ **2.** $5 \times 59(250)$ **3.** $7 \times 34(200)$ **4.** $6 \times 68(420)$

Estimate the product. Is it more or less than the reference point in parentheses?

**1.** $3 \times 58(200)$     **2.** $7 \times 23(200)$     **3.** $5 \times 47(250)$     **4.** $8 \times 19(160)$

**5.** $4 \times 79(350)$     **6.** $6 \times 82(480)$     **7.** $9 \times 27(300)$     **8.** $8 \times 53(400)$

Estimate each product by rounding to the nearest ten. Tell whether the exact product is over or under your estimate.

**9.** $7 \times 29$    **10.** $4 \times 49$    **11.** $6 \times 88$    **12.** $6 \times 42$    **13.** $8 \times 52$

**14.** $6 \times 23$    **15.** $7 \times 56$    **16.** $8 \times 32$    **17.** $2 \times 59$    **18.** $7 \times 28$

**19.** $\begin{array}{r} 18 \\ \times\ 5 \\ \hline \end{array}$    **20.** $\begin{array}{r} 27 \\ \times\ 8 \\ \hline \end{array}$    **21.** $\begin{array}{r} 44 \\ \times\ 3 \\ \hline \end{array}$    **22.** $\begin{array}{r} 78 \\ \times\ 6 \\ \hline \end{array}$    **23.** $\begin{array}{r} 19 \\ \times\ 9 \\ \hline \end{array}$

## APPLY

**MATH REASONING** These are estimates. What could the exact products be? Be ready to support your choices.

**24.** 120     **25.** 150     **26.** 180     **27.** 200     **28.** 240

### PROBLEM SOLVING

**29.** Most third graders spend about 6 hours in school each school day. How many hours do they spend in school in 30 school days?

**30.** Manny can hardly wait for his friend Pablo to arrive from Mexico in 17 weeks. Estimate how many days Manny must wait.

 ALGEBRA

Choose 3, 4, 5, or 6 to fill in the missing factor.

**31.** $\square \times 40 = 120$      **32.** $\square \times 40 = 200$      **33.** $\square \times 70 = 280$

**34.** $\square \times 70 = 350$      **35.** $\square \times 30 = 180$      **36.** $\square \times 30 = 150$

# Multiplying
## Making the Connection

**LEARN ABOUT IT**

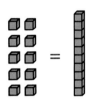

**EXPLORE  Use a Place Value Model**
Work with a partner.

■ Choose a one-digit number from the number
bank. Have your partner choose a two-digit
number from the number bank. Write the two
numbers in a table like the one below.

■ Use blocks to show the two-digit number.
Repeat it as many times as the one-digit
number you have chosen. For example, if you
choose 3 and your partner chooses 16, you
would show 3 sets of 16 with blocks.

■ Push the blocks together. Make as many trades
as you can. On the last line of the table, write
the number you have after trading.

■ Do this five times.

| Number Bank |
| --- |
| 19  13  18  2 |
| 3 |
| 16  22 |
| 23  17 |
| 4  14 |

**TALK ABOUT IT**

1. Did you have to trade for each pair of
   numbers?

2. Can you pick two numbers from the
   bank so that you do not have to trade?

3. Can you give a rule for deciding when you
   will need to trade?

|  | tens | ones |
| --- | --- | --- |
| 2-digit number → |  |  |
| 1-digit number → |  |  |
| number in all → |  |  |

354

You have pushed blocks together, traded, and figured out how many blocks you have in all. Now you will learn a way to record what you have done.

Use blocks to show $4 \times 17$.

**What You Do**                        **What You Record**

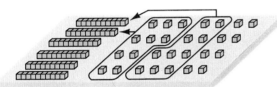

$$\begin{array}{r} 17 \\ \times\ 4 \\ \hline \end{array}$$

1. Are there enough ones on the table for some trading?

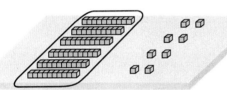

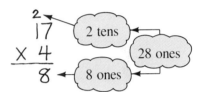

2. How many ones are there after the trade?

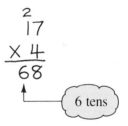

3. How many tens are there after the trading?

4. What is the product of $4 \times 17$?

5. Suppose you use blocks to show $4 \times 12$. Will there be enough ones for a trade? Explain.

---

**TRY IT OUT**

Show with place value blocks. Record the total.

1. Three 18s        2. Four 17s        3. Three 13s        4. Two 16s

5. Use blocks to solve your own multiplication problem.

355

# Multiplying
## Trading Ones

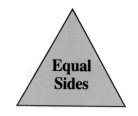

**Equal Sides**

**EXPLORE  Think About the Process**

The table shows the perimeters of some triangles with sides of equal length. What is the perimeter of a triangle whose sides are 18 cm long?

| Side | Perimeter |
|------|-----------|
| 7    | 21        |
| 8    | 24        |
| 9    | 27        |
| 10   | 30        |

**You multiply because you need to put together groups of equal amounts to find the total.**

| Multiply the ones. | Trade if you can. | Multiply the tens. Add any extra tens. |
|---|---|---|
| $\begin{array}{r} 18 \\ \times\ 3 \end{array}$  (24) | $\begin{array}{r} \scriptstyle 2 \\ 18 \\ \times\ 3 \\ \hline 4 \end{array}$  (Trade) | $\begin{array}{r} \scriptstyle 2 \\ 18 \\ \times\ 3 \\ \hline 54 \end{array}$  ((3 × 1) plus 2) |

**TALK ABOUT IT**

1. Why do you need to trade in the second step?

2. Explain why the product has 5 tens.

3. How could you have estimated the product?

4. Use a sentence to answer the problem.

**Other Examples**

$\begin{array}{r} \scriptstyle 1 \\ 25 \\ \times\ 3 \\ \hline 75 \end{array}$
$\begin{array}{r} 12 \\ \times\ 4 \\ \hline 48 \end{array}$ (No Trade)
$\begin{array}{r} \scriptstyle 3 \\ 18 \\ \times\ 4 \\ \hline 72 \end{array}$
$\begin{array}{r} \scriptstyle 1 \\ 35 \\ \times\ 2 \\ \hline 70 \end{array}$

Decide if you need to trade. Then find the products.

**1.** $16 \times 3$      **2.** $23 \times 4$      **3.** $37 \times 3$      **4.** $14 \times 2$      **5.** $41 \times 4$

Decide if you need to trade. Then find the products.

**1.** 15
$\times$ 4

**2.** 33
$\times$ 3

**3.** 28
$\times$ 2

**4.** 46
$\times$ 2

**5.** 11
$\times$ 9

**6.** 13
$\times$ 5

Multiply.

**7.** 19
$\times$ 5

**8.** 15
$\times$ 6

**9.** 12
$\times$ 7

**10.** 43
$\times$ 2

**11.** 16
$\times$ 5

**12.** 19
$\times$ 4

**13.** $2 \times 34$

**14.** $2 \times 14$

**15.** $4 \times 22$

**16.** $2 \times 18$

**17.** $5 \times 12$

**18.** $3 \times 30$

**19.** $5 \times 15$

**20.** $2 \times 42$

**21.** Find the product of 17 and 3.

**22.** What is 24 multiplied by 4?

**APPLY**

**MATH REASONING**

**23.** If $18 + 18 + 18 + 18$ goes with $4 \times 18$, what goes with $3 \times 27$?

**PROBLEM SOLVING**

**24. Understanding the Question** Baby elephants at the zoo like to chase each other around the edge of their square park. One side is 48 yards long. How far must the elephants run to go around the square once?

**25.** The baby gazelles live in a park whose sides are each 52 yards long. How far will the gazelles run if they go halfway around the square?

▶ **CALCULATOR**

**26.** Think of two ways to find the product $5 \times 15$ on a calculator. Try both ways. Did you get the same answer? Explain.

# More About Multiplying

**LEARN ABOUT IT**

**EXPLORE  Think About the Process**

In the book *Henry Huggins* by Beverly Cleary, Henry buys 2 guppies. His guppies have babies. Then the babies have babies. Soon Henry has so many guppies that he has to return them all.

Suppose 37 people each buy 4 pairs of guppies. How many pairs of guppies is that in all?

**You multiply because you need to put together same-size groups to find the total.**

| Multiply the ones. | Trade if you can. | Multiply the tens. Add any extra tens. |
|---|---|---|
| 37<br>× 4   (28) | 2<br>37<br>× 4  (Trade)<br>8 | 2<br>37<br>× 4  (4 × 3) plus 2<br>148 |

**TALK ABOUT IT**

1. How many tens do you get in the third step? How do you show this number in the answer?

2. How could you have estimated the product?

3. Use a sentence that answers the problem.

**Other Examples**

$$21 \times 6 = 126 \quad \text{(No Trade)}$$

$$\overset{1}{46} \times 3 = 138$$

$$\overset{4}{35} \times 8 = 280$$

$$\overset{2}{25} \times 4 = 100$$

**TRY IT OUT**

Multiply.

**1.** $62 \times 4$  **2.** $58 \times 3$  **3.** $65 \times 8$  **4.** $84 \times 6$  **5.** $21 \times 7$

Multiply.

| | | | | | |
|---|---|---|---|---|---|
| **1.** 62 <br> × 4 | **2.** 42 <br> × 8 | **3.** 36 <br> × 5 | **4.** 19 <br> × 4 | **5.** 52 <br> × 6 | **6.** 66 <br> × 3 |
| **7.** 25 <br> × 7 | **8.** 35 <br> × 6 | **9.** 13 <br> ×7 | **10.** 94 <br> × 2 | **11.** 23 <br> × 7 | **12.** 62 <br> × 5 |

**13.** $9 \times 23$    **14.** $8 \times 25$    **15.** $3 \times 83$    **16.** $4 \times 49$

**17.** Find the product of 2 and 75.    **18.** What is 53 multiplied by 6?

**MATH REASONING**

**19.** Without finding these products, tell how much they differ from each other. $3 \times 87$ and $3 \times 88$.

**PROBLEM SOLVING**

**20.** If 15 people returned their 4 pairs of guppies to the pet store, how many pairs of guppies did they return in all?

**DATA BANK**

**21. Language Arts Data Bank** Suppose Henry Huggins's allowance was the same as Mona's in *The Saturdays*. If they paid him in quarters, how many quarters would he save in two weeks? See page 480.

Find the quotients.

| | ÷ 2 | |
|---|---|---|
| | 24 | 12 |
| **22.** | 10 | |
| **23.** | 14 | |
| **24.** | 8 | |

| | ÷ 3 | |
|---|---|---|
| | 24 | 8 |
| **25.** | 12 | |
| **26.** | 36 | |
| **27.** | 27 | |

| | ÷ 4 | |
|---|---|---|
| | 24 | 6 |
| **28.** | 36 | |
| **29.** | 32 | |
| **30.** | 28 | |

| | ÷ 5 | |
|---|---|---|
| | 25 | 5 |
| **31.** | 15 | |
| **32.** | 30 | |
| **33.** | 45 | |

*More Practice, page 509, set D*

# Using Critical Thinking

Jill and Chuck emptied packages of peanuts into a bowl to fill it. Then they estimated how many peanuts were in the bowl.

"Each package has over 50 peanuts," said Jill. "Now we can multiply that number by the number of bags to estimate how many are in the bowl."

But then Jill spilled paint on her multiplication problem when she was painting the contest poster.

"Oh no, I can't remember the numbers," she said. "We'll have to count all the peanuts."

"Wait," said Chuck. "I just thought of a way to figure it out."

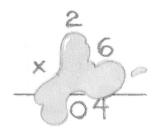

**Contest**

Guess how many peanuts are in the jar. The person with the closest guess wins the peanuts!

## TALK ABOUT IT

1. Why did Jill and Chuck multiply?

2. How did Chuck find the number of packages?

3. What clue on the paper helped Chuck finish the problem?

4. About how many peanuts were in the jar?

Find the missing digits.

1.
```
   ■
   3■
 ×  4
 ■■0
```

2.
```
     2
   ■8
 ×  ■
 ■74
```

3.
```
    6
   59
 ×  ■
 ■■3
```

4.
```
    4
   ■7
 ×  6
 52■
```

**360**

# MIDCHAPTER REVIEW/QUIZ

Multiply the first two numbers. Then add the third number. Write answers only.

**1.** 3, 5, 4    **2.** 2, 7, 3    **3.** 6, 8, 3    **4.** 9, 4, 2    **5.** 8, 7, 6

Multiply.

**6.** $7 \times 60$    **7.** $9 \times 80$    **8.** $8 \times 70$    **9.** $4 \times 60$    **10.** $6 \times 9 \times 5$

Estimate. Tell if the exact product is over or under your estimate.

**11.** $5 \times 23$    **12.** $3 \times 64$    **13.** $8 \times 47$    **14.** $9 \times 58$    **15.** $4 \times 75$

Multiply.

**16.**    15    **17.**    27    **18.**    52    **19.**    86    **20.**    43    **21.**    20
      $\times\ 6$        $\times\ 3$        $\times\ 4$        $\times\ 9$        $\times\ 2$        $\times\ 4$

**22.**    24    **23.**    75    **24.**    16    **25.**    19    **26.**    34    **27.**    54
      $\times\ 7$        $\times\ 8$        $\times\ 6$        $\times\ 5$        $\times\ 3$        $\times\ 6$

## PROBLEM SOLVING

**28.** Max used place value blocks to find these products. For which products did he trade ones for tens?

| | |
|---|---|
| **A.** $5 \times 16$ | **B.** $3 \times 23$ |
| **C.** $2 \times 18$ | **D.** $4 \times 12$ |
| **E.** $3 \times 30$ | **F.** $8 \times 7$ |

**29.** When Meghan multiplied two digits and added 7, her answer was 19. Which digits did she multiply? Can you find another way to get the same answer?

**30.** What different answers can you get when you multiply two of these digits and then add the third?

| 2 | 8 | 3 |
|---|---|---|

# Problem Solving
## Choosing a Calculation Method

| |
|---|
| **UNDERSTAND** |
| **FIND DATA** |
| **PLAN** |
| **ESTIMATE** |
| **SOLVE** |
| **CHECK** |

## LEARN ABOUT IT

One decision you must make when you solve problems is to choose a calculation method.

MO'S VIDEO
Monday
Rentals

Science Fiction  20
Comedies  37
Action  42
Classics  17
Drama  31
Westerns  16

> Customers at Mo's Video rented the same number of science fiction movies every day last week. How many did they rent in all during the 7 days?

$7 \times 20$
That's easy to do in my head.
I can use mental math.

How many comedies did customers rent in 6 days if they rented the same number every day?

$6 \times 37$
I'll need to trade. I could use pencil and paper, but a calculator might be better.

When you choose a calculation method:

- **First try mental math.** Look for numbers you can work with easily in your head.

- **If you can't use mental math, choose between pencil and paper and a calculator.**
  A calculator is often better if you need to trade or do many steps to solve a problem.

| Calculation Methods |
|---|
| ■ Mental Math |
| ■ Paper and Pencil |
| ■ Calculator |

## TRY IT OUT

Tell which calculation method you would use and why. Then solve.

**1.** $4 \times 36$  **2.** $2 \times 41$  **3.** $6 \times \$7.98$  **4.** $3 \times 80$

362

Choose a calculation method. Then solve. Use
each method at least twice.

**1.** $4 \times 80$    **2.** $4 \times 24$    **3.** $7 \times 688$    **4.** $2 \times 34$    **5.** $82 - 77$

**6.** $\$8.95 \times 4$    **7.** $9 \times 6$    **8.** $78 - 25$    **9.** $36 \times 4$    **10.** $6 \times 20$

**11.** $8 \times 11$    **12.** $3 \times 32$    **13.** $5 \times 75$    **14.** $5 \times 8$    **15.** $54 + 21$

**16.** $\begin{array}{r} 70 \\ \times\ 5 \\ \hline \end{array}$    **17.** $\begin{array}{r} 36 \\ -\ 6 \\ \hline \end{array}$    **18.** $\begin{array}{r} 54 \\ \times\ 6 \\ \hline \end{array}$    **19.** $\begin{array}{r} 21 \\ \times\ 3 \\ \hline \end{array}$    **20.** $\begin{array}{r} \$8.12 \\ -\$5.78 \\ \hline \end{array}$

Choose a calculation method. Then solve. You
may use the list on page 362.

**21.** How many more action movies
than westerns did Mo's
customers rent on Monday?

**22.** Suppose customers rented 5
fewer movies on Tuesday than
they did on Monday. How many
movies did they rent on
Tuesday?

**23.** Mo charges $2.95 a day to rent
comedies. How much money
did he earn from comedies on
Monday?

**24.** How many movies did
customers rent in all on
Monday?

**25.** If customers rented the same
number of westerns on Monday,
Tuesday, Wednesday, and
Thursday, how many westerns
did they rent during the four
days?

**26.** Jerome rented a comedy for
$2.95 and a drama for $2.49.
He gave Mo a ten dollar bill.
How much change did he get?

**27. Write Your Own Problem**
Write and solve a problem that
you can solve with mental math.
Use data from the list on page
362.

# Estimating and Finding Money Products

**EXPLORE** **Think About the Process**

The chart shows ticket prices for the school play. Jamie wants to buy adult tickets for her mother and father, her aunt and uncle, and two neighbors. She has $6. Does Jamie have enough money to buy tickets on the main floor? How much will she pay?

Tickets for Sale

|  | Main Floor | Balcony |
|---|---|---|
| Child | $0.45 | $0.35 |
| Adult | $0.95 | $0.50 |
| Senior | $0.35 | $0.25 |

**TALK ABOUT IT**

1. How many tickets does Jamie need?
2. Can Jamie estimate, or must she know the exact product to figure out if she has enough money? Explain.

**Estimate the product.**

$$\begin{array}{r} \$0.95 \\ \times \quad 6 \\ \hline \end{array} \qquad \begin{array}{r} \$1.00 \\ \times \quad 6 \\ \hline \$6.00 \end{array}$$

Round to the nearest ten cents.

**Find the exact product.**

$$\begin{array}{r} \$0.95 \\ \times \quad 6 \\ \hline \$5.70 \end{array}$$

Multiply the same way that you multiply whole numbers. Show dollars and cents.

## Other Examples

**Estimates**

$$\begin{array}{r} \$0.45 \\ \times \quad 5 \\ \hline \end{array} \qquad \begin{array}{r} \$0.50 \\ \times \quad 5 \\ \hline \$2.50 \end{array} \qquad \begin{array}{r} \$0.34 \\ \times \quad 4 \\ \hline \end{array} \qquad \begin{array}{r} \$0.30 \\ \times \quad 4 \\ \hline \$1.20 \end{array}$$

**Exact Answer**

$$\begin{array}{r} \$0.45 \\ \times \quad 5 \\ \hline \$2.25 \end{array} \qquad \begin{array}{r} \$0.34 \\ \times \quad 4 \\ \hline \$1.36 \end{array}$$

Estimate the products. Then find the exact answers.

| 1. | 2. | 3. | 4. | 5. |
|---|---|---|---|---|
| $0.26 | $0.32 | $0.43 | $0.78 | $0.23 |
| × 4 | × 5 | × 6 | × 3 | × 9 |

Estimate the products. Then find the exact answers.

**1.**  $0.51     **2.**  $0.77     **3.**  $0.55     **4.**  $0.26     **5.**  $0.34
    $\times$ 5          $\times$ 3          $\times$ 9          $\times$ 4          $\times$ 7

**6.**  $0.98     **7.**  $0.48     **8.**  $0.15     **9.**  $0.42     **10.**  $0.81
    $\times$ 4          $\times$ 7          $\times$ 6          $\times$ 9           $\times$ 5

**11.** $8 \times \$0.25$     **12.** $9 \times \$0.13$     **13.** $4 \times \$0.49$     **14.** $5 \times \$0.78$

**15.** $6 \times \$0.41$     **16.** $3 \times \$0.38$     **17.** $6 \times \$0.24$     **18.** $4 \times \$0.56$

**19.** 4 at $0.67 each        **20.** 8 at $0.19 each        **21.** 7 at $0.32 each

**MATH REASONING** Use mental math to find these
   products. Remember that 4 quarters equal $1.00.

**22.** $4 \times \$0.25$        **23.** $5 \times \$0.25$        **24.** $8 \times \$0.25$        **25.** $9 \times \$0.25$

**PROBLEM SOLVING**

**26. Data Hunt** Find the price of
   something you would like to
   buy that costs less than a dollar.
   How much money would you
   spend if you bought one for
   yourself and three friends?

**27.** Jared bought 5 children's tickets
   in the balcony and 2 adult
   tickets on the main floor. How
   much money did he spend? Use
   the chart on page 364.

Is the dashed line a line of symmetry? Check by tracing and folding.

**28.**      **29.**      **30.**      **31.**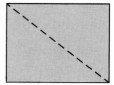

*More Practice, page 509, set E*

# Problem Solving
## Using a Calculator

| UNDERSTAND |
|---|
| FIND DATA |
| PLAN |
| ESTIMATE |
| SOLVE |
| CHECK |

**LEARN ABOUT IT**

Sometimes you need to do several steps to solve a problem. A calculator can help.

In her sporting goods store, Ms. Rogers has 27 packages of baseballs that have 3 balls in each package and 19 boxes of baseballs with 12 balls in each box. She also has 28 single baseballs. How many baseballs does Ms. Rogers have?

Mrs. Rogers wrote a plan for solving the problem. Then she found the answer using a calculator.

First I'll find how many balls are in the 27 packages.

$\boxed{\text{ON/AC}}$ 27 $\boxed{\times}$ 3 $\boxed{=}$ $\boxed{81}$

Then I'll find how many balls are in the boxes.

$\boxed{\text{ON/AC}}$ 19 $\boxed{\times}$ 12 $\boxed{=}$ $\boxed{228}$

Now I can find the sum of these products and the single balls.

$\boxed{\text{ON/AC}}$ 81 $\boxed{+}$ 228 $\boxed{+}$ 28 $\boxed{=}$

$\boxed{337}$

Ms. Rogers has 337 baseballs in her store.

**TRY IT OUT**

Use a calculator to solve these problems.

1. Ms. Rogers has 7 boxes with 18 tennis balls in each box and 43 cans with 3 balls in each can. She also has 2 single tennis balls. How many tennis balls does Ms. Rogers have?

2. Ms. Rogers sold 44 packs of fishhooks with 6 hooks in each pack, 22 packs with 25 hooks in each pack, and 7 boxes with 100 hooks in each box. How many fishhooks did she sell?

366

Solve. You may use a calculator.

1. Raul hung up 36 tennis rackets at Tina's Sports Depot. He could fit 6 rackets on one hook. How many hooks did he use?

2. Large skateboard stickers cost $0.79 and small stickers cost $0.59. How much did Carna pay for 4 small stickers?

3. Juanita counted golf balls for Ms. Rogers after school. She found 16 boxes with 4 balls in each box, 22 boxes with 2 balls in each box, and 32 single golf balls. How many golf balls did Juanita count in all?

4. Carlos works at Tina's Sports Depot. He counted 27 cans of racquetballs with 3 balls in each can, 32 boxes with 6 racquetballs in each box, and 42 single balls. How many racquetballs did he count in all?

5. Yan wants to buy a silver-colored baseball bat that is in the middle of a row of bats. There are 4 bats to the left of the silver one. How many bats are in the whole row?

**Some Strategies**

Act It Out
Choose an Operation
Make an Organized List
Look for a Pattern
Use Logical Reasoning
Use Objects
Draw a Picture
Guess and Check
Make a Table
Work Backward

6. Johnna paid $5.95 for a handball and $9.95 for a soccer ball. How much change did she get back from her twenty dollar bill?

7. Vic saw swimming goggles on sale for $7.98. The regular price was $9.89. How much money did he save by buying the goggles on sale?

8. Li wants to buy a baseball mitt. She can choose from three different brands: Compete, Rally, and Fox. Each brand offers mitts at two prices: $20 and $40. How many types of mitts can Li choose from?

*More Practice, page 524, set C*

**367**

# Applied Problem Solving
## Group Decision Making

UNDERSTAND
FIND DATA
PLAN
ESTIMATE
SOLVE
CHECK

**Group Skill:**
Disagree in an Agreeable Way

Your teacher, Mr. Glue, likes to give stickers to students who do special jobs or show good behavior. He has asked you to order this year's supply of stickers.

**Facts to Consider**

1. Mr. Glue gives out about 200 stickers every year.

2. He likes to have several types of stickers.

3. He can buy no more than 4 packages of the most expensive stickers.

4. He buys stickers from a catalog that sells the stickers shown in the table.

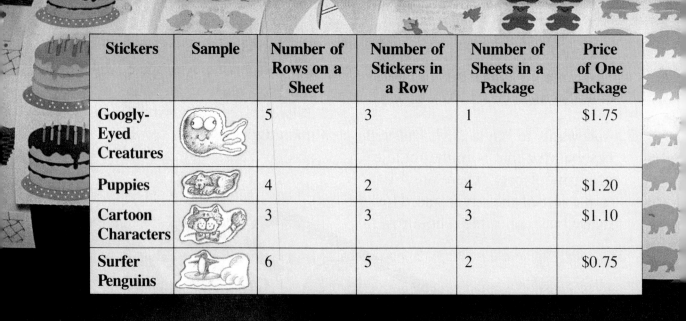

| Stickers | Sample | Number of Rows on a Sheet | Number of Stickers in a Row | Number of Sheets in a Package | Price of One Package |
|---|---|---|---|---|---|
| Googly-Eyed Creatures | | 5 | 3 | 1 | $1.75 |
| Puppies | | 4 | 2 | 4 | $1.20 |
| Cartoon Characters | | 3 | 3 | 3 | $1.10 |
| Surfer Penguins | | 6 | 5 | 2 | $0.75 |

1. How many stickers are in a package of googly-eyed creatures?

2. How many puppy stickers are on 1 sheet?

3. How many puppy stickers are in a package?

4. What is a good way to keep track of how many stickers of each type you are ordering?

5. What is the total cost of the stickers?

How many packages of each type of sticker did you decide to order? Make a table to show how many stickers of each type Mr. Glue will have.

369

# Wrap Up

## Producing Products

Choose the number in the box that best completes each sentence.

1. To multiply 6 × 70, think of 70 as _____ tens.

2. Multiply 7 tens by 6 to get _____ tens.

3. 42 tens is _____.

4. The estimated product for 6 × 47 is _____, an overestimate.

5. To estimate the product of $0.23 × 7, round $0.23 to _____ and multiply by 7.

> 42   40   47
>
> 360   420
>
> 25  6  7  25
>
> 300   0.20

## Sometimes, Always, Never

Decide which word should go in the blank, *sometimes*, *always*, or *never*. Explain your choices.

6. When you multiply a 2-digit number by a 1-digit number, you _____ add the extra tens before you multiply the tens.

7. Rounding down in multiplication _____ gives an answer close to the exact answer.

8. Rounding up in multiplication _____ gives an answer greater than the exact answer.

## Project

Work with a partner. Each of you measure one of your normal steps to the nearest inch. Make a table to show how far you would walk in 2 through 9 steps. Compare results. Would one of you walk farther in fewer steps? You may want to use a calculator to make your table.

# CHAPTER REVIEW/TEST

## Part 1  Understanding

**1.** $8 \times 5$ tens is the same as $8 \times 50$. Do you agree or disagree? Explain your answer.

**2.** How would you estimate the cost of 4 melons that each cost $0.89?

Tell when you would choose the calculation method.

**3.** using mental math

**4.** using a calculator

## Part 2  Skills

**5.** Add 5 to the product of $7 \times 9$.

**6.** Add 3 to the product of $6 \times 4$.

Is the estimate more or less than the reference point in parentheses?

**7.** $8 \times 66$ (550)

**8.** $4 \times 72$ (300)

**9.** $5 \times 57$ (270)

Multiply.

**10.** $60 \times 9$

**11.** $26 \times 3$

**12.** $19 \times 5$

**13.** $38 \times 5$

**14.** $28 \times 7$

**15.**  $\begin{array}{r} 14 \\ \times\ 6 \\ \hline \end{array}$

**16.**  $\begin{array}{r} 78 \\ \times\ 4 \\ \hline \end{array}$

**17.**  $\begin{array}{r} 43 \\ \times\ 8 \\ \hline \end{array}$

**18.**  $\begin{array}{r} \$0.56 \\ \times\ \ \ 7 \\ \hline \end{array}$

**19.**  $\begin{array}{r} \$0.81 \\ \times\ \ \ 6 \\ \hline \end{array}$

## Part 3  Applications

**20.** Larry earns $0.95 each day doing chores. If he saves his money for a week, about how much does he save? Can he buy a dictionary for $6.39?

**21.** Linda has $4.00 to buy 6 marking pens. Some pens cost $0.64 each. Other pens cost $0.67 each. What can Linda buy? Explain.

**22.** Ms. Wells has 6 display cases with 24 seashells in each, 8 cases with 33 shells in each, and 7 cases with 29 shells each. How many seashells are in her collection?

**23.** **Challenge** Students sold 8 cases of 19 notebooks each and 5 cases of 31 book covers each. A notebook costs $1.32. A book cover costs $0.39. How much did they earn?

# ENRICHMENT
## Estimating by Clustering

Wanda helped the librarian by checking out books to students. She checked out 33 books on Monday, 27 on Tuesday, 29 on Wednesday, and 34 on Thursday. About how many books did Wanda check out?

Each addend is close to 30, so you can estimate by using **clustering.** Multiply $4 \times 30$ to estimate the sum.

Wanda checked out about 120 books.

Estimate each sum. Use clustering.

**1.** Each addend is close to what number?
THINK: $3 \times 20$

$17 + 24 + 19$

**2.** Each addend is close to what number?
THINK: $4 \times 40$

$37 + 41 + 35 + 43$

**3.** $37 + 36 + 40 + 38$

**4.** $52 + 45 + 50 + 52$

**5.** $32 + 29 + 34 + 27$

**6.** $44 + 41 + 43 + 42$

**7.** $73 + 69 + 70$

**8.** $85 + 93 + 87 + 94$

**9.** $61 + 57 + 64 + 62 + 56$

**10.** $67 + 71 + 69 + 70 + 74$

**11.** Matt traveled with his family. They drove 74 miles the first day, 68 miles the second day, 71 miles the third day, and 65 miles the fourth day. About how many miles did they travel in all?

**12.** Carol practiced hopping on one foot. She hopped 27 meters without stopping. Then she hopped 31 meters, 29 meters, and 34 meters. About how many meters did she hop all together?

**13.** A store has stickers in boxes that give the price of each sticker. About how much would it cost to buy one weather sticker from each box?

# CUMULATIVE REVIEW

**1.** Give the best description of this angle.

A. greater than a right angle
B. less than a right angle
C. a triangle
D. a right angle

**2.** Give the best description of these shapes.

A. congruent  B. polygons
C. space figures  D. intersecting

**3.** Give the best description of these lines.

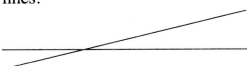

A. congruent  B. symmetrical
C. intersecting  D. parallel

**4.** Which number sentence shows 12 items shared 3 ways?

A. $12 \div 4 = 3$  B. $12 \div 3 = 4$
C. $3 \times 4 = 12$  D. $12 - 3 = 9$

**5.** Find the quotient. $4\overline{)32}$

A. 4  B. 6
C. 7  D. 8

**6.** Divide. $7\overline{)49}$

A. 5  B. 7
C. 8  D. 9

**7.** Find the quotient. $56 \div 8$

A. 5  B. 6
C. 7  D. 8

**8.** $9\overline{)81}$

A. 6  B. 7
C. 8  D. 9

**9.** Give the number pair for X.

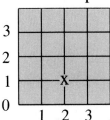

A. (0, 2)  B. (1, 2)
C. (2, 1)  D. (2, 2)

**10.** Which two problems are related?
**1.** Nat earns $1 for every 3 magazines he sells. How much does he earn for 9 magazines?
**2.** How much does Ernie pay for 3 magazines at $2 each?
**3.** Cindy can buy a bag of 6 marbles for $2. How much do 24 marbles cost?

A. 1 and 3  B. 1 and 2
C. 2 and 3  D. 3 and 2

# 14

## METRIC MEASUREMENT

**M**ATH AND
HEALTH AND FITNESS

DATA BANK

Use the Health and
Fitness Data Bank on
page 483 to answer

1 If you eat 1 cup of
whole wheat cereal
with 1 cup of milk
for breakfast, how many
calories are in the food?

**2** Five children shared one serving of grapes. How many grapes did they each eat? How many calories were in each child's share?

**3** Which uses more calories, sleeping for 9 hours or running for 1 hour?

**4** **Use Critical Thinking** Compare the activities. Which would use more calories in 15 minutes, roller skating or writing? Explain.

# Estimating and Measuring Length in Centimeters

**LEARN ABOUT IT**

**EXPLORE** Use a Benchmark

This paper clip is about 3 **centimeters** (cm) long. How can you use the paper clip as a benchmark to estimate the length of the pencil?

**About 3 cm**

**TALK ABOUT IT**

1. About how many paper clips long is the pencil? How did you find out?

2. How can you use multiplication with this benchmark to estimate the pencil's length?

You can use what you already know about measuring to measure to the nearest centimeter.

**10 cm to the nearest cm**

**9 cm to the nearest cm**

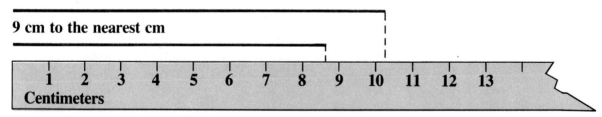

Centimeters

You can use a known length as a benchmark to help you estimate length.

**TRY IT OUT**

Use the paper clip as a benchmark to make these estimates. Check your estimates by measuring to the nearest centimeter.

1. Length of your shoe

2. Width of your shoe

376

Use your hand span as a benchmark to estimate these lengths. Check your estimates by measuring to the nearest centimeter. You may wish to use string to help you with your measuring.

**1.** Your height

**2.** Your neck

**3.** Your waist

**APPLY**

**MATH REASONING** Suppose a footprint is 12 cm to the nearest centimeter. Of which statement can you be sure?

**4.** The footprint is more than 12 and less than 13 cm.

**5.** It is more than 11 and less than 13 cm.

**6.** It is more than 11 and less than 12 cm.

**PROBLEM SOLVING**

**7.** Su Lin's footprint is about 17 cm long. How long are 8 of her prints measured end to end?

**8.** Chaco's suitcase is 23 cm long and 15 cm wide. What is the perimeter of his suitcase?

**MIXED REVIEW**

Divide.

**9.** $16 \div 8$  **10.** $5 \div 5$  **11.** $54 \div 9$  **12.** $18 \div 6$  **13.** $63 \div 7$

**14.** $28 \div 7$  **15.** $30 \div 30$  **16.** $0 \div 7$  **17.** $72 \div 8$  **18.** $32 \div 8$

**19.** $9\overline{)81}$  **20.** $8\overline{)56}$  **21.** $6\overline{)24}$  **22.** $6\overline{)42}$  **23.** $9\overline{)36}$

**24.** $6\overline{)54}$  **25.** $9\overline{)45}$  **26.** $1\overline{)8}$  **27.** $7\overline{)14}$  **28.** $9\overline{)9}$

# Decimeter, Meter, and Kilometer

## EXPLORE Use Paper Strips

Work in groups. Cut strips of paper 10 cm long.
Tape them together to make a strip 1 meter long.
Use a meter stick to be sure. Make one extra
10-cm strip. Estimate to find objects that are as
long as each of your strips.

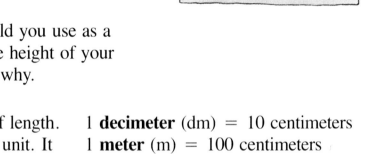

## TALK ABOUT IT

1. How many 10-cm strips did you need to make
   a 1-meter strip?

2. Which of your strips would you use as a
   benchmark to estimate the height of your
   desk? of a door? Explain why.

Here are three metric units of length.
The kilometer is a very long unit. It
takes about 10 minutes to walk
1 km.

1 **decimeter** (dm) = 10 centimeters
1 **meter** (m) = 100 centimeters
1 **kilometer** (km) = 1,000 meters

## TRY IT OUT

Is it longer than 1 decimeter?

| | | | |
|---|---|---|---|
| **1.** 15 cm | **2.** 1 m | **3.** 1 cm | **4.** 10 cm |

Is it longer than 1 meter?

| | | | |
|---|---|---|---|
| **5.** 100 cm | **6.** 5 dm | **7.** 120 cm | **8.** 15 dm |

Is it longer than 1 kilometer?

| | | | |
|---|---|---|---|
| **9.** 1,500 m | **10.** 500 m | **11.** 1,000 m | **12.** 5,000 m |

1. Which object is shorter than 1 dm?

2. Which objects are longer than 1 dm?

3. Which objects are shorter than 1 m?

4. Which object is longer than 1 m?

**60 cm**

**9 cm**

**125 cm**

APPLY

**MATH REASONING** Is the distance more than, less than, or the same as 1 km?

5. from the park to the store

6. from the store to the gas station

7. from the gas station to the park

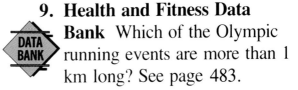

Store

1,254 m        1,000 m

Park

Gas station    952 m

**PROBLEM SOLVING**

8. If a table is 78 cm tall and 160 cm long, how much less than 1 m is its height?

9. **Health and Fitness Data Bank** Which of the Olympic running events are more than 1 km long? See page 483.

DATA BANK

▶ **ESTIMATION**

10. Work with a partner. Plan a way to estimate time in seconds. Practice until you are able to estimate 10 seconds. Use your method to estimate how long it takes you to hop a distance of 10 meters. Take turns checking each other's estimates against a clock.

*More Practice, page 525, set A*

# Area

LEARN ABOUT IT

## EXPLORE  Use Objects

Cover the inside of these shapes with pennies.
Count the number of pennies on each shape.

## TALK ABOUT IT

**1.** What is the measure of the blue rectangle in pennies? of the red rectangle?

**2.** What object could you use to get more exact measurements of these rectangles? Explain.

The **area** of a shape is the number of square units needed to cover the inside of the shape.

You can measure area using a **square centimeter** unit.

**Square centimeter**          **The area is 12 square centimeters.**

TRY IT OUT

Find the area of each shape in square centimeters.

**1.**          **2.**          **3.**          **4.**

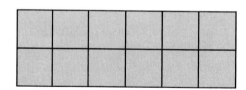

Give the area of each shape in square centimeters.

**1.**                    **2.**                    **3.**

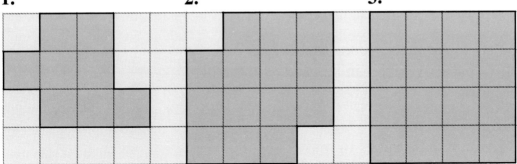

Use graph paper. Draw shapes with these areas.

**4.** 12 square units     **5.** 20 square units     **6.** 24 square units

**MATH REASONING**

**7.** Rectangle A and Rectangle B have the same area. A is wider than B. Which rectangle is longer?

**PROBLEM SOLVING**

**8.** A counter top is covered with three rows of square tiles. There are 8 tiles in each of the first two rows and 7 tiles in the third row. What is the area of the counter top in tiles?

**9. Draw a Picture** Use graph paper. Find two shapes with the same area but different perimeters.

▶ **USING CRITICAL THINKING**

**10.** Give the area for this figure.

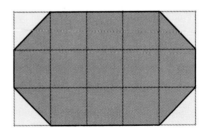

# Volume

**Volume** is the number of units that fills a geometric space. A cube can be used as the unit for measuring volume.

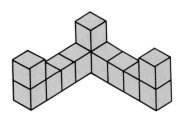

### EXPLORE  Use Cubes

Work in groups. Use cubes to make 3 or more figures. Each figure should have a volume of 12 cubic units and look different from the one shown here.

### TALK ABOUT IT

**1.** How many cubes tall is your tallest figure?

**2.** How many cubes wide is your widest figure?

**3.** Do figures with the same volume have to have the same shape?

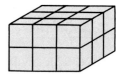

**The volume is
18 cubic units.**

When measuring volume, you sometimes have to count cubic units that you cannot see.

Give the volume of each figure in cubic units.

**1.**

**2.**

**3.**

**4.**

**5.**

**6.**

Find the volume of each figure in cubic units.

**1.**

**2.**

**3.**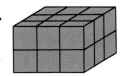

**4.** Which figure has the greatest volume? What is its volume?

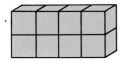 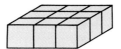

**MATH REASONING**

**5.** The volume of two figures is the same. One base is smaller than the other. Which one is taller?

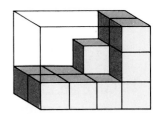

**PROBLEM SOLVING**

**6.** Inside each large box, Raul packed 18 boxes of shoes. If he packed 9 large boxes, how many boxes of shoes did he pack in all?

**7. Use Objects** Jenny used 27 small cubes to build a large cube. How many cubes were on the bottom layer? (Use the hint in Exercise 8.)

▶ **CALCULATOR**

**8.** The distance around this cube is 500 cm. What is the length of each side of the cube in centimeters?

HINT: The height, length, and width of a cube are all the same.

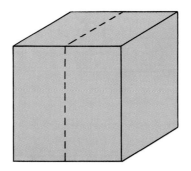

# Temperature

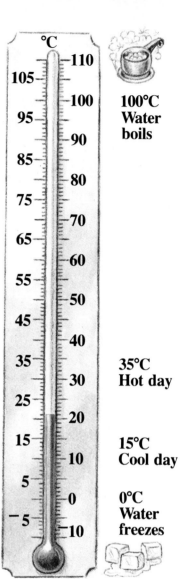

°C

110
105
100
95
90
85
80
75
70
65
60
55
50
45
40
35
30
25
20
15
10
5
0
-5
-10

100°C
Water
boils

35°C
Hot day

15°C
Cool day

0°C
Water
freezes

**LEARN ABOUT IT**

**EXPLORE  Study the Information**

The **degree Celsius** (°C) is the metric unit for measuring temperature. This thermometer reads 21°C.

**TALK ABOUT IT**

1. What happens to the liquid in the thermometer when the temperature gets warmer? What happens to the liquid when the temperature gets cooler?

2. Give an example of a very hot temperature and an example of a very cold temperature.

**Cold lemonade**

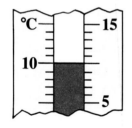

°C  15
10
5

**Hot soup**

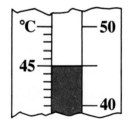

°C  50
45
40

**PRACTICE**

Record each temperature.

**1.**

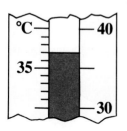

°C  40
35
30

**2.**

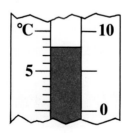

°C  10
5
0

**3.**

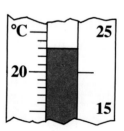

°C  25
20
15

**4.**

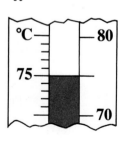

°C  80
75
70

**384**

*More Practice, page 525, set D*

# MIDCHAPTER REVIEW/QUIZ

Use the length of your little finger as a benchmark to make these estimates in centimeters. Check your estimates by measuring to the nearest centimeter.

**1.** Length of your scissors

**2.** Length of your favorite pencil

**3.** Length of a sheet of paper

**4.** Width of a sheet of paper

Is it longer than 1 decimeter?

**5.** 80 cm

**6.** 1 km

**7.** 8 m

Is it the same as 1 meter?

**8.** 10 dm

**9.** 100 cm

**10.** 1,000 km

Give the area of each shape in square centimeters.

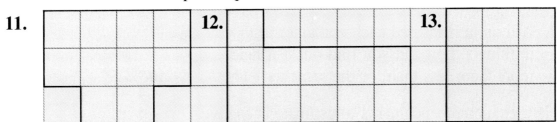

**11.**

**12.**

**13.**

Give the volume of each figure in cubic units.

**14.**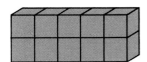

**15.**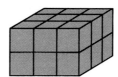

## PROBLEM SOLVING

**16.** It takes Barbie about 23 minutes to walk to school. Choose the best estimate for that distance.

    **A** about 1 km
    **B** between 3 km and 4 km
    **C** about 2 km

**17.** Friday the temperature was 29°C. Saturday the temperature rose 8 degrees. Give the temperature on Saturday.

**18.** If there are about 5 city blocks in 1 km, how many city blocks are there in 6 km?

# Liters and Milliliters

The **liter** and the **milliliter** are metric
units used to measure capacity.

### EXPLORE Estimate and Measure

Work in groups. Get some large containers.
Estimate which of them would hold about 1 liter.
Use water to check your estimates. Measure each
container to the nearest 100 milliliters.

**1 liter (L) =
1,000 milliliters (mL)**

### TALK ABOUT IT

1. How did you estimate which containers would
   hold about 1 liter?

2. After you measured each container in
   milliliters, how did you find out if it held
   more than, less than, or the same as 1 liter?

Here are some benchmarks to estimate capacity.

**A medicine dropper holds about 1 mL.**

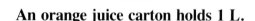

**An orange juice carton holds 1 L.**

### PRACTICE

Use a benchmark. Estimate whether the object
holds more than, less than, or about the same as 1 liter.

**1.** Pitcher

**2.** Cereal bowl

**3.** Water pail

# Problem Solving

UNDERSTAND
FIND DATA
PLAN
ESTIMATE
SOLVE
CHECK

## MIXED PRACTICE

Solve. Use any problem solving strategy.

1. The 14 girls in the third grade exercise 45 minutes a day, 5 days a week. How many minutes a week do the girls exercise?

2. The boys jog 3 days a week. If they each want to cover 6 km a week, how many kilometers should they plan to jog each day?

3. Darnell brought 5 liters of water for the 10 boys to drink when they finished jogging. If each boy gets the same amount of water, how many milliliters will they each have?

4. Joshua wants to jog 6 km a week. How can he do it in 3 days if he jogs a different number of kilometers each day?

5. Kin Lee did 2 push-ups the first week, 4 the second week, 6 the third week, and so on. At that rate, how many will she do the tenth week?

6. Melissa decided to jog 4 km a week. How many weeks will it take Melissa to cover 96 km?

7. One group of students goes to a dance class 2 times a week. They dance the same amount of time at each class. They dance 256 minutes in 4 weeks. How many minutes do they dance at each class?

8. Rosalia swims to stay fit. The pool she uses is 40 m long and 20 m wide. How many meters does she cover if she swims the length 8 times?

9. **More Than One Answer** Plan a schedule for Rosalia so that she swims 50 laps of the pool a week. Make sure she swims at least 3 days a week but no more than 5 days a week.

# Grams and Kilograms

**LEARN ABOUT IT**

**EXPLORE  Use a Benchmark**

Work in groups. Use your math book as a benchmark. Compare it to objects in your classroom. Record whether each object is heavier than, lighter than, or about the same as your math book.

**TALK ABOUT IT**

**1.** Can you always tell how heavy something is by its size? Why or why not?

**2.** How can you check on your estimates?

The **gram** (g) and the **kilogram** (kg) are metric units used to measure how heavy objects are. 1 kg = 1,000 g. By using benchmarks, you can estimate objects in these units.

**A paper clip is about 1 g.**

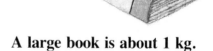

**A large book is about 1 kg.**

**TRY IT OUT**

Use a benchmark to estimate these objects in grams or kilograms. Check each object on a metric scale.

**1.** Penny

**2.** Brick

**3.** Apple

Use benchmarks to estimate these objects in grams or kilograms. Check each object on a metric scale.

**1.** Box of crayons

**2.** Typewriter

**3.** Softball

**4.** Safety pin

**5.** Dictionary

**6.** Paper cup

**APPLY**

**MATH REASONING**

**7.** Malcolm weighs 40 kg on the bathroom scale. How much will he weigh if he picks up one foot and stands on one leg instead of two?

**PROBLEM SOLVING**

**8.** Ann's apple weighs 80 g. She cut it into 4 pieces. Each piece weighs the same. How much does each piece weigh?

**9. Health and Fitness Data Bank**

Which of the food servings weighs the most? Which weighs the least? See page 483.

**MIXED REVIEW**

**A** 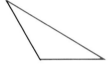    **B**     **C**     **D**

**10.** How many segments does each polygon have?

**11.** How many points does each polygon have?

**12.** Which polygons have red parallel segments?

**13.** Which polygons have red intersecting segments?

# Problem Solving
## Deciding When to Estimate

| UNDERSTAND |
| FIND DATA |
| PLAN |
| ESTIMATE |
| SOLVE |
| CHECK |

### LEARN ABOUT IT

It is important to know when you can use an estimate or when you have to give an exact measurement. Knowing how the measurement will be used can help you decide.

> Ron is cutting strips of paper about 5 cm long to make a chain. Heather is cutting strips to match the heights of baby animals on a poster.

The strips of paper for the chain do not have to be exactly 5 cm long.

Ron can estimate the length of each strip he cuts.

The strips of paper for the poster must be the same height as the baby animals.

Heather needs to measure each strip to match the height of each animal on the poster.

### TRY IT OUT

1. You are buying rope for a new jump rope. Should you estimate or measure the length of rope?

2. Your pet parakeet is supposed to drink 1 milliliter of medicine each day. Should you estimate or measure the amount of medicine you give her?

3. You are building a gate. Should you estimate or measure the distance between the posts?

4. You need to cut a post for your school's car wash sign. Should you estimate or measure the length of the post?

Choose a strategy from the strategies list or other strategies you know to solve these problems.

**Some Strategies**

Act it Out
Use Objects
Choose an Operation
Draw a Picture
Make an Organized List
Guess and Check
Make a Table
Look for a Pattern
Use Logical Reasoning
Work Backward

1. Therese is a mail carrier. She said, "I have between 10 and 25 letters for your family. I have fewer than $3 \times 5$. I have an odd number of letters. I don't have 11 letters. How many do I have?"

2. To keep in shape for her job, Therese swims. The first week, she swam 250 m. The second week, she swam 300 m, and the third week, 350 m. If this pattern continued, how far did she swim the sixth week?

3. Leah is going to the post office. She needs to buy 5 stamps. Should she estimate or count the amount of money she needs to take with her?

4. There are 5,393 workers at the Bay City post office. They handle 9 million pieces of mail a day. There are 1,586 workers at the Midtown post office. How many more workers are at the Bay City office than the Midtown office?

5. Clark has 28 blocks on his mail route. He usually finishes 4 blocks in 1 hour. How long does it usually take for him to finish his route?

6. There are about 24 houses on a block. Clark delivers about 6 pieces of mail per house. How many pieces of mail does Clark deliver on a block?

7. **Determining Reasonable Answers**
Tell whether the calculator answer is reasonable. Explain why or why not.

On the coldest day that Clark delivered the mail, it was 3°C. On the warmest day, it was 2 degrees warmer than 11 times the temperature on the coldest day. What was the temperature on Clark's warmest day?

*More Practice, page 526, set B*

391

# Data Collection and Analysis
## Group Decision Making

UNDERSTAND
FIND DATA
PLAN
ESTIMATE
SOLVE
CHECK

**Doing an Investigation**

**Group Skill:**
Explain and Summarize

To do this investigation, your group will need a paper bag and three small pieces of paper that are all the same size. Draw a triangle on two of the pieces of paper and draw a square on the other one. Put the pieces in the paper bag.

Your group should predict whether more triangles or squares will be drawn. Record your predictions.

**Collecting Data**

1. Draw a piece of paper out of the bag. Record if the paper has a triangle or a square on it. Return the paper to the bag. Take turns drawing a piece of paper out of the bag. Your group should do a total of at least 40 draws.

| Shape | Number Drawn |
|-------|--------------|
| ■ | |
| ▲ | |

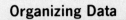

**2.** Make a picture graph using the data that you have collected.

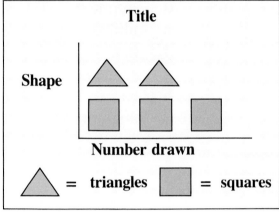

**3.** Were your group's predictions correct? Explain how you made your predictions.

**4.** Write at least three sentences about the information in your graph.

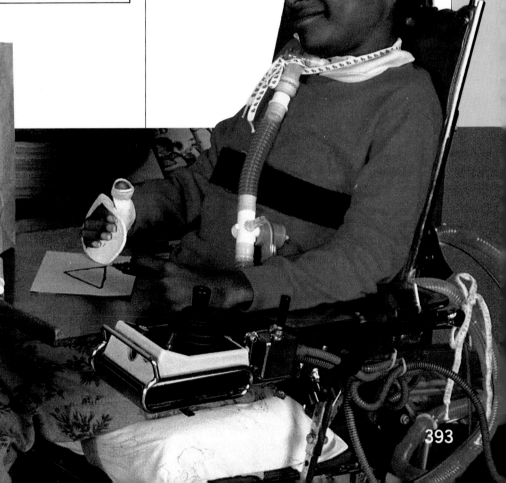

393

# WRAP UP

## Measurement Match

Match each description on the left with the most likely unit of measure on the right.

1. length of a rubber band
2. amount of soup in a spoon
3. weight of a chair
4. width of a poster
5. distance to the next town
6. heat of liquid in a pot
7. height of a diving board
8. weight of a ring
9. amount of water in a fishbowl

a. kilometer
b. degree Celsius
c. meter
d. centimeter
e. milliliter
f. gram
g. decimeter
h. liter
i. kilogram

## Sometimes, Always, Never

Decide which word should go in the blank, *sometimes*, *always*, or *never*. Explain your choices.

10. Units in the metric system _____ are related by multiples of 10.

11. Benchmarks _____ are used to give estimates in measurement.

12. The measure of an object using a large unit _____ will be a greater number than the measure of that object using a smaller unit.

13. The volume of a space figure is _____ the number of cubic units the figure will hold.

## Project

Build this figure with centimeter cubes. What is the perimeter of the base? Give the area of the base of the figure in square centimeters. Give the volume in cubic centimeters.

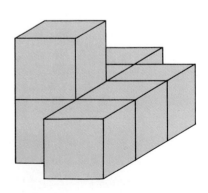

# CHAPTER REVIEW/TEST

## Part 1   Understanding

1. How would you use an eraser about 5 cm long to estimate the width of your desk?

2. Would you use an eraser to estimate the length of a school hallway? Explain.

3. How do the areas of these two shapes compare?

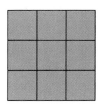 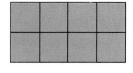

4. Can different shapes have the same area? Explain.

5. The temperature is 5°C. Would you wear a warm coat or swim at the beach? Why?

6. Would you use liters or milliliters to find the amount of juice in a cup? Tell why.

## Part 2   Skills

7. Give the volume of the figure in cubic units.

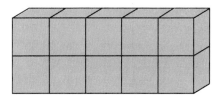

Choose centimeters, decimeters, meters, or kilometers to measure.

8. the distance from school to home

9. the length of a book shelf

10. the length of a necklace

11. a lap in a swimming pool

Choose grams or kilograms to measure how heavy each object is.

12. a strawberry

13. a bag of flour

14. a slice of bread

## Part 3   Applications

15. The hardware store clerk tells you that 2 or 3 mL of oil can loosen a rusty lock. Should you measure or estimate the amount of oil you need? Why?

16. **Challenge** Your class is planning a 500-meter obstacle race course. Do you measure or estimate the distances in your course? Why?

# ENRICHMENT
## Surface Area Applications

Suppose it costs 5¢ to paint each face of this 1 centimeter cube. How much would it cost to paint the whole cube?

6 square centimeters × 5¢ = 30¢

Use centimeter cubes to build each figure to match the picture. Count the square centimeters to find the cost of painting the outside of each stack of cubes. You can use your calculator to help find the answer.

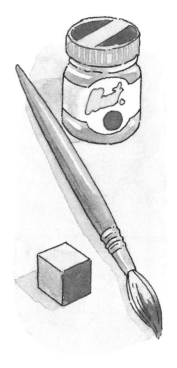

**1.**

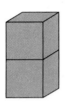

**2.**

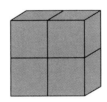

**3.**

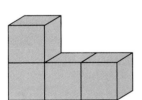

**4.**

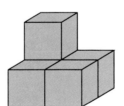

**5.**

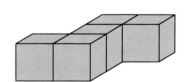

**6.**

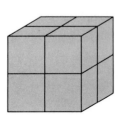

**7.**

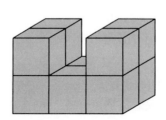

**8.**

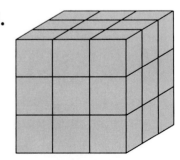

# CUMULATIVE REVIEW

1. Which shows how 15 apples are shared equally by 5?

    **A.** $15 \div 3 = 5$     **B.** $15 \div 5 = 3$

    **C.** $3 + 5 = 8$     **D.** $15 - 5 = 10$

2. Which is from a fact family for the factors 4 and 6?

    **A.** $24 \div 6 = 4$     **B.** $24 \div 8 = 3$

    **C.** $24 \div 3 = 8$     **D.** $24 \div 2 = 12$

3. Divide. $5\overline{)0}$

    **A.** 5     **B.** 10

    **C.** 1     **D.** 0

4. Find the quotient. $36 \div 6$

    **A.** 5     **B.** 6

    **C.** 7     **D.** 8

5. Find the quotient. $9\overline{)54}$

    **A.** 4     **B.** 5

    **C.** 6     **D.** 7

6. What is $63 \div 7$?

    **A.** 6     **B.** 9

    **C.** 8     **D.** 10

7. Multiply. $19 \times 4$

    **A.** 56     **B.** 66

    **C.** 76     **D.** 86

8. Give the product. $4 \times 40$

    **A.** 120     **B.** 200

    **C.** 160     **D.** 400

9. Find the estimate and the exact product. $\$0.76 \times 7$

    **A.** \$4.90, \$5.32    **B.** \$5.60, \$5.32

    **C.** \$5.60, \$5.42    **D.** \$6.30, \$5.42

10. One book costs \$3.27. Which calculation method would you choose to find the exact cost of 29 books?

    **A.** mental math
    **B.** paper and pencil
    **C.** calculator
    **D.** addition

11. Virginia bought special drawing paper on sale for \$1.63 a sheet. How much did 25 sheets cost?

    **A.** \$40.75     **B.** \$407.50

    **C.** \$40.00     **D.** \$84.76

12. What would it cost to buy 1 marker, 2 pens, and 1 eraser?

    | Write Stuff Catalog | | | |
    |---|---|---|---|
    | paper | \$3.19 | notebook | \$1.19 |
    | pencil | \$0.25 | marker | \$0.89 |
    | pen | \$0.55 | eraser | \$0.39 |

    **A.** \$1.78     **B.** \$2.68

    **C.** \$2.38     **D.** \$1.83

# 15

## FRACTIONS AND DECIMALS

**M**ATH AND SCIENCE

DATA BANK

Use the Science Data Bank on page 474 to answer the questions.

1 What is the total number of shells that are shown in the data bank? How many of the shells come from animals that live on land?

2 How many of the total number of shells in the data bank are from snails? Are they all land snails?

3 How many of the shells are from limpets? Do they all live in salt water?

4 **Use Critical Thinking** The shells are grouped by where they are found. What other ways could you sort these shells into groups? Explain.

# Naming Parts of a Whole

**EXPLORE  Use Fraction Pieces**

Work in groups. Look at the chart. Pretend your fraction pieces are pizzas. Find a whole pizza that can be shared equally among the members of your group. Record the size of your group and the name of the pieces you shared.

**TALK ABOUT IT**

1. If you were in a group of 3, what is the name of the pieces you shared?

2. If your group shared sixths, how many were in the group?

Here are some other ways to divide a whole into equal parts.

**Sharing Equally**

2 shares

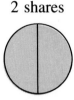

**Halves**

3 shares

**Thirds**

4 shares

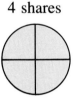

**Fourths**

6 shares

**Sixths**

**5 Equal Parts**

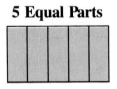

**Fifths**

**8 Equal Parts**

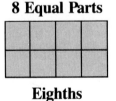

**Eighths**

**10 Equal Parts**

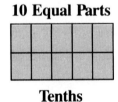

**Tenths**

**12 Equal Parts**

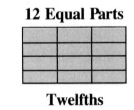

**Twelfths**

**TRY IT OUT**

Show these with your fraction pieces. Give the missing word.

1. If a group of 8 people share something equally, each person gets one _____ of the whole.

2. A group of _____ people shared something equally. Each of them got one tenth.

These graph-paper figures are divided into equal parts. Give the name of the parts of each whole.

**1.**    **2.** 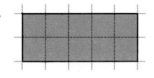   **3.** 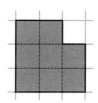   **4.**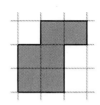

Draw a graph-paper figure for each of the following.

**5.** 6 equal parts          **6.** 10 equal parts          **7.** Fourths

**8.** A square divided into ninths          **9.** A square divided into sixteenths

**MATH REASONING**

**10.** Start with a paper strip. Fold it once. Fold it again. Fold it once more. Into how many parts do the folds divide the paper? First guess the answer. Then do the folding to check your guess.

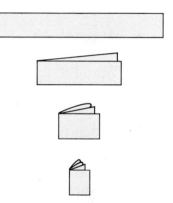

**PROBLEM SOLVING**

Show the answers to the questions using fraction pieces or graph paper.

**11.** April cut her pizza into thirds. She ate 2 of the pieces. Did April eat more or less than a half?

**12.** Suppose April had cut her pizza into fifths instead of thirds. She still ate 2 of the pieces. Did she eat more or less than half of the pizza?

▶ **USING CRITICAL THINKING  Number Sense**

**13.** Andrew and Erica each bought a foot-long submarine sandwich. Andrew divided his into sixths. Erica divided hers into eighths. Who had larger pieces?

*More Practice, page 526, set C*                                                    **401**

# Understanding Fractions

You can use the names you learned
in the last lesson to name fractions.

**Cory's Garden**

Tomatoes                        Corn

**EXPLORE Think About the Process**

Cory divided his garden into fifths.
He planted tomatoes in 2 of the
fifths and corn in 3 of the fifths.
What can you say about Cory's
garden?

**TALK ABOUT IT**

**1.** Did Cory plant all of his garden?

**2.** Were the parts of the garden the same size?

Fractions name parts of a whole. This fraction
tells what part of the garden was planted in
tomatoes.

Parts planted in tomatoes $\longrightarrow$ $\dfrac{2}{5}$
Number of same-size parts $\longrightarrow$

We read "two fifths" for the fraction $\dfrac{2}{5}$.

We say, "Two fifths of the garden is planted in tomatoes."

**TRY IT OUT**

Write a fraction that tells what part is planted.

**1.**                         **2.** 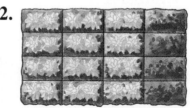     **3.**

Only the green part of each garden is planted. Write a fraction that tells what part of each garden is planted.

**1.**  **2.** 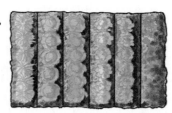 **3.**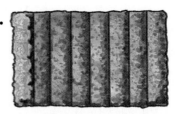

Use graph paper to show each garden. Use color to show what part is planted.

**4.** $\frac{1}{4}$ planted      **5.** $\frac{2}{3}$ planted      **6.** $\frac{7}{8}$ planted      **7.** $\frac{3}{10}$ planted

**APPLY**

**MATH REASONING**

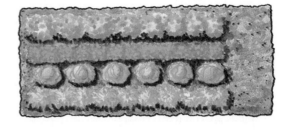

**8.** Write a fraction to estimate what part of this garden is planted.

**PROBLEM SOLVING**

**9.** Carrie and Derek have gardens the same size. Carrie has planted $\frac{3}{8}$ of her garden. Derek has planted $\frac{3}{4}$ of his. Who has more of the garden planted?

**10.** Shannon has planted $\frac{7}{10}$ of her garden. How much of it is not planted?

▶ **COMMUNICATION  Write to Learn**

**11.** Lacie said, "$\frac{1}{4}$ of something is more than $\frac{1}{6}$ of it, because each of 4 equal parts is larger than each of 6 equal parts." Write a sentence explaining what Lacie's idea tells you about $\frac{1}{10}$ and $\frac{1}{12}$.

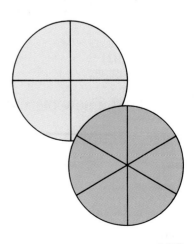

# Equivalent Fractions

**LEARN ABOUT IT**

The same amount can have more than one name.

**EXPLORE  Use Paper and Crayons**
Fold a sheet of paper into 2 equal parts. Unfold and color one of the parts. Now fold the paper twice so that you have 4 equal parts.

**TALK ABOUT IT**

1. What fraction of the paper did you color after the first fold?

2. After the second fold, what fraction does the colored part show?

3. If you fold the paper a third time, what fraction could you show? Try it.

Fractions that name the same amount are called equivalent fractions.

**We write:**  $\frac{1}{2} = \frac{2}{4}$

On the third fold, you showed that

$$\frac{1}{2} = \frac{4}{8}$$

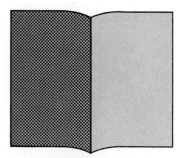

**Folded once**

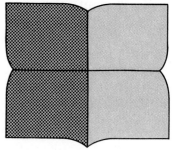

**Folded twice**

**TRY IT OUT**

- Fold a new piece of paper to show fourths.

- Color $\frac{1}{4}$ of the paper.

- Fold your paper to show a fraction that is equivalent to $\frac{1}{4}$.    $\frac{1}{4} = \frac{||||}{||||}$

**404**

These pictures show other fractions that are equivalent to $\frac{1}{2}$. What are they?

**1.**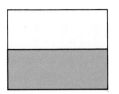

$\frac{1}{2} = \frac{||||}{||||}$

**2.**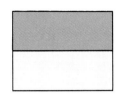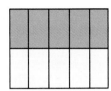

$\frac{1}{2} = \frac{||||}{||||}$

Give the missing fractions.

**3.**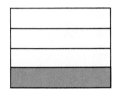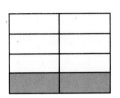

$\frac{1}{4} = \frac{||||}{||||}$

**4.**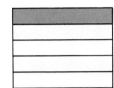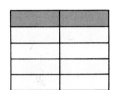

$\frac{1}{5} = \frac{||||}{||||}$

**APPLY**

**MATH REASONING**

**5.** How many times would you have to fold a paper to get 32 equal parts? Guess first. Then try it.

**PROBLEM SOLVING**

**6.** Janet's sandwich was cut into fourths. She ate half of it. How many pieces did she eat?

**7.** Mac cut a pan of corn bread into sixteenths. The family ate 7 pieces at lunch and 6 at dinner. What fraction of the bread is left?

▶ **ESTIMATION**

Give a fraction to tell about how much gasoline is in the tank.

**Example**

Answer $\frac{1}{4}$

**8.**

**9.**

**10.**

# Comparing Fractions

**LEARN ABOUT IT**

**EXPLORE** **Use Fraction Pieces**

Using just the fraction pieces shown here, how many comparisons can you make like the examples below?

$\frac{1}{2}$ is greater than $\frac{1}{4}$     $\frac{1}{4}$ is less than $\frac{1}{2}$

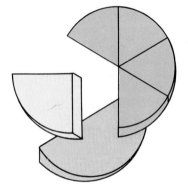

**TALK ABOUT IT**

1. When you find that $\frac{1}{3}$ is greater than $\frac{1}{6}$, what else do you know about these fractions?

2. How do two of the $\frac{1}{3}$ pieces together compare in size to the $\frac{1}{2}$ piece? What does that tell you about $\frac{2}{3}$ and $\frac{1}{2}$?

3. How does $\frac{2}{6}$ compare to $\frac{1}{2}$?

**Remember:** The symbol $>$ points to the smaller number.

**We say:** $\frac{1}{2}$ is greater than $\frac{1}{4}$          $\frac{1}{4}$ is less than $\frac{1}{2}$

**We write:** $\frac{1}{2} > \frac{1}{4}$          $\frac{1}{4} < \frac{1}{2}$

**TRY IT OUT**

Write the sign $>$, $<$, or $=$ for each ⦀. Use your fraction pieces when you need help.

**1.** $\frac{1}{2}$ ⦀ $\frac{1}{3}$          **2.** $\frac{1}{5}$ ⦀ $\frac{1}{4}$          **3.** $\frac{1}{10}$ ⦀ $\frac{1}{5}$          **4.** $\frac{2}{4}$ ⦀ $\frac{3}{6}$

Write the sign $>$, $<$, or $=$ for each ⫴.

**1.** $\frac{1}{6}$ ⫴ $\frac{1}{2}$    **2.** $\frac{1}{8}$ ⫴ $\frac{1}{6}$    **3.** $\frac{1}{5}$ ⫴ $\frac{1}{10}$    **4.** $\frac{1}{2}$ ⫴ $\frac{1}{3}$    **5.** $\frac{2}{4}$ ⫴ $\frac{1}{2}$

Give the comparison shown by each picture of fraction pieces.

**Example**

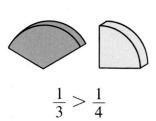

$\frac{1}{3} > \frac{1}{4}$

**6.**

$\frac{1}{6}$ ⫴ $\frac{1}{4}$

**7.**

$\frac{2}{6}$ ⫴ $\frac{1}{3}$

**8.**

$\frac{1}{2}$ ⫴ $\frac{2}{5}$

**MATH REASONING**

**9.** As you read the list of fractions, are they getting larger or smaller?

$\frac{1}{2}$, $\frac{1}{3}$, $\frac{1}{4}$, $\frac{1}{5}$, $\frac{1}{8}$, $\frac{1}{10}$, $\frac{1}{12}$

**PROBLEM SOLVING**

**10. Extra Data** Leo has 8 cheese wheels in his shop. He cut 1 of them into twentieths. After selling 5 pieces, he split the rest of the cut wheel equally among his 5 brothers. What fraction of the wheel did each brother get?

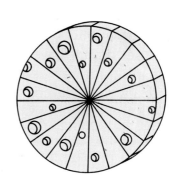

**MIXED REVIEW**

Multiply.

**11.** $6 \times 80$    **12.** $9 \times 70$    **13.** $25 \times 10$    **14.** $8 \times 50$    **15.** $80 \times 7$

Estimate. Tell if the exact product is over or under your estimate.

**16.** $6 \times 48$    **17.** $8 \times 23$    **18.** $5 \times 74$    **19.** $4 \times 19$    **20.** $9 \times 87$

# Fractional Part of a Set

## LEARN ABOUT IT

### EXPLORE  Understand the Process

The post office uses wavy lines to cancel stamps.
$\frac{3}{8}$ of this strip of stamps has been cancelled.
When you cut them apart, you can see that $\frac{3}{8}$ of
the set of stamps has been cancelled too.

Draw a strip of stamps on graph paper. Cancel
some of them. Then cut out all of your stamps.

### TALK ABOUT IT

**1.** What fraction of your strip did you cancel?

**2.** How many stamps do you have in all?

**3.** What fraction of your stamps are cancelled?

The fractional part of a set is much like the
fractional part of an object except the objects in
the set do not have to be the same size.

$\frac{3}{8}$ **of the stamps are cancelled.**

Cancelled stamps  $\longrightarrow$  $\dfrac{2}{6}$

Total number of stamps  $\longrightarrow$

$\frac{2}{6}$ of the stamps are cancelled.

### TRY IT OUT

Tell what fraction of the stamps are cancelled.

**1.**

**2.**

**408**

What fraction of the stamps are cancelled?

**1.**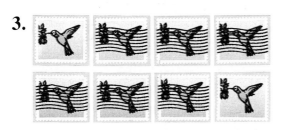

**2.**

**3.**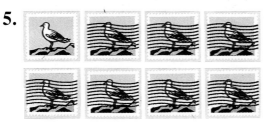

**4.**

**5.**

**6.**

**MATH REASONING** Draw a picture for each statement.

**7.** $\frac{3}{4}$ of the stamps are cancelled.

**8.** $\frac{1}{3}$ of the stamps are cancelled.

**PROBLEM SOLVING**

**9.** Kevin had 12 stamps. He used 2 of them for letters and 3 for a package. What fraction of the stamps did he use?

**10.** Cindy has 8 stamps. She used 5 of them on a package. What fraction of the stamps does she have left?

**USING CRITICAL THINKING  Take a Look**

**11.** How many different ways can you tear 3 stamps from a large sheet of stamps?

Draw pictures on graph paper to show your answers. Remember: These stamps are not square.

**Example:**

*More Practice, page 526, set E*

# Finding a Fraction of a Number

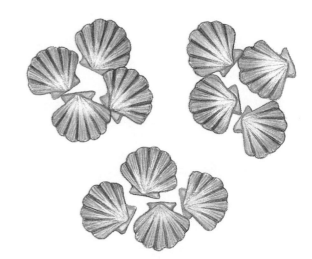

## LEARN ABOUT IT

You can use division to find a fractional part of a number.

### EXPLORE  Use Counters

Alan collected 12 seashells. He decided to give $\frac{1}{3}$ of them to his friends. Use your counters to show how he decided how many to give away.

### TALK ABOUT IT

**1.** How did Alan find $\frac{1}{3}$ of the seashells?

**2.** How would you use groups to find $\frac{1}{4}$ of 12 seashells?

**3.** What if Alan had 13 seashells and wanted to give away $\frac{1}{3}$ of them. Could he do it?

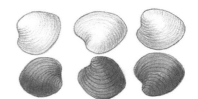

$\frac{1}{2}$ of 6 is 3.

Here is a way to find a fraction of a number.

- To find $\frac{1}{2}$ of a number, divide by 2.

- To find $\frac{1}{3}$ of a number, divide by 3.

How would you find $\frac{1}{4}$ of a number? $\frac{1}{5}$ of a number?

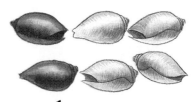

$\frac{1}{3}$ of 6 is 2.

## TRY IT OUT

Find each of the following.

**1.** $\frac{1}{2}$ of 8      **2.** $\frac{1}{3}$ of 15      **3.** $\frac{1}{4}$ of 20      **4.** $\frac{1}{5}$ of 15

Find each of the following.

**1.** $\frac{1}{3}$ of 21      **2.** $\frac{1}{5}$ of 40      **3.** $\frac{1}{2}$ of 10      **4.** $\frac{1}{4}$ of 24

**5.** $\frac{1}{6}$ of 42      **6.** $\frac{1}{8}$ of 8      **7.** $\frac{1}{3}$ of 27      **8.** $\frac{1}{8}$ of 40

**9.** $\frac{1}{5}$ of 30      **10.** $\frac{1}{2}$ of 18      **11.** $\frac{1}{4}$ of 32      **12.** $\frac{1}{3}$ of 24

| APPLY |

## MATH REASONING

**13.** Is this statement sometimes true, always true, or never true?

$\frac{1}{2}$ of a set of shells is more than $\frac{1}{3}$ of the same set of shells.

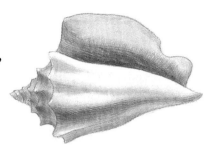

## PROBLEM SOLVING

**14.** Sixteen students went to the beach to collect shells. Half of them were girls. How many girls were in the group?

**15.** **Science Data Bank** What  fraction of all the shells on the chart are snail shells? See page 474.

## ▶ CALCULATOR

How much would you save on each sale item?

**16.**

$24.96
Sale $\frac{1}{3}$ off
| ON/AC | 24.96 | ÷ | 3 | = |

**17.**

$36.98
Sale $\frac{1}{2}$ off
| ON/AC | 36.98 | ÷ | 2 | = |

**18.**

$31.80
Sale $\frac{1}{4}$ off
| ON/AC | 31.80 | ÷ | 4 | = |

# Exploring Algebra

**LEARN ABOUT IT**

What will the next three figures will look like?

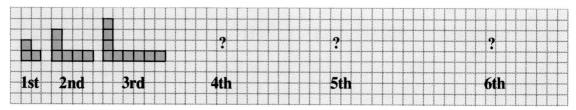

## TALK ABOUT IT

1. Describe the pattern in the figures.
2. What is the number pattern in the table?
3. Use a calculator to find the number of blocks in the twentieth design.

| Design Number | 1st | 2nd | 3rd | 4th | 5th | 6th |
|---|---|---|---|---|---|---|
| Number of Blocks | 3 | 6 | 9 | ? | ? | ? |

**TRY IT OUT**

Use blocks or draw pictures to make the next three designs. Copy and complete the table.

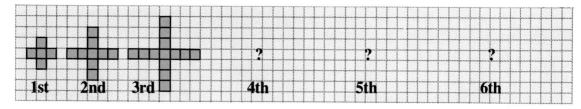

1. Describe the pattern in the figures.

2. What is the number pattern in the table?

3. Use a calculator to find the number of blocks in the twentieth design.

| Design Number | 1st | 2nd | 3rd | 4th | 5th | 6th |
|---|---|---|---|---|---|---|
| Number of Blocks | 5 | 9 | 13 | ? | ? | ? |

412

# MIDCHAPTER REVIEW/QUIZ

Draw and shade a graph-paper figure for each.

**1.** 8 equal parts.
Shade $\frac{5}{8}$.

**2.** Halves.
Shade $\frac{1}{2}$.

**3.** Fifths.
Shade $\frac{5}{5}$.

Use the rectangles to write equations showing equivalent fractions.

**4.** $\frac{1}{4} = \square$

**5.** $\frac{2}{4} = \square$

**6.** $\frac{2}{3} = \square$

Write the sign $>$, $<$, or $=$ for each ▥.

**7.** $\frac{1}{2}$ ▥ $\frac{6}{12}$

**8.** $\frac{2}{8}$ ▥ $\frac{2}{4}$

**9.** $\frac{1}{3}$ ▥ $\frac{1}{4}$

**10.** $\frac{5}{8}$ ▥ $\frac{3}{4}$

Find each of the following.

**11.** $\frac{1}{2}$ of 16

**12.** $\frac{1}{3}$ of 24

**13.** $\frac{1}{6}$ of 42

**14.** $\frac{1}{4}$ of 36

## PROBLEM SOLVING

**15.** Greg planted $\frac{2}{5}$ of his garden with tomatoes. What fraction of his garden was not tomatoes?

**16.** Alice planted $\frac{1}{3}$ of her garden with corn and $\frac{1}{4}$ of her garden with peas. Did she plant more of her garden in corn or in peas?

**17.** Christian bought 8 postcards. He mailed 2 to friends in France and 3 to friends in America. What fraction of the postcards did he mail?

**18.** Draw a figure with 7 equal parts. Color some parts red and the others blue. Use $<$ and $>$ to write two statements comparing the fractions for the red and the blue parts.

413

# Estimating Fractional Parts

## LEARN ABOUT IT

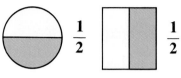

**Benchmarks**

### EXPLORE Use a Benchmark

The picture shows some examples of the fraction $\frac{1}{2}$. Can you use these pictures as benchmarks to help you decide if more or less than $\frac{1}{2}$ of the food is left?

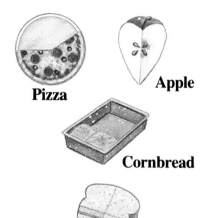

### TALK ABOUT IT

1. Which of the pictures show more than $\frac{1}{2}$? How can you tell?

2. Which show less than $\frac{1}{2}$? How can you tell?

The examples show two circles that you can use as benchmarks to help you estimate fractional parts.

### Examples

$\frac{1}{2}$

$\frac{1}{4}$

**More than** $\frac{1}{2}$

**More than** $\frac{1}{4}$

**Less than** $\frac{1}{2}$

**Less than** $\frac{1}{4}$

## TRY IT OUT

Estimate the shaded part of the figure. Compare to the benchmarks.

1.

2.

3.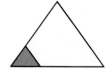

414

Estimate how much orange juice or toast is left.
Compare to the benchmarks.

**1.**   **2.**   **3.**   **4.**

Choose the fraction that is the best estimate of the shaded part.

**5.** 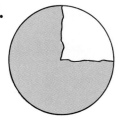   $\frac{1}{2}$   **6.** 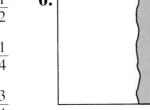   $\frac{1}{2}$   **7.** 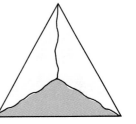   $\frac{1}{3}$

$\frac{1}{4}$      $\frac{1}{4}$      $\frac{1}{2}$

$\frac{3}{4}$      $\frac{3}{4}$      $\frac{3}{4}$

**APPLY**

**MATH REASONING**

**8.** Larry estimated that he had eaten about $\frac{1}{4}$ of his spaghetti. About how much does he have left?

**PROBLEM SOLVING**

**9.** Victoria bought $1.98 worth of mixed nuts that cost $3.95 a pound. Give a fraction to estimate what part of a pound she bought.

**10.** Trudy is mowing a lawn that is 930 sq. yd. By lunch time she estimated that she was $\frac{2}{3}$ done. About how many square yards does she have left to mow?

▶ **MENTAL MATH**

**11.** Suppose each container holds 24 ounces. Estimate the fractional amount that is in each one. Use mental math to calculate the number of ounces in each jar.

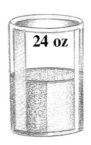

*More Practice, page 527, set B*

# Mixed Numbers

**EXPLORE** **Think About It**

Brent said, "I'm really full. I had 1 whole waffle and then $\frac{3}{4}$ of another waffle for breakfast." Is there a single number to tell how many waffles Brent had for breakfast?

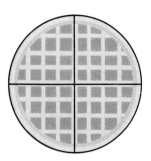

**TALK ABOUT IT**

1. Did Brent have more than 1 waffle? How much more?

2. Did he have as many as 2 waffles?

You write **mixed numbers** using a whole number and a fraction.

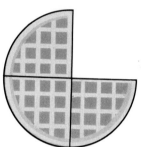

**Brent's Breakfast**

For 1 and $\frac{3}{4}$ more:   We write $1\frac{3}{4}$.

We read, "one and three fourths."

Mixed numbers are often used to give measures.

### Examples

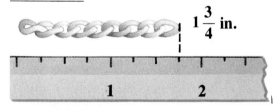 $1\frac{3}{4}$ in.

 $2\frac{1}{2}$ in.

Write a mixed number for each picture.

1.

2.

Write a mixed number for each picture.

**1.**      **2.**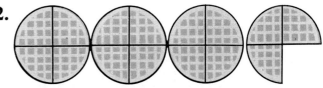

Give each length to the nearest quarter of an inch.

**3.**      **4.**

**MATH REASONING**

**5.** A sandwich shop cut their submarine sandwiches into fourths. They sold 15 of these fourths the first day. Did they sell more or less than 4 whole sandwiches?

**PROBLEM SOLVING**

**Understand the Operation** Tell what operation you would use to solve the following problems. Do not solve the problems.

**6.** Jean measured her corn plant on Friday and it was $6\frac{1}{4}$ in. tall. It grew $\frac{1}{2}$ in. overnight. How tall was it then?

**7.** Tom had a ribbon that was $12\frac{1}{4}$ in. long. He cut off pieces that were each $1\frac{3}{4}$ in. long. How many of these pieces could he get?

▶ **ESTIMATION**

Do not add. Just tell if the sum is more or less than 8.

**8.** $5\frac{1}{4} + 2\frac{1}{4}$     **9.** $5\frac{1}{2} + 2\frac{3}{4}$     **10.** $4\frac{3}{4} + 2\frac{3}{4}$     **11.** $5\frac{3}{4} + 2\frac{1}{2}$

# Adding and Subtracting Fractions

**EXPLORE** **Model with Fraction Pieces**
Work in groups. Use your fraction pieces
to show each of these actions.

- Put 2 eighths with 3 eighths.

- Take 3 eighths away from 7 eighths.

- Find how many more 7 eighths is than
  4 eighths.

Can you match each action you showed with one
of these fraction equations?

**TALK ABOUT IT**

1. Which action suggests addition? subtraction?
   Tell why.

2. Use fraction pieces to explain why this
   equation is false. $\frac{4}{8} + \frac{3}{8} = \frac{7}{16}$

When you add or subtract fractions with the same
denominator, add or subtract the numerators and
use the same denominator.

### Examples

$$\frac{3}{6} + \frac{2}{6} = \frac{5}{6} \qquad \frac{7}{10} - \frac{4}{10} = \frac{3}{10} \qquad \frac{4}{5} - \frac{1}{5} = \frac{3}{5}$$

| Key Actions |
| --- |
| ■ **Addition**<br>Put Together |
| ■ **Subtraction**<br>Take Away<br>Compare |

A. $\dfrac{7}{8} - \dfrac{4}{8} = \dfrac{3}{8}$

B. $\dfrac{3}{8} + \dfrac{2}{8} = \dfrac{5}{8}$

C. $\dfrac{7}{8} - \dfrac{3}{8} = \dfrac{4}{8}$

Find each sum and difference. Use fraction pieces
if you need help.

1. $\dfrac{3}{8} + \dfrac{1}{8}$     2. $\dfrac{6}{8} - \dfrac{4}{8}$     3. $\dfrac{7}{8} - \dfrac{5}{8}$     4. $\dfrac{2}{8} + \dfrac{2}{8}$

Find each sum and difference.

**1.** $\dfrac{1}{6} + \dfrac{3}{6}$     **2.** $\dfrac{7}{8} - \dfrac{6}{8}$     **3.** $\dfrac{5}{10} + \dfrac{4}{10}$     **4.** $\dfrac{6}{12} - \dfrac{2}{12}$

**5.** $\dfrac{6}{8} + \dfrac{1}{8}$     **6.** $\dfrac{9}{10} - \dfrac{6}{10}$     **7.** $\dfrac{11}{12} - \dfrac{6}{12}$     **8.** $\dfrac{3}{5} + \dfrac{1}{5}$

Write a fraction equation for each action.

**9.** Take 5 eighths from 7 eighths.     **10.** Put 2 eighths with 5 eighths.

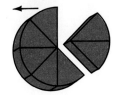           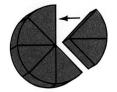

**APPLY**

**MATH REASONING** Give a number for each ||||| to make a true equation.

**11.** $\dfrac{|||||}{12} + \dfrac{|||||}{12} = \dfrac{|||||}{|||||}$         **12.** $\dfrac{|||||}{10} + \dfrac{|||||}{10} = \dfrac{|||||}{|||||}$

**PROBLEM SOLVING**

**13.** Jeremy cut a pan of corn bread into eighths. He ate $\dfrac{3}{8}$ and his sister ate $\dfrac{2}{8}$. What part of the corn bread did they eat?

**14. Be a Problem Finder**
What if Jeremy had cut the corn bread into tenths? Write a fraction problem about this pan of bread.

**MIXED REVIEW**

**15.** $3 \times 15$     **16.** $54 \div 6$     **17.** $48 \div 8$     **18.** $8 \times 12$     **19.** $18 \div 6$

**20.** $54 \div 9$     **21.** $3 \times 54$     **22.** $8 \times 27$     **23.** $27 \div 9$     **24.** $56 \div 8$

**25.** $2 \times 46$     **26.** $35 \div 7$     **27.** $3 \times 33$     **28.** $4 \times 24$     **29.** $45 \div 9$

# Reading and Writing Decimals

<u>EXPLORE</u> **Use Graph-Paper Models**
Draw four rectangles on graph paper so that each one is divided into **tenths.** Color different numbers of tenths in each one. Record how much of each rectangle you colored.

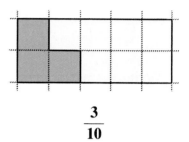

$$\frac{3}{10}$$

**TALK ABOUT IT**

1. Did you color more than half of any of your rectangles?

2. What are the different fractions that are less than half?

3. How many squares would you color if you colored exactly half of a rectangle?

You can use **decimals** to show fractional parts in tenths.

| **We Show** | **We Say** | **We Write** | |
|---|---|---|---|
| | | fraction | decimal |

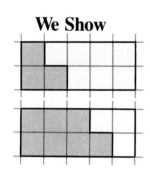

"three tenths"   $\frac{3}{10}$   0.3

"seven tenths"   $\frac{7}{10}$   0.7

decimal point

**TRY IT OUT**

Read each decimal.

**1.** 0.6 **2.** 0.4 **3.** 0.9 **4.** 0.1 **5.** 0.8 **6.** 0.5

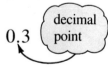

**7.** Write a fraction and a decimal for the shaded part of this rectangle.

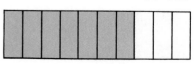

Write a decimal for the shaded part of each region.

**1.** 

**2.** 

**3.** 

**4.** 

**5.** 

**6.** 

**7.** Draw and shade 0.8 of a figure.

**8.** Write a decimal to estimate what part of this figure is shaded.

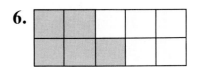

**MATH REASONING**

**9.** Brett said, "0.5 is the same amount as one half." Color a graph-paper model to show that he is correct.

**PROBLEM SOLVING**

**10.** Jamie said "I'm thinking of a decimal. It's more than half and less than 1. It's an even number of tenths. It's not 0.6." What is Jamie's decimal?

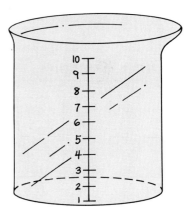

▶ **ALGEBRA**

Tell how to find △ when you know ☐.

**11.**

| ☐ | 1 | 2 | 3 | 4 | 5 | 6 | 7 | 8 |
|---|---|---|---|---|---|---|---|---|
| △ | 3 | 5 | 7 | 9 | | | | |

**12.**

| ☐ | 1 | 2 | 3 | 4 | 5 | 6 | 7 | 8 |
|---|---|---|---|---|---|---|---|---|
| △ | 2 | 5 | 10 | 17 | | | | |

*More Practice, page 510, set C*

**421**

# Larger Decimals

**EXPLORE  Use Graph Paper**
These bulletin boards hold 10 pictures each. How much of the two bulletin boards are filled?

**TALK ABOUT IT**

**1.** How many bulletin boards are completely filled?

**2.** What decimal part of the second board is filled?

**3.** What mixed number tells how many bulletin boards are filled?

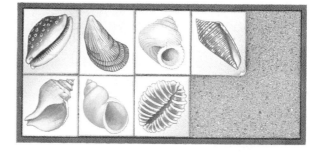

You can write either a mixed number or a decimal to show one and seven tenths.

| Mixed Number | Decimal |
|:---:|:---:|
| $1\dfrac{7}{10}$ | 1.7 |

Read, "one and seven tenths"

## Other Examples

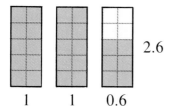

2.6

1   1   0.6

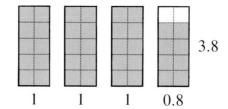

3.8

1   1   1   0.8

**TRY IT OUT**

Read and write a decimal for each model.

**1.**

**2.**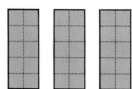

**3.**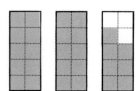

422

Read and write a decimal for each graph-paper model.

**1.**       **2.**

Draw a model to show each of the following.

**3.** 2.6          **4.** 3.4          **5.** 1.7

**APPLY**

**MATH REASONING**

**6.** Put these numbers in order from least to greatest. 4, 2.7, 3.2, 3

**PROBLEM SOLVING**

**7.** One group of students in Ryan's class collected 16 shell pictures. Another group collected 18 shell pictures. How much of the bulletin boards did the class use if they put 10 pictures on each bulletin board?

**8. Science Data Bank** Write the size of the Marsh Snail as a decimal. See page 474.

 DATA BANK

▶ **ESTIMATION**

Estimate each length using the red unit  as 1. Give your estimate using a decimal.

**9.**

**10.**

# Extending Decimal Concepts

**EXPLORE** **Use Decimal Models**

Cut out graph paper models to show some units, tenths, and hundredths.

**When this is 1,** **this is 1 tenth,** **and this is 1 hundredth.**

**TALK ABOUT IT**

1. How many 1-tenth strips can you cut from the unit?
2. How many 1-hundredth squares can you cut from a 1-tenth strip? from the full unit?

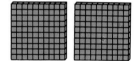

**We see:**

**We write:**     2          0.3 or 0.30     0.07          2.37

**We read:**      two      three tenths or     seven        two and
                            thirty hundredths  hundredths   thirty seven hundredths

A calculator can help you understand how these decimals are put together.

|  | **By Adding** |  |  | **Without Adding** |
|---|---|---|---|---|
| | Show 2 | Add 0.3 | Add 0.07 | |

Write a decimal to show how much is shaded.
Show the decimal on a calculator.

1.

2.

3.

424

Write a decimal to tell how much is shaded.
Show the decimal on a calculator.

1. 2. 3.

4. 5. 6.

Write a decimal for each of these.

**7.** Three and seventy-nine hundredths  **8.** Four and fifty-six hundredths

<u>MATH REASONING</u>  Study the pattern. Show the pattern
with models. Use the models to find the next
3 decimals.

**9.** 2.9, 2.8, 2.7, 2.6 . . .          **10.** 0.25, 0.35, 0.45, 0.55 . . .

<u>PROBLEM SOLVING</u>

**11.** Suppose stamps come in sheets
of 100. How many stamps
would you buy if you bought
2.46 sheets of stamps?

**12. Data Hunt**  Find 3 places in a
newspaper that use decimals.

▶ <u>ALGEBRA</u>

Rewrite each equation giving the missing decimal.

<u>**Example**</u>   $4.28 = 4 + ||||| + 0.08$
      $4.28 = 4 + 0.2 + 0.08$

**13.** $7.64 = 7.60 + |||||$          **14.** $1.73 = 1 + ||||| + 0.03$

**15.** $8.42 = 8 + ||||| + 0.02$          **16.** $5.19 = 5 + 0.1 + |||||$

# Adding and Subtracting Decimals
## Making the Connection

**EXPLORE** Use Place-Value Blocks

Work with a partner. Think of the small cubes as tenths.

- Each partner shows the starting number 5.4 using place-value blocks. Work together to choose one number from the number bank.

- Decide who will show adding the two numbers and who will show subtracting the two numbers. Check each other's work.

**Adding**

- Show the number you chose with the blocks. Push the two piles of blocks together. Trade 10 tenths for a 1 if you can. Record the numbers in a table.

**Subtracting**

- Take away from the pile of 5.4 the number of blocks that shows the number you chose. How many do you have left? You may have to trade a 1 for 10 tenths. Record the numbers in a table.

- Repeat this exercise three times, each time trading operations with your partner.

**TALK ABOUT IT**

1. What numbers in the bank will not require a trade when adding? when subtracting?

2. How are adding and subtracting decimals like adding and subtracting whole numbers?

0.1      1

**Trade
10 tenths = 1 one**

=

| Number Bank | | |
|---|---|---|
| 2.3 | 1.9 | 3.6 |
| | 2.7 | 2.8 |
| 3.8 | 2.5 | |
| 3.9 | 1.4 | 3.7 |
| | 2.9 | 1.7 |

| | ones | tenths |
|---|---|---|
| Starting Number → | 5 | 4 |
| Add-on Number → | | |
| Sum → | | |

| | ones | tenths |
|---|---|---|
| Starting Number → | 5 | 4 |
| Take-away Number → | | |
| Difference→ | | |

426

You have used blocks to show adding and
subtracting decimals. Now you will learn a way
to record what you have done.

Add 1.5 and 2.8.

**What You Do**

**Trace**

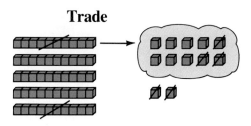

**What You Record**

$$\begin{array}{r} 1.5 \\ +2.8 \\ \hline 4.3 \end{array}$$

**1.** How many ones are there before the trade?
after the trade?

**2.** How many tenths are there after the trade?

Subtract 1.5 from 5.2.

**What You Do**

**Trade**

**What You Record**

$$\begin{array}{r} 5.2 \\ -1.5 \\ \hline 3.7 \end{array}$$

**3.** Why do you need to trade before you can take
away 1.5?

**4.** After the trade, how many tenths are left?

---

**TRY IT OUT**

Use place-value blocks to add or subtract.

**1.** Add 1.6 and 3.9.

**2.** Add 2.4 and 1.3.

**3.** Subtract 1.8 from 4.3.

**4.** Subtract 2.4 from 5.9.

# Problem Solving
## Estimating the Answer

UNDERSTAND
FIND DATA
PLAN
ESTIMATE
SOLVE
CHECK

Before solving a problem, it is important to decide what would be a reasonable answer. You can do this by estimating the answer.

Food at the Petting Zoo costs $0.39 for 1 bag. How much would 4 bags of food cost?

First I'll round the cost of 1 bag of food.

$0.39 → $0.40

Then I'll use the rounded number to estimate the cost of 4 bags.

$0.40 × 4 = $1.60

Now I'll solve the problem.

$0.39 × 4 = $1.56

$1.56 is a reasonable answer because it is close to $1.60.

Before solving each problem, estimate the answer. Then solve the problem and decide if your answer is reasonable.

1. Joe weighs 96 pounds. The gorilla at the zoo weighs 392 pounds. How much heavier is the gorilla than Joe?

2. Three trainers fed the seals at the zoo. Each trainer had 28 fish. How many fish were there in all?

3. There were 787 visitors to the zoo on Saturday and 823 on Sunday. How many people in all came for the two days?

4. Doug's family bought 3 small bags of popcorn for $0.89 each. How much did they pay all together?

Choose a strategy from the strategies list.

| Some Strategies |
| :---: |
| Act It Out |
| Choose an Operation |
| Make an Organized List |
| Look for a Pattern |
| Logical Reasoning |
| Use Objects |
| Draw a Picture |
| Guess and Check |
| Make a Table |
| Work Backward |

**1.** Joseph has fed $\frac{5}{9}$ of the lions. What fraction of the lions have not been fed?

**2.** At the zoo, Kristin's mother gave her a choice for lunch. She could have a hamburger or a fishburger to eat. She could have apple juice or lemonade to drink. How many different choices did Kristin have?

**3.** There are 36 insects in the poisonous section of the insect zoo. There are 4 insects per cage. How many cages are there?

**4.** In the line to buy fish for the seals, Mike was behind Lia. Lia was between Jeff and Mike. Bea was ahead of Jeff. Who was third in line?

**5.** The elephant weighs 9,865 pounds. His tusk weighs 3 times the weight of the female leopard. The leopard weighs 76 pounds. How much does the tusk weigh?

**6.** Which zoo has 94 more kinds of animals than 3 times the number of animals at the Riverbanks Zoo?

**7.** What is the total number of acres in all the zoos in the table?

| Zoo | Acres | Kinds of Animals |
| --- | --- | --- |
| Bronx | 265 | 674 |
| Cleveland | 164 | 460 |
| Houston | 55 | 649 |
| Miami Metrozoo | 280 | 283 |
| Milwaukee | 185 | 538 |
| Riverbanks (S.C.) | 153 | 185 |
| St. Louis | 83 | 660 |
| San Diego | 100 | 780 |
| San Diego (Wild Animal Park) | 1800 | 225 |

**8. Determining Reasonable Answers** Joey used his calculator to figure out how many more acres there are in the San Diego Wild Animal Park than there are in the Bronx Zoo. Is his answer reasonable? Explain why or why not.

*More Practice, page 527, set D*

# Group Decision Making
## Applied Problem Solving

UNDERSTAND
FIND DATA
PLAN
ESTIMATE
SOLVE
CHECK

**Group Skill:**

Disagree in an Agreeable Way

You and your cousins are having a picnic. You are planning the menu with your parents. Decide how many servings you need and how much of each kind of food to buy.

**Facts to Consider**

- You and your 7 cousins will be at the picnic.

- Your oldest cousin, Steven, eats at least 2 servings of most things.

- You can choose from these foods:

| | |
|---|---|
| barbequed chicken | 1 chicken has 8 servings |
| barbequed beef kabobs | 1 pound has 4 servings |
| corn on the cob | $\frac{1}{2}$ corn ear for each serving |
| foot-long sandwiches | 1 sandwich has 4 servings |
| potato salad | $\frac{1}{2}$ potato for each serving |
| sliced tomatoes | 1 tomato for 4 servings |
| chips | 1 bag has 6 servings |
| strawberries | $\frac{1}{2}$ basket for each serving |
| watermelon | 1 melon has 12 servings |

1. How many chickens would you need to buy if everyone eats 2 servings?
2. How many ears of corn would you need to buy if Steven eats 2 servings and everyone else eats 1 serving?
3. How many bags of chips would you need to buy if everyone eats 1 serving?
4. How many potatoes would you buy if you think you will need 10 servings?
5. How many baskets of strawberries would you buy if everyone ate 1 serving?

Choose at least 4 foods for the picnic. Tell how many servings you think you will need of each. Tell how much of each food you will need to buy.

# WRAP UP

## Name or Number, Please

Write word names for the fractions or decimals.

**1.** $\frac{1}{7}$      **2.** 0.5      **3.** 0.3      **4.** $\frac{3}{8}$      **5.** $1\frac{2}{3}$

Write fractions or decimals for the word names.

**6.** four tenths      **7.** seven tenths      **8.** five ninths

**9.** one third      **10.** two and one half      **11.** one and two tenths

## Sometimes, Always, Never

Decide which word should go in the blank, *sometimes, always,* or *never.* Explain your choices.

**12.** The number for three tenths _____ can be written as either 0.3 or $\frac{3}{10}$.

**13.** When you add or subtract numbers with decimals, you _____ make trades.

**14.** If two fractions have the same numerator, the fraction with the larger denominator _____ is larger.

**15.** Fractions with different numerators and denominators _____ can show the same amount.

## Project

Use play money dollars, dimes, and pennies to show these amounts of money.

- 3 hundredths of a dollar
- 35 hundredths of a dollar
- 5 tenths of a dollar
- 2 and 35 hundredths of a dollar

Is the dot in money notation the same as a decimal point?

# CHAPTER REVIEW/TEST

## Part 1    Understanding

1. How many friends are sharing if each gets an equal share that is $\frac{1}{9}$ of a whole pizza?

2. If 6 people share a pizza, what is each one's equal share?

3. Draw a picture using circles to show $1\frac{1}{4}$.

4. How many sixths are equivalent to $\frac{1}{2}$? Draw a picture to explain.

5. Describe how you could show 0.11 on graph paper.

6. How does a benchmark for $\frac{1}{2}$ help you estimate a fraction?

## Part 2    Skills

7. Find $\frac{1}{5}$ of 15 marbles.

8. Find $\frac{1}{3}$ of 24 students.

Write $<$ or $>$ for each .

9. $\frac{4}{5}$  $\frac{4}{9}$

10. $\frac{1}{8}$  $\frac{1}{7}$

11. $\frac{2}{3}$  $\frac{2}{10}$

Add or subtract.

12. $\frac{2}{5} + \frac{1}{5}$

13. $\frac{7}{8} - \frac{4}{8}$

14. $\frac{3}{4} - \frac{1}{4}$

15. $\frac{3}{10} + \frac{5}{10}$

Write a decimal for each of these.

16. $3\frac{3}{10}$

17. seven tenths

18. forty-one hundredths

Write a fraction for the shaded part.

19.

20.

21.

## Part 3    Applications

22. Follow the pattern. How many blocks are in the fourth and fifth designs? Explain.

23. **Challenge** Silly Sticky Stamps come in sheets of 10 stamps. Can 5 friends divide 3 sheets of stamps equally? What part of a sheet will each friend get? Tell how you solved the problem.

**433**

# ENRICHMENT
## Magic Squares

Look at the square at the right. It is a **magic square.** Can you tell why?

The sum along each
ROW →
COLUMN ↓
DIAGONAL ↗ ↘
is the same.

In the square, the magic sum is 15. Check one row, one column, and one diagonal to be sure.

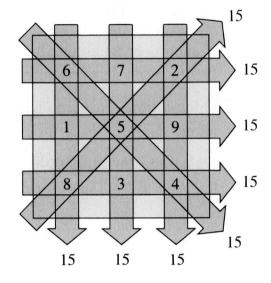

Copy this square. Fill in the missing numbers to make it a magic square.

**Hint 1:** Add the numbers in the top row to find the magic sum.

**Hint 2:** Then find the number in the middle box. The sum for the middle column should be 18.

Now you should have enough clues to complete the magic square!

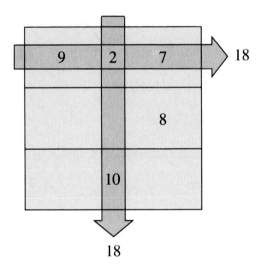

Use a calculator to make these magic squares.

**1.**

| 160 | 105 | 116 |
|-----|-----|-----|
|     |     | 171 |
|     | 149 |     |

**2.**

| 7.3  | 15.0 |      |
|------|------|------|
| 12.8 |      |      |
| 11.7 |      | 13.9 |

**434**

# CUMULATIVE REVIEW

**1.** Divide. $7\overline{)56}$

    **A.** 6    **B.** 7

    **C.** 8    **D.** 9

**2.** Find the quotient. $64 \div 8$

    **A.** 6    **B.** 7

    **C.** 8    **D.** 9

**3.** Find $72 \div 9$.

    **A.** 6    **B.** 7

    **C.** 8    **D.** 9

**4.** What is 5 added to the product of $4 \times 6$?

    **A.** 25    **B.** 27

    **C.** 28    **D.** 29

**5.** Multiply. $40 \times 8$

    **A.** 300    **B.** 320

    **C.** 340    **D.** 400

**6.** How does the estimated product of $4 \times 76$ compare to 300?

    **A.** equal to    **B.** greater than

    **C.** less than    **D.** same as

**7.** Which is longer than a meter?

    **A.** 50 centimeters
    **B.** 5 decimeters
    **C.** 75 centimeters
    **D.** 15 decimeters

**8.** What is the volume?

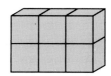

    **A.** 2 cubic units   **B.** 3 cubic units

    **C.** 5 cubic units   **D.** 6 cubic units

**9.** What is the temperature in degrees Celsius?

    **A.** 9°

    **B.** 19°

    **C.** 20

    **D.** 29°

**10.** A basket of apples produces 3 liters of juice. How much juice would 10 baskets of apples produce?

    **A.** 30 mL      **B.** 300 L

    **C.** 3,000 mL   **D.** 30 L

**11.** Ms. Kio is getting carpet to cover the floor of a room from wall to wall. What does she need to know about the room to make sure the carpet fits?

    **A.** exact length and width
    **B.** estimated area
    **C.** exact perimeter
    **D.** estimated width

# 16

# DIVISION

## Math and Social Studies

### DATA BANK

Use the Social Studies Data Bank on page 477 to answer the questions.

1 Galveston, Texas, is on an island called Treasure Island. Is $10 enough money to buy 3 children's tickets for the Treasure Island tour?

**2** St. Augustine, Florida, is the oldest city in the United States. What is the cost of 12 adult tickets for the train tour of St. Augustine?

**3** Denver, Colorado, is called the Mile-High City. How many children's tickets for the City of Denver tour can you buy for $49?

**4 Use Critical Thinking** If a group paid $36 for tickets on the Galveston boat, there must be at least one child in the group. How can you tell?

# Estimating Quotients Using Compatible Numbers

**EXPLORE** **Think About the Situation**

Cindy, Beth, and Jan found 25 shells and put them in a bucket. They decided to share them equally. Then they discussed how many shells each girl should get. Cindy said, "We need to divide 25 by 3, but $25 \div 3$ is not a basic division fact."

**TALK ABOUT IT**

1. What if the girls had found 24 shells? How many would each girl get?

2. How would the girls have shared 27 shells?

3. Would you use $24 \div 3$ or $27 \div 3$ to estimate $25 \div 3$? Why?

- In division, pairs of numbers such as 24 and 3 or 30 and 5 are called **compatible numbers.** When you choose a close basic fact to estimate a quotient, you are using compatible numbers.

## Other Examples

Estimate     $33 \div 4$
Close fact    $32 \div 4 = 8$
The estimated quotient is 8.

Estimate     $29 \div 5$
Close fact    $30 \div 5 = 6$
The estimated quotient is 6.

Estimate each quotient using compatible numbers.

**1.** $23 \div 3$      **2.** $29 \div 4$      **3.** $43 \div 5$      **4.** $30 \div 4$      **5.** $19 \div 3$

Find the quotients.

**1.** $2\overline{)12}$    **2.** $4\overline{)36}$    **3.** $3\overline{)12}$    **4.** $2\overline{)18}$    **5.** $4\overline{)32}$    **6.** $5\overline{)20}$

**7.** $5\overline{)30}$    **8.** $3\overline{)21}$    **9.** $4\overline{)20}$    **10.** $2\overline{)10}$    **11.** $3\overline{)27}$    **12.** $5\overline{)35}$

**13.** $5\overline{)45}$    **14.** $4\overline{)28}$    **15.** $2\overline{)14}$    **16.** $3\overline{)18}$    **17.** $2\overline{)8}$    **18.** $4\overline{)16}$

Estimate each quotient using compatible numbers.

**19.** $4\overline{)17}$    **20.** $2\overline{)9}$    **21.** $3\overline{)17}$    **22.** $2\overline{)15}$    **23.** $4\overline{)29}$    **24.** $5\overline{)47}$

**25.** $5\overline{)33}$    **26.** $3\overline{)26}$    **27.** $2\overline{)11}$    **28.** $4\overline{)22}$    **29.** $3\overline{)22}$    **30.** $5\overline{)31}$

**31.** $2\overline{)13}$    **32.** $4\overline{)34}$    **33.** $3\overline{)11}$    **34.** $2\overline{)17}$    **35.** $4\overline{)33}$    **36.** $5\overline{)24}$

## APPLY

**MATH REASONING** Which pairs are compatible numbers for division?

**37.** 20 and 5      **38.** 20 and 3      **39.** 20 and 4

### PROBLEM SOLVING

**40.** Chad and 3 friends worked together to earn 25 dollars. Estimate the number of dollars each boy would get if they shared equally.

**41.** Don and his 3 friends worked on a job. They each made $2 an hour. How much did the 4 friends earn if they worked 3 hours each?

### COMMUNICATION

**42.** The word compatible means to get along together in an agreeable manner.

Write a sentence telling why you think a pair of numbers such as 24 and 3 are called compatible when you are dividing.

# Finding Quotients and Remainders
## Making the Connection

**EXPLORE** **Model with Counters
and Cups**
Work in groups. Use 28 counters
and 7 paper cups.

- Think of the counters as raisins
  that you might like to share.
  Show how you can share the
  "raisins" equally among
  different numbers by using
  different numbers of cups.

- Complete a table like this one.

| Number of cups | Number in each | Number left over |
|:---:|:---:|:---:|
| 2 | | |
| 3 | | |
| 4 | | |
| 5 | | |
| 6 | | |
| 7 | | |

**TALK ABOUT IT**

1. For which numbers of cups did you have
   counters left over?

2. Why should the number left over be less than
   the number of cups?

3. If you wanted to share 28 counters so that
   there are 2 in each cup with none left over,
   how many cups would you need?

You have explored ways of sharing a number of counters equally among different sizes of groups. Now you will see a convenient way to record what you have done. Use counters and cups to help you understand a division problem such as $23 \div 4$.

**What You Do**                    **What You Record**

**1.** How many cups will you use for the sharing?

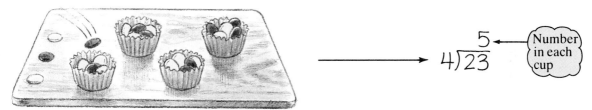

**2.** How many are in each cup when you have shared equally as many as possible?

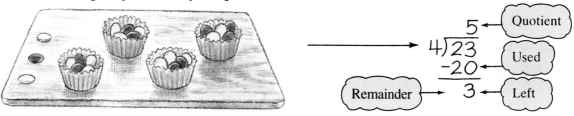

**3.** How many of the counters were you able to use? How many are left?

**4.** What is the quotient? What is the remainder?

---

**TRY IT OUT**

Use counters and cups. Record what you did as before. Find the quotient and remainder.

**1.** Divide 28 by 5.                    **2.** Divide 22 by 3.

# Dividing
## Finding Quotients and Remainders

EXPLORE **Think About the Process**

Amber took a roll of 32 pictures on her vacation. If she puts 6 pictures on each page of her photo book, how many pages can Amber fill? How many pictures will be left?

You divide because you are sharing equally.

| Estimate the quotient. | Multiply. | Subtract. | Compare. Write the remainder beside the quotient. |
|---|---|---|---|
| $6\overline{)30}$ $6\overline{)32}$ | $5 \times 6$ $\quad$ 5 $6\overline{)32}$ $\quad$ 30 | 5 $6\overline{)32}$ $-30$ $\quad$ 2 | 5 R 2 $6\overline{)32}$ $-30$ $\quad$ 2 $\quad$ 2 < 6 |

**TALK ABOUT IT**

1. Why is $30 \div 6$ a good way to estimate?

2. Is the remainder less than the divisor?

3. Use complete sentences to answer the problem.

**Other Examples**

$$\begin{array}{r} 4 \text{ R } 2 \\ 3\overline{)14} \\ -12 \\ \hline 2 \end{array} \qquad \begin{array}{r} 4 \text{ R } 4 \\ 5\overline{)24} \\ -20 \\ \hline 4 \end{array} \qquad \begin{array}{r} 5 \\ 4\overline{)20} \\ -20 \\ \hline 0 \end{array}$$

Find the quotients and remainders.

1. $4\overline{)29}$ 　　　　　2. $5\overline{)44}$ 　　　　　3. $3\overline{)23}$ 　　　　　4. $2\overline{)18}$

442

Find the quotients and remainders.

**1.** $5\overline{)31}$    **2.** $3\overline{)22}$    **3.** $2\overline{)16}$    **4.** $4\overline{)22}$    **5.** $2\overline{)11}$    **6.** $3\overline{)26}$

**7.** $4\overline{)28}$    **8.** $5\overline{)33}$    **9.** $4\overline{)17}$    **10.** $2\overline{)9}$    **11.** $5\overline{)40}$    **12.** $3\overline{)17}$

**13.** $41 \div 5$      **14.** $26 \div 4$      **15.** $25 \div 3$      **16.** $19 \div 4$

**17.** Divide 27 by 5.     **18.** Divide 22 by 4.     **19.** Divide 26 by 3.

**APPLY**

**MATH REASONING**

**20.** What are the possible remainders when you divide by 2? What can these remainders tell you about odd and even numbers?

**PROBLEM SOLVING**

**21. Social Studies Data Bank** Amber's mother paid for the horse and carriage tour of St. Augustine with a ten-dollar bill. How many children's tickets will that buy? Is there any money left over? See page 477.

 **MIXED REVIEW**

Decide if each is longer than, shorter than, or the same as 1 meter.

**22.** 1cm     **23.** 1km     **24.** 75cm     **25.** 10dm     **26.** 17dm

Give the area of each shape in square centimeters.

**27.** 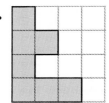     **28.** 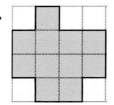     **29.**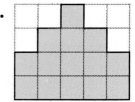

*More Practice, page 511, set B*

# Checking Division

Your understanding of quotients and remainders will help you learn to check division calculations.

**EXPLORE** **Use Mental Math**

Betsy knew that there were 42 school days until summer. She wondered how many weeks that would be. There are 5 school days in each week.

How can you check Betsy's work to be sure it is correct?

**TALK ABOUT IT**

1. How many school days are in 8 weeks?

2. How many school days are in 8 weeks and 2 days?

3. Does your second answer match the number of days until the end of school?

You can check division by multiplying the quotient by the divisor and adding the remainder. Use mental math to check these answers.

$$\begin{array}{r} 7 \\ 5\overline{)39} \\ -35 \\ \hline 4 \end{array}$$  Think $7 \times 5$ plus 4

$$\begin{array}{r} 8 \\ 4\overline{)34} \\ -32 \\ \hline 2 \end{array}$$  Think $8 \times 4$ plus 2

**PRACTICE**

Find the quotients and remainders. Check your answers.

1. $3\overline{)28}$  2. $5\overline{)38}$  3. $4\overline{)35}$  4. $2\overline{)15}$  5. $5\overline{)40}$  6. $3\overline{)26}$

7. $4\overline{)15}$  8. $5\overline{)12}$  9. $3\overline{)17}$  10. $4\overline{)19}$  11. $5\overline{)31}$  12. $2\overline{)17}$

**444**

*More Practice, page 511, set C*

# MIDCHAPTER REVIEW/QUIZ

Estimate each quotient using a compatible number.

**1.** $5\overline{)24}$      **2.** $4\overline{)27}$      **3.** $2\overline{)13}$      **4.** $3\overline{)22}$      **5.** $4\overline{)37}$

**6.** $2\overline{)11}$      **7.** $3\overline{)28}$      **8.** $5\overline{)41}$      **9.** $4\overline{)31}$      **10.** $3\overline{)17}$

**11.** $3\overline{)14}$      **12.** $5\overline{)29}$      **13.** $2\overline{)19}$      **14.** $4\overline{)17}$      **15.** $5\overline{)21}$

Ely has 26 counters. Draw pictures to find quotients and remainders if he divides them equally in 3 cups, 4 cups, and 5 cups.

**16.** 3 cups      **17.** 4 cups      **18.** 5 cups

Find the quotients and remainders.

**19.** $2\overline{)18}$      **20.** $3\overline{)21}$      **21.** $4\overline{)18}$      **22.** $5\overline{)19}$      **23.** $5\overline{)33}$

**24.** $4\overline{)34}$      **25.** $3\overline{)25}$      **26.** $2\overline{)9}$      **27.** $3\overline{)19}$      **28.** $4\overline{)30}$

**29.** $30 \div 5$      **30.** $28 \div 3$      **31.** $15 \div 4$      **32.** $17 \div 2$

## PROBLEM SOLVING

**33.** Write yes or no. Which of these can be shared equally among Claire and 2 friends?
**A.** 12 books      **B.** 16 counters      **C.** 15 pencils

**34.** Steve took a roll of 24 pictures. He gave 2 to his aunt and 1 to his uncle. He put the rest in an album. If 4 pictures fit on one page, how many pages did he fill? How many pictures were left over?

**35.** When Christina divided a number by 2 she had remainder 1. When she divided the same number by 3 she had remainder 2. When she divided it by 4 the remainder was 3. What is the smallest number it could be?

# Problem Solving
## Interpreting Remainders

UNDERSTAND
FIND DATA
PLAN
ESTIMATE
SOLVE
CHECK

When you solve a division problem that has a remainder, you have to think what the remainder means.

| | |
|---|---|
| 33 students are going on a field trip to the Science Exploratorium. Each car can hold 5 students. How many cars do they need? | Mr. Asahi needs 3 pieces of ribbon to make each prize for the science fair. How many prizes can he make from 29 pieces? |

$$\begin{array}{r} 6 \text{ R } 3 \\ 5\overline{)33} \\ 30 \\ \hline 3 \end{array}$$

$$\begin{array}{r} 9 \text{ R } 2 \\ 3\overline{)29} \\ 27 \\ \hline 2 \end{array}$$

They need 7 cars.

He can make 9 prizes.

> 6 of the cars will be full. They need a 7th car for the 3 remaining students.

> The remaining 2 ribbons aren't enough to make a prize.

1. Mrs. Hernandez bought cans of juice for the 40 students on the field trip. If the juice came in 6-packs, how many packs did she buy?

2. At the science museum, groups of 7 students at a time can do a laser experiment. In a class of 32, how many groups can do the experiment?

3. T-shirts cost $10. Mr. Jonah bought as many Einstein T-shirts as he could. If he started with $21, how much money did he have left over?

4. At the science fair, there were 43 projects on display. Each table can hold 9 projects. How many tables did the fair need?

Choose a strategy from the strategies list or other strategies you know to solve these problems.

**Some Strategies**

Act It Out
Use Objects
Choose an Operation
Draw a Picture
Make an Organized List
Guess and Check
Make a Table
Look for a Pattern
Use Logical Reasoning
Work Backward

1. At the Optical Illusions Booth, there were between 30 and 40 people lined up. Jim counted an even number of people less than 33. How many people were in line?

2. Mrs. Lu wanted 35 students to work on science projects at tables of 4. How many students were left over?

3. The first 2 weeks the Exploratorium was open, the admission was 1/2 off the regular price. The first week, 3,527 people came. The next week, 2,987 people came. How many more people came the first week than the second?

4. At most, 4 students can use 1 microscope in Ms. Lopez's class. How many microscopes does she need for 37 students?

5. Judges at the science fair gave 3 prizes, first, second, and third. Milly, Erin, and Jim each got one of the prizes. How many different ways could the prizes have been awarded?

6. Aquarium posters were on sale at $25 for 6 posters. About how much did each poster cost?

7. The Bionic Dinosaur Display was on exhibit in 4 different cities for 65 days each. It was on exhibit in another 3 cities for 90 days each. How long was the display on exhibit all together?

8. **Think About Your Solution**
   On the day of the planetarium field trip, 3 out of 51 students were absent. If 9 students can go in each van, how many vans did the school need?

■ Write your answer in a complete sentence.

■ Write a description of how you solved the problem.

*More Practice, page 527, set E*

# Finding 2-Digit Quotients
## Making the Connection

**LEARN ABOUT IT**

**EXPLORE** **Model with Play Money**
Work in groups. Use dimes and pennies in play money.

- Show 56 cents using 5 dimes and 6 pennies

**1 dime = 10 pennies**

- Show how to share the money equally among different numbers of people in your group. Trade dimes for pennies as needed for the sharing.

- Complete a table like this one.

| Number sharing | Amount for each person | | Amount left over |
|---|---|---|---|
| | dimes | pennies | |
| 2 | 2 | 8 | 0 |
| 3 | | | |
| 4 | | | |
| 5 | | | |
| 6 | | | |
| 7 | | | |

**TALK ABOUT IT**

1. Use play money. Tell about some of the trades you made as you shared the money equally among different numbers of people.

2. How many times did you have money left over? Was it less than the number of people? Why?

3. Does the amount for each person get greater or less as the number of people increases? Why?

You have found ways of sharing money equally among different numbers of people. Now you will see a convenient way to record what you have done. Use play money dimes for tens and pennies for ones to help you understand a problem such as 57 ÷ 2.

**What You Do**                                                    **What You Record**

$$2\overline{)57}$$

1. How many tens can you put in each of the 2 sets?

$$\begin{array}{r} 2\phantom{7} \\ 2\overline{)57} \\ -4\phantom{} \\ \hline 17 \end{array}$$ ← 17 ones

2. How many ones do you have left to divide after you trade the extra ten?

$$\begin{array}{r} 28 \\ 2\overline{)57} \\ -4\phantom{} \\ \hline 17 \\ -16 \\ \hline 1 \end{array}$$

3. How many ones were you able to put into each of the two sets? How many extra ones were there?

4. How many are in each set? What is the quotient for 57 ÷ 2? What is the remainder?

**TRY IT OUT**

Solve. Use play money. Record what you did.

1. $2\overline{)35}$                                    2. $3\overline{)74}$

449

# Estimating Money Quotients

**EXPLORE** **Understand the Situation**

Nancy wanted to estimate how much each map would cost her if she bought three. Paul said, "I'm sure it's more than $2 and not as much as $5."

**TALK ABOUT IT**

1. Why do you think Paul was so sure it was more than $2?

2. Can you prove the maps are less than $5?

3. Does each map cost closer to $3 or $4?

**State Map Special
3 for $10.95**

Here is how to estimate money quotients.

- Round to the nearest dollar.

- If needed, choose compatible numbers so that you have a close basic fact.

### Examples

| | Round | Close Basic Fact |
|---|---|---|
| $19.28 \div 4$ | $19 \div 4$ | $20 \div 4 = 5$ |
| $19.28 \div 4$ is about $5. | | |
| $29.95 \div 7$ | $30 \div 7$ | $28 \div 7 = 4$ |
| $29.95 \div 7$ is about $4. | | |

Estimate each quotient. Decide if your estimate is over or under the exact answer.

1. $17.25 \div 3$     2. $22.95 \div 6$     3. $9.15 \div 4$     4. $49.50 \div 8$

450

Estimate. Decide if your estimate is over or under the exact answer.

**1.** $34.95 ÷ 6     **2.** $15.75 ÷ 4     **3.** $46.75 ÷ 9     **4.** $18.95 ÷ 3

**5.** $19.95 ÷ 5     **6.** $31.20 ÷ 8     **7.** $14.88 ÷ 2     **8.** $34.95 ÷ 7

Estimate the cost of one of each. Is your estimate over or under the exact answer?

**9.** Travel books
2 for $15.75

**10.** National park tapes
$22.25 for 3

**11.** State park maps
4 for $12.75

**APPLY**

**MATH REASONING**

**12.** Use estimation to decide which is the better buy.

4 for $29.15     5 for $34.75

**PROBLEM SOLVING**

**13.** The Tylers bought 3 travel tapes for their trip to Florida. The tapes were $19.95. They could have got them for $7 each. Did they save or lose money by buying 3?

**14. Social Studies Data Bank**
Mr. and Mrs. Tyler bought tickets for themselves and their 3 children for the train ride in St. Augustine, Florida. How much were the tickets? See page 477.

**MIXED REVIEW**

**A.**      **B.** 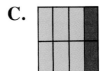     **C.**

**15.** What fraction of each shape is shaded red?
**16.** What fraction of each shape is shaded green?
**17.** Which shape shows the most red?
**18.** Which shape shows the most green?

*More Practice, page 511, set D*

# Data Collection and Analysis
## Group Decision Making

UNDERSTAND
FIND DATA
PLAN
ESTIMATE
SOLVE
CHECK

**Doing an Investigation**

**Group Skill:**
Explain and Summarize

It is important for children to think about being safe. Do you think children in your school are careful? Make a questionnaire by writing down some questions asking for information that you would like to find out.

**Collecting Data**

1. Talk with your group. List some things that children can do to be safe.

   Cross the street at the corner. Don't get in the car with a stranger. Know your home address.

2. Make up two questions to ask like this one.

   Do you cross the street at the corner?

   ___ always ___ sometimes ___ never

3. Have your group make at least 10 copies of your questionnaire. Give your questionnaire to children to mark their answers.

**4.** Make a table to record the information from your questionnaire. Make a tally for each person's answer.

| Question 1 | Question 2 |
|------------|------------|
| always | always |
| sometimes | sometimes |
| never | never |

**5.** Make a bar graph from the table. Use even numbers along the side of the graph.

**6.** How many children answered your questionnaire?

**7.** Do you think most children do things that are safe?

**8.** Write at least three true statements about what your graph shows.

# WRAP UP

## Follow a Good Example

Use this example to answer the questions.

1. Which number is the quotient? What does it tell you?

2. Which number is the remainder? What does it tell you?

3. Which number is the divisor? What does it tell you?

4. Which number is the dividend? What does it tell you?

29 tomatoes

6 in a box

How many boxes?

$$\begin{array}{r} 4\ \text{R5} \\ 6)\overline{\phantom{0}29\phantom{0}} \\ -\ 24 \\ \hline 5 \end{array}$$

## Sometimes, Always, Never

Decide which word should go in the blank, *sometimes, always,* or *never.* Explain your choices.

5. To check a division answer, you _____ multiply the quotient by the remainder.

6. The remainder in division _____ is greater than the divisor.

7. When the divisor is 2, the remainder _____ is 0 or 1.

8. When the divisor is 3, the remainder _____ is 2.

## Project

Make a list of 20 numbers. Mix odd and even numbers. Trade lists with a partner.

Divide each number by 2. What do you notice?
Write a rule about an odd number divided by 2.
Write a rule about an even number divided by 2.

454

# CHAPTER REVIEW/TEST

## Part 1    Understanding

1. Tell how you would estimate $17 \div 4$.

2. Tell how you would estimate $24 \div 5$.

3. How can you check a division problem such as $30 \div 7$? Show all of your steps.

4. Tell how you would estimate the cost of one pair of socks if 4 pairs cost $7.75.

5. Draw a picture to show the result when you divide 19 by 2.

6. Explain what the quotient 5 R3 means if you are telling about dividing some students into work groups.

## Part 2    Skills

Estimate the quotients.

7. $17 \div 3$

8. $14.82 \div 5$

9. $35.29 \div 9$

Divide.

10. $22 \div 7$

11. $28 \div 8$

12. $31 \div 6$

13. $5 \overline{)34}$

14. $9 \overline{)57}$

15. $8 \overline{)40}$

16. $7 \overline{)69}$

17. $8 \overline{)74}$

18. $7 \overline{)39}$

19. $18 \div 4$

20. $27 \div 9$

21. $44 \div 6$

## Part 3    Applications

22. Jerry wants to buy some posters that cost $4.00 each. He has $11.00 to spend. How many posters can Jerry buy?

23. Caryl is mailing posters, putting no more than 5 in each mailing tube. How many tubes does she need to mail 27 posters?

24. For every $6.00 spent in Poster World, a customer receives a coupon. For a $53.98 purchase, how many coupons will you get?

25. **Challenge** The cost to frame a team picture is $8.95. Can the Little League have 7 pictures framed for $57.00?

# ENRICHMENT
## Using a Calculator to Find Quotients and Remainders

Vincent said, "I can find the quotient and remainder for this problem by subtracting." This is how he showed his work.

**The quotient is 5.**
**The remainder is 25.**

Crystal answered, "I can use the subtraction constant on a calculator to do that."

Here is the key code she used to divide 200 by 35.

| ON/AC | 200 | − | 35 | = | = | = | = | = |

"One, two, three, four, five." Crystal counted the number of times she pressed the $=$ sign until the answer was less than 35. Her count is the quotient. The number less than 35 is the remainder.

$$
\begin{array}{r}
200 \\
-35 \quad ① \\
\hline
165 \\
-35 \quad ② \\
\hline
130 \\
-35 \quad ③ \\
\hline
95 \\
-35 \quad ④ \\
\hline
60 \\
-35 \quad ⑤ \\
\hline
25
\end{array}
$$

Use the constant on a calculator to help you find the quotient and remainder for each problem.

1. $45\overline{)400}$  2. $38\overline{)276}$

3. $56\overline{)395}$  4. $29\overline{)470}$

5. $61\overline{)872}$  6. $126\overline{)789}$

7. $624\overline{)4,287}$  8. $968\overline{)6,000}$

9. $927\overline{)4,280}$  10. $348\overline{)696}$

11. $94\overline{)1,629}$  12. $400\overline{)2,431}$

13. $789\overline{)9,468}$  14. $808\overline{)7,325}$

# Cumulative Review

**1.** Multiply. $13 \times 5$

    **A.** 55    **B.** 60

    **C.** 63    **D.** 65

**2.** Find the product. $34 \times 4$

    **A.** 116    **B.** 126

    **C.** 136    **D.** 134

**3.** Students piled 30 math books on a scale. The scale showed _____.

    **A.** 30 grams    **B.** 30 kilograms

    **C.** 3 grams    **D.** 3 kilograms

**4.** If 7 people share a pizza equally, each person gets one _____ of the whole pizza.

    **A.** three    **B.** third

    **C.** seventh    **D.** seventeen

**5.** Which fraction describes the shaded part?

    **A.** $\frac{1}{4}$    **B.** $\frac{2}{3}$

    **C.** $\frac{1}{2}$    **D.** $\frac{3}{4}$

**6.** Which fraction is equal to $\frac{1}{2}$?

    **A.** $\frac{3}{6}$    **B.** $\frac{2}{3}$

    **C.** $\frac{5}{8}$    **D.** $\frac{4}{9}$

**7.** Find $\frac{1}{5}$ of 20.

    **A.** $\frac{1}{2}$    **B.** 4

    **C.** 5    **D.** $\frac{5}{20}$

**8.** What number is shown in this picture?

    **A.** $\frac{2}{2}$    **B.** $1\frac{1}{2}$

    **C.** $\frac{1}{2}$    **D.** $2\frac{1}{2}$

**9.** Add. $\frac{2}{9} + \frac{3}{9}$

    **A.** $\frac{1}{9}$    **B.** $\frac{5}{9}$

    **C.** $\frac{1}{5}$    **D.** 5

**10.** Give the decimal number.

    **A.** 3.2    **B.** 2.3

    **C.** $2\frac{1}{3}$    **D.** 3.3

**11.** Malcolm and 3 friends washed cars to earn $25. About how many dollars does each one get if they share the money equally?

    **A.** $5    **B.** $6

    **C.** $7    **D.** $8

# RESOURCE BANK
# AND
# APPENDIX

## RESOURCE BANK

# APPENDIX

# Place Value: Tens and Ones

Erica is putting 10 cubes together to show that <u>ten ones</u> are equal to <u>one ten</u>.

**ten ones**       **one ten**

We use these <u>tens</u> and <u>ones</u> to show numbers from 1 to 99.

<u>**Examples**</u>

**3 tens and 4 ones**

**34**

**1 ten and 5 ones**

**15 ·**

Write the number for each picture.

**1.**

_____ tens and _____ ones

_____

**2.**

_____ ten and _____ ones

_____

**3.**

_____ tens and _____ ones

_____

**4.**

_____ tens and _____ ones

_____

Write the number.

**5.** 9 tens and 6 ones _____

**6.** 1 ten and 7 ones _____

**7.** 5 tens and 0 ones _____

**8.** 8 tens and 9 ones _____

# Counting and Order (1 to 100)

Earl is reading *Superfudge*. The pages of his book are numbered in order. He is on page 63.

A number line shows the order of numbers.

1  2  3  4  5  6  7  8          1, 2, 3, 4, 5, 6, 7, 8

15 16 17 18 19 20 21 22          15, 16, 17, 18, 19, 20, 21, 22

48 49 50 51 52 53 54 55          48, 49, 50, 51, 52, 53, 54, 55

## Example

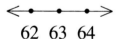
62 63 64
62 comes **before** 63
64 comes **after** 63
63 is **between** 62 and 64

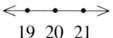
19 20 21
19 comes **before** 20
21 comes **after** 20
20 is **between** 19 and 21

Write the number that comes after.

**1.** 6 _____      **2.** 49 _____      **3.** 40 _____      **4.** 87 _____

Write the number that comes before.

**5.** _____ 9      **6.** _____ 19      **7.** _____ 50      **8.** _____ 76

Write the number that is between.

**9.** 1 _____ 3      **10.** 17 _____ 19    **11.** 53 _____ 55    **12.** 89 _____ 91

**461**

# Comparing Numbers (1 to 100)

Jenny and Maria are reading the same book. Jenny is on page 34. Maria is on page 29.

**34 = 3 tens and 4 ones**          **29 = 2 tens and 9 ones**

**Think:**   3 tens are greater than 2 tens.          **Write:**   34 > 29
        34 is greater than 29.

The next day Jenny is on page 63 and Maria is on page 68.

**63 = 6 tens and 3 ones**          **68 = 6 tens and 8 ones**

**Think:**   Tens are the same.          **Write:**   63 < 68
        Compare the ones.
        3 ones are less than 8 ones.
        63 is less than 68.

Compare each pair of numbers. Use > or <.

**1.** 29 ☐ 54          **2.** 90 ☐ 71          **3.** 16 ☐ 61          **4.** 84 ☐ 87

**5.** 52 ☐ 48          **6.** 98 ☐ 96          **7.** 27 ☐ 31          **8.** 77 ☐ 88

# Addition: Counting on from a 2-Digit Number

Kim has just finished counting her stickers. She has 21 cat stickers, 39 heart stickers, and 57 dog stickers. Then Shari gives her 3 more of each kind. Kim uses **counting on** to find how many stickers she now has.

21 cat stickers and 3 more
**Think:** 21 → 22, 23, 24          24 cat stickers

39 heart stickers and 3 more
**Think:** 39 → 40, 41, 42          42 heart stickers

57 dog stickers and 3 more
**Think:** 57 → 58, 59, 60          60 dog stickers

Add. Use counting on.

**1.** 16 + 3 = _____      **2.** 28 + 3 = _____      **3.** 49 + 2 = _____

**4.** 55 + 2 = _____      **5.** 43 + 4 = _____      **6.** 58 + 4 = _____

**7.** 79 + 1 = _____      **8.** 18 + 4 = _____      **9.** 30 + 2 = _____

| | | | |
|---|---|---|---|
| **10.** 83<br>+ 3 | **11.** 78<br>+ 4 | **12.** 37<br>+ 3 | **13.** 97<br>+ 3 |
| **14.** 74<br>+ 3 | **15.** 88<br>+ 4 | **16.** 64<br>+ 2 | **17.** 57<br>+ 3 |
| **18.** 99<br>+ 3 | **19.** 33<br>+ 2 | **20.** 46<br>+ 4 | **21.** 77<br>+ 2 |

# Subtraction: Counting Back from a 2-Digit Number

Felipe has 37 baseball cards. Oscar has 29 baseball cards. Margo has 90 baseball cards. Since Nikki does not have any cards, they each give her 3 cards.

Felipe **counts back** to see how many cards he has left.

37 cards → 36, 35, 34      34 cards left

Oscar counts back to see how many cards he has left.

29 cards → 28, 27, 26      26 cards left

Margo counts back to see how many cards she has left.

90 cards → 89, 88, 87      87 cards left

Subtract. Use counting back.

**1.** $49 - 3 =$ _____   **2.** $81 - 3 =$ _____   **3.** $21 - 2 =$ _____

**4.** $37 - 2 =$ _____   **5.** $55 - 4 =$ _____   **6.** $62 - 4 =$ _____

**7.**   31       **8.**   45       **9.**   92       **10.**   83
     $-\ 3$           $-\ 4$           $-\ 3$            $-\ 4$

**11.**   81       **12.**   64       **13.**   19       **14.**   42
     $-\ 2$           $-\ 3$           $-\ 3$            $-\ 4$

**15.**   51       **16.**   94       **17.**   75       **18.**   30
     $-\ 3$           $-\ 2$           $-\ 4$            $-\ 2$

# Hour and Half Hour

Shaun has to phone home every half hour, starting at 2 o'clock.

An hour has 60 minutes.

A half hour has 30 minutes.

**30 minutes later** →

**30 minutes later** →

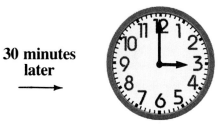

**2:00**
two o'clock

**2:30**
two-thirty

**3:00**
three o'clock

Write each time two ways.

**1.**

**2.**

**3.**

**4.**

_____

_____

_____

_____

**5.**

**6.**

**7.**

**8.**

_____

_____

_____

_____

# Quarter Hour

Martha is trying to call her friend on the telephone. She phones every quarter hour starting at 8 o'clock.

8:00

An hour has 60 minutes.

A quarter hour has 15 minutes.

**8:00**
**eight o'clock**

15 minutes later →

**8:15**
**eight-fifteen**

15 minutes later →

**8:30**
**eight-thirty**

15 minutes later →

**8:45**
**eight-forty-five**

Write each time two ways.

**1.**
_____

**2.**
_____

**3.**
_____

**4.**
_____

**5.**
_____

**6.**
_____

**7.**
_____

**8.**
_____

# Skip-Counting

Reggie is counting his money. First
he finds the value of his nickels. He
**skip-counts** by 5s.

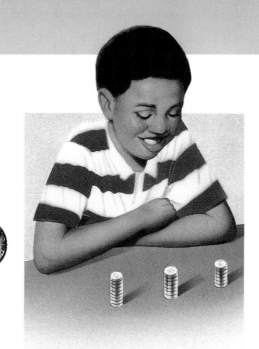

| 5¢ | 10¢ | 15¢ | 20¢ | 25¢ | 30¢ |

**He has 30¢ in nickels.**

Then he finds the value of his
dimes. He skip-counts by 10s.

 **He has $1.00
in dimes.**

| 10¢ | 20¢ | 30¢ | 40¢ | 50¢ | 60¢ | 70¢ | 80¢ | 90¢ | 100¢ or
$1.00 |

Skip-count to tell how much.

**1.** _____ _____ _____

**2.** _____ _____ _____ _____ _____ _____ _____

**3.** _____ _____ _____ _____ _____ _____ _____

**4.** _____ _____ _____ _____

**467**

# Math and Fine Arts Data Bank

## The Life of Ludwig van Beethoven

Beethoven is thought to be one of the greatest composers that ever lived. This time line shows some of the things he did during his life.

4 years old
**1774**

Birth
**1770**

12 years old
**1782**

28 years old
**1798**

54 years old
**1824**

Beethoven was born in Germany.

Some of his music was published.

He started to become deaf.

He finished writing his last symphony, the Ninth Symphony.

He began taking piano lessons.

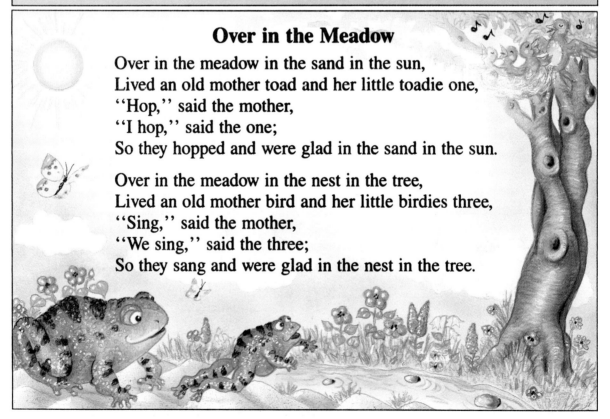

## Over in the Meadow

Over in the meadow in the sand in the sun,
Lived an old mother toad and her little toadie one,
''Hop,'' said the mother,
''I hop,'' said the one;
So they hopped and were glad in the sand in the sun.

Over in the meadow in the nest in the tree,
Lived an old mother bird and her little birdies three,
''Sing,'' said the mother,
''We sing,'' said the three;
So they sang and were glad in the nest in the tree.

# Math and Fine Arts Data Bank

## The Statue of Liberty

### Some Measurements of Length

| | |
|---|---|
| Torch | 21 feet |
| Right arm | 42 feet |
| Hand | 16 feet, 5 inches |
| Index finger | 7 feet, 11 inches |
| Fingernail | 13 inches |

### More About the Statue

There are 171 steps in the statue.

There are 167 steps in the pedestal.

There is room for 40 people in the statue's crown.

There is room for 12 people on the torch's balcony.

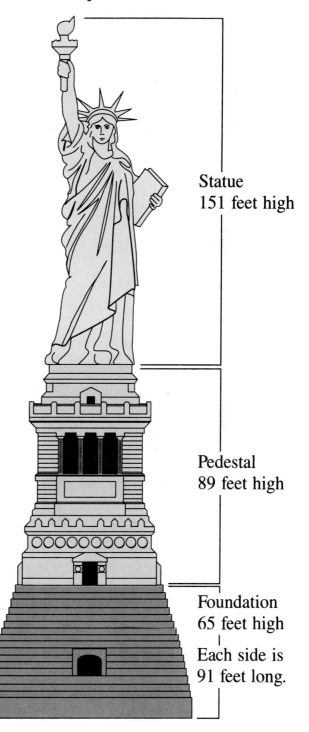

Statue 151 feet high

Pedestal 89 feet high

Foundation 65 feet high

Each side is 91 feet long.

# Math and Fine Arts Data Bank

*Tableau II* was painted by
Piet Mondrian in 1925.

*The Boating
Party* was
painted by
Mary Cassatt
in 1894.

# Math and Science Data Bank

## How Many Seeds in a Pound?

| Seed | Number of seeds in a pound |
|------|---------------------------|
| Cucumber | 17,600 |
| Lettuce | 400,000 |
| Lima Bean | 800 |
| Onion | 152,000 |
| Radish | 48,000 |
| Watermelon | 4,000 |

## World's Largest Fruits and Vegetables

| Fruit or Vegetable | Weight in Pounds |
|--------------------|------------------|
| Apple | 2 |
| Beet | 22 |
| Carrot | 15 |
| Lemon | 8 |
| Onion | 8 |
| Pear | 3 |
| Pumpkin | 612 |
| Radish | 27 |
| Squash | 513 |

# Math and Science Data Bank

## Lengths of Dinosaurs

| Name | Length |
|------|--------|
| Alamosaurus | 69 ft |
| Aristosaurus | 5 ft |
| Brachiosaurus | 80 ft |
| Coloradisaurus | 13 ft |
| Elmisaurus | 7 ft |
| Kakuru | 8 ft |
| Megalosaurus | 30 ft |
| Stegosaurus | 30 ft |
| Triceratops | 30 ft |
| Ultrasaurus | 100 ft |

## Heights of Dinosaurs

| Name | Height |
|------|--------|
| Brachiosaurus | 40 ft |
| Saltasaurus | 17 ft |
| Iguanodon | 16 ft |

# Math and Science Data Bank

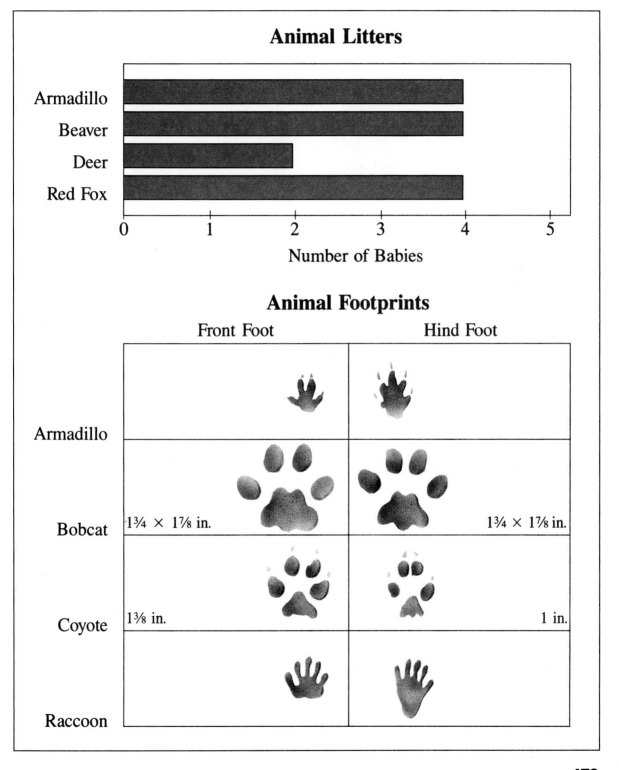

## Animal Litters

| | Number of Babies |
| Armadillo | |
| Beaver | |
| Deer | |
| Red Fox | |

## Animal Footprints

| | Front Foot | Hind Foot |
| --- | --- | --- |
| Armadillo | | |
| Bobcat | 1¾ × 1⅞ in. | 1¾ × 1⅞ in. |
| Coyote | 1⅜ in. | 1 in. |
| Raccoon | | |

**473**

# Math and Science Data Bank

## Shells Found in Salt Water, Fresh Water, or on Land
### Fresh Water

**Ram's Horn Snail**

$\frac{1}{2}$ to $1\frac{1}{2}$ in.

**Little Freshwater Clam**

1 to $2\frac{1}{2}$ in.

**Pearl Mussel**

3 to 6 in.

***Lanx* Limpet**

$\frac{3}{8}$ to $\frac{3}{4}$ in.

### Salt Water

**Marsh Snail**

$\frac{1}{2}$ in.

**Atlantic Surf Clam**

2 to 4 in.

**Blue Mussel**

1 to 3 in.

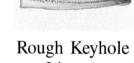

**Rough Keyhole Limpet**

$1\frac{1}{2}$ to 2 in.

### Land

**Banded Forest Snail**

1 to $1\frac{3}{4}$ in.

**Striped Forest Snail**

$\frac{3}{4}$ to $1\frac{1}{4}$ in.

**Speckled Garden Snail**

$1\frac{1}{2}$ to $1\frac{3}{4}$ in.

**White-lipped Forest Snail**

$\frac{3}{4}$ to $1\frac{3}{4}$ in.

# Math and Social Studies Data Bank

## The History of Kites

Kites were invented around 3000 years ago in China. This time line shows some of the history of kites after they were invented.

1752             1826             1901 1903             1969

| Benjamin Franklin used kites to study lightning. | An Englishman used a kite to pull a carriage. | A kite was used for the first long-distance radio. | The Wright brothers put an engine in a kite and flew it. This was the first airplane flight. | Children in Indiana flew a string of 19 kites and broke the record height. |

## Alexander Graham Bell's Kites

Alexander Graham Bell is most famous for inventing the telephone. He also made inventions with kites, using them to lift people.

Bell's kites had sections called cells. His kite called Frost King had 1,300 cells. His kite called Cygnet had 3,393 cells. Frost King lifted one of Bell's helpers 30 feet in the air.

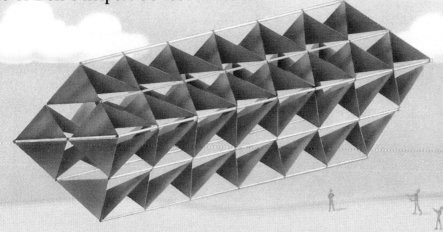

# Math and Social Sciences Data Bank

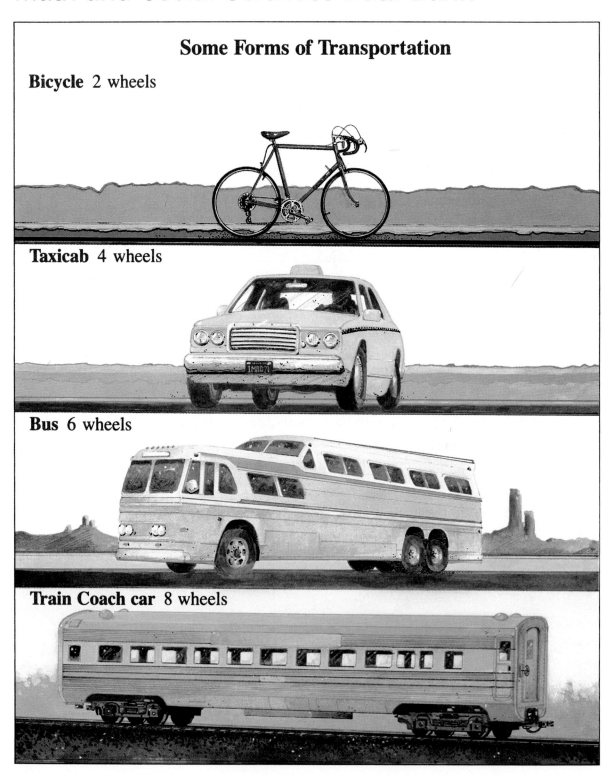

**Some Forms of Transportation**

**Bicycle** 2 wheels

**Taxicab** 4 wheels

**Bus** 6 wheels

**Train Coach car** 8 wheels

# Math and Social Studies Data Bank

### Galveston, Texas

| Tour | Transportation | Adult | Child |
|---|---|---|---|
| All About Town | bus | $5.00 | $2.50 |
| Harbor | paddlewheel boat | $8.00 | $4.00 |
| Treasure Island | train | $4.30 | $2.50 |

### Denver, Colorado

| Tour | Transportation | Adult | Child |
|---|---|---|---|
| City of Denver | bus | $10.00 | $7.00 |
| Denver and Mountains | bus | $25.00 | $16.50 |
| Old Denver | walking | $7.00 | $5.00 |
| Old Denver | van | $12.00 | $10.00 |

### St. Augustine, Florida

| Tour | Transportation | Adult | Child |
|---|---|---|---|
| City History | train | $7.00 | $2.00 |
| City History | horse and carriage | $7.00 | $3.00 |

# Math and Language Arts Data Bank

**Nonsense Poem** by Edward Lear

There was an Old Man with a beard
Who said, ''It is just as I feared!''
   Two owls and a hen,
   four larks and a wren
Have all built their nests in my beard.''

There was an Old Person of Sparta
Who had twenty-five sons and one ''darter'';
   He fed them on snails,
   and weighed them on scales,
That wonderful person of Sparta.

**The Story of the Four Little Children
Who Went Round the World**
   by Edward Lear

Edward Lear wrote a nonsense story about four
small children who went around the world in a
boat. These are some of the things the children saw
during their trip.

65 red parrots with blue tails
260 tail feathers lost by the parrots
A tree that is 503 feet high
Millions and millions of oranges
A crowd of kangaroos and gigantic cranes
About 600 or 700 crabs and crawfish

# Math and Language Arts Data Bank

### Tongue Twisters

Four fat frogs frying fritters.
Fanny fried five fish for Francis's father.
Six sand castles sitting in the sand.
Silly Sally swiftly shooed seven silly sheep.
Eight gray geese gazing into Greece.
Nine nimble noblemen nibbling nuts.

# Math and Language Arts Data Bank

## Weekly Allowances

In *Henry Huggins* by Beverly Cleary

Henry Huggins    25 cents

In *The Saturdays* by Elizabeth Enright

| Mona | 50 cents |
|------|----------|
| Oliver | 10 cents |
| Randy | 50 cents |
| Rush | 50 cents |

## Rainboots in Ramona's Class

In *Ramona Quimby* by Beverly Cleary

| 16 girls | |
|----------|--------|
| White Boots | 6 pairs |
| Red Boots | 10 pairs |

| 13 boys | |
|---------|----------|
| Brown Boots | 13 pairs |

# Math and Health and Fitness Data Bank

| Home Run Leaders | | |
|---|---|---|
| Year | Hitter | Number of Home Runs |
| 1927 | Babe Ruth | ⚾⚾⚾⚾⚾⚾⚾⚾⚾⚾⚾⚾ |
| 1969 | Willie McCovey | ⚾⚾⚾⚾⚾⚾⚾⚾ |
| 1985 | Darrell Evans | ⚾⚾⚾⚾⚾⚾⚾ |

⚾ stands for 5 home runs

| Prices at a World Series Baseball Game | |
|---|---|
| Item | Price |
| Peanuts | $0.80 |
| Popcorn | $0.80 |
| Team pennants | $2.50 |
| World Series pins | $4.00 |
| World Series programs | $5.00 |

# Math and Health and Fitness Data Bank

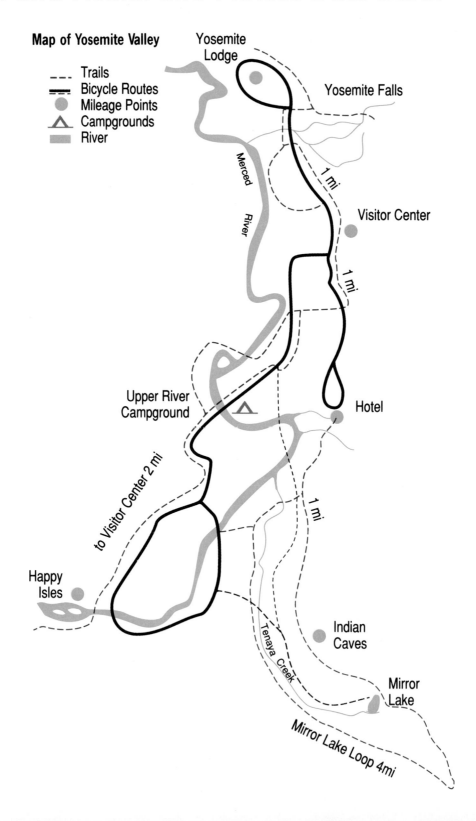

**Map of Yosemite Valley**

- - - Trails
━━━ Bicycle Routes
● Mileage Points
△ Campgrounds
▨ River

Yosemite Lodge

Yosemite Falls

Merced River

1 mi

Visitor Center

1 mi

Upper River Campground

Hotel

to Visitor Center 2 mi

1 mi

Happy Isles

Tenaya Creek

Indian Caves

Mirror Lake

Mirror Lake Loop 4mi

# Math and Health and Fitness Data Bank

## The Number of Calories Activities Use

| Activity | Calories used per hour |
|---|---|
| Sleeping | 60 |
| Standing | 100 |
| Walking (medium pace) | 300 |
| Running | 480 |

## The Number of Calories in Food

| Food | Serving | Weight | Calories |
|---|---|---|---|
| Butter | 1 tablespoon | 14 g | 100 |
| Cereal, whole wheat | 1 cup | 30 g | 100 |
| Fish | 1 small serving | 85 g | 135 |
| Grapes | 10 grapes | 50 g | 35 |
| Milk, whole | 1 cup | 244 g | 160 |
| Potato | 1 medium | 160 g | 145 |

## Olympic Running Events

100-meter run
200-meter run
400-meter run
800-meter run
1,000-meter run
1,500-meter run
3,000-meter run
5,000-meter run
10,000-meter run

**483**

# Counting and Counting Patterns

Try these key codes to count on by twos.

ON/AC $+$ 2 $=$ $=$ $=$ $=$ . . .

ON/AC $+$ 2 Cons Cons Cons Cons . . .

Entering $+$ 2 or $+$ 2 Cons sets up a **constant,** a number that stays the same. Each time you press $=$ or Cons the calculator will add 2 to the number in the display.

Start at 2. Count on by 5s.

Enter  ON/AC 2 $+$ 5 $=$ $=$ $=$ $=$ . . .
or      ON/AC $+$ 5 Cons 2 Cons Cons Cons . . .

Your display should show 7, 12, 17, 22, 27 . . .

Try **counting backward** by 4s. Start at 40.

Enter  ON/AC 40 $-$ 4 $=$ $=$ $=$ $=$ . . .
or      ON/AC $-$ 4 Cons 40 Cons Cons Cons Cons . . .

The display should show 36, 32, 28, 24 . . .

Enter as many 9s as you can on your calculator. What happens if you enter $+$ 1 $=$ ? The number is too large for many calculators. The display shows an **Overflow Error.**

## Activity

For each pattern, find its rule and write the next 5 numbers in the pattern.

1. 20 23 26 29
2. 412 408 404 400
3. 75000 74500 74000 73500
4. 155 153 151 149

| Rules |
| --- |
| Count back by 4 |
| Count back by 500 |
| Count back by 2 |
| Count on by 3 |
| Count on by 10 |

# Whole Number Addition and Subtraction

To solve an addition or subtraction problem, enter
the key code just the way you say the problem. Try these.

| Say | Enter | Display |
|---|---|---|
| Five plus seven equals | ON/AC 5 + 7 = | 12 |
| Twelve minus four equals | ON/AC 12 − 4 = | 8 |
| Twenty minus fourteen equals | ON/AC 20 − 14 = | 6 |
| Fifty-five plus eleven equals | ON/AC 55 + 11 = | 66 |
| Two hundred two minus nine equals | ON/AC 202 − 9 = | 193 |
| Five thousand eight plus ninety equals | ON/AC 5008 + 90 = | 5098 |

Addition and subtraction problems with more than
two numbers work the same way. Try these.

| Problem | Enter | Display |
|---|---|---|
| $16 + 83 + 46 =$ | ON/AC 16 + 83 + 46 = | 145 |
| $77 + 41 + 32 - 15 =$ | ON/AC 77 + 41 + 32 − 15 = | 135 |
| $85 - 17 - 19 - 21 =$ | ON/AC 85 − 17 − 19 − 21 = | 28 |
| $109 - 42 + 68 - 9 =$ | ON/AC 109 − 42 + 68 − 9 = | 126 |

## Activity

This is a game for two or more
players. Each person enters 21 on
his or her calculator. Take turns
rolling a pair of dice. If you roll a
double, subtract the number rolled.
If you do not roll a double, add the
number rolled. The person with the
largest display after 12 turns is the
winner.

# Whole Number Multiplication

Here are some ways you can find the total of 5 nines.

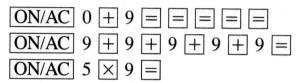

ON/AC  0  +  9  =  =  =  =  =
ON/AC  9  +  9  +  9  +  9  +  9  =
ON/AC  5  ×  9  =

Use your calculator to find these totals.

|  | **Display** |
|---|---|
| 7 eights | 56 |
| 14 twelves | 168 |

Suppose you want to multiply 9, 15, 46, and 119 each by 12. If your calculator has a multiplication constant, you do not have to enter an equation for each multiplication. You can enter the constant, ×  12, and then multiply each number by the constant. Try these key codes.

ON/AC  9  ×  12  =  15  =  46  =  119  =
ON/AC  ×  12  Cons  9  Cons  15  Cons
46  Cons  119  Cons

Your display should show these products:
108, 180, 552, 1428.

## Activity

Use your calculator to find the first four products. Guess what the fifth product will be, then check.

**1.** $9 \times 9 =$
$9 \times 98 =$
$9 \times 987 =$
$9 \times 9876 =$
$9 \times 98765 =$

**2.** $202 \times 22 =$
$202 \times 222 =$
$202 \times 2222 =$
$202 \times 22222 =$
$202 \times 222222 =$

**3.** $111 \times 12 =$
$111 \times 13 =$
$111 \times 14 =$
$111 \times 15 =$
$111 \times 16 =$

# Whole Number Division Without Remainders

How many 8s are there in 40? You can count the number of times you can subtract 8 from 40.

[ON/AC] 40 [−] 8 [=] [=] [=] [=] [=]

Or you can divide 40 by 8.

Enter [ON/AC] 40 [÷] 8 [=]. The display will show 5.

| Find: | Display |
|---|---|
| $90 \div 15$ | 6 |
| $330 \div 33$ | 10 |

Suppose you want to divide 72, 54, 36, and 24 by 6. If your calculator has a division constant, you do not have to enter an equation for each division. You can enter the constant, [÷] 6, and then divide each number by that constant.

[ON/AC] 72 [÷] 6 [=] 54 [=] 36 [=] 24 [=] 18 [=]
[ON/AC] [÷] 6 [Cons] 72 [Cons] 54 [Cons] 36 [Cons] 24 [Cons]

Your display should show 12, 9, 6, and 4.

Divide 8, 22, 119, and 2406 by 0. Dividing by zero is not logical. The calculator will display a **Logic Error.**

## Activity

How many of each flower would there be in a bunch if you divided this spring bouquet into 6 bunches? 1 bunch? 4 bunches? 12 bunches?

96 tulips       144 daffodils
48 roses        252 violets
180 lilies

# Whole Number Division with Remainders

Mr. Franklin has 135 stamps to divide among the 12 members of the stamp club. How many can he give to each member? How many will be left over? Use your calculator to divide 135 by 12.

| **Enter** | **Display** |
|---|---|
| ON/AC 135 ÷ 12 = | 11.25 |

Mr. Franklin can give each member 11 stamps. To find the number left over, multiply $11 \times 12$ to find the number of stamps Mr. Franklin gave away: $11 \times 12 = 132$. Then subtract 132 from the original number of stamps: $135 - 132 = 3$. There are 3 stamps left over.

How many stamps would Mr. Franklin have left over if he divided 78 stamps among the 12 club members?

With the Math Explorer calculator, you can use the INT ÷ key for division with remainders. Divide 97 by 5.

**Enter**

ON/AC 97 INT ÷ 5 =

The quotient is 19. The remainder is 2.

**Display**

Q 19     R 2

## Activity

Find the remainder for each division problem.

| | | | |
|---|---|---|---|
| **1.** $87 \div 8$ | **2.** $91 \div 14$ | **3.** $73 \div 5$ | **4.** $124 \div 10$ |
| **5.** $82 \div 7$ | **6.** $51 \div 4$ | **7.** $103 \div 9$ | **8.** $149 \div 12$ |

# Order of Operations: Memory Keys

In problems involving more than one operation,
you must do the operations in a certain order.
First multiply and divide, then add and subtract.

To find $5 \times 14 + 7 \times 8$ you first find $5 \times 14$. Then
find $7 \times 8$. Then add the two products.

| Enter | Display |
|---|---|
| ON/AC 5 × 14 = | 70 |
| ON/AC 7 × 8 = | 56 |
| ON/AC 70 + 56 = | 126 |

You can find $5 \times 14 + 7 \times 8$ using memory keys.

| M+ | Adds the display to the calculator's memory |
|---|---|
| M− | Subtracts the display from the calculator's memory |
| MR | Recalls the total in memory |
| ON/AC | Clears the memory |

| Enter | Display |
|---|---|
| ON/AC 5 × 14 = M+ | 70 |
| 7 × 8 = M+ | 56 |
| MR | 126 |

Unlike some calculators, the Math Explorer does
operations in the correct order. You can enter the
problem just as it is written. Find $15 \times 4 - 42 \div 7$.

ON/AC 15 × 4 − 42 ÷ 7 =

## Activity

Solve.

**1.** $12 \times 3 - 20$    **2.** $5 + 30 \div 5 + 5$    **3.** $46 - 5 \times 5$    **4.** $19 - 18 \div 3 + 8$

**489**

# Decimals

When entering decimal numbers on your calculator, remember to enter the decimal point. Notice that the calculator places a zero in front of the decimal point if you do not enter a whole number. It drops extra zeros on the right side after $=$ or another operation sign is entered.

Count by .05. Start at 1.

ON/AC 1 + . 05 = = = . . .

or ON/AC + . 05 Cons 1 Cons Cons Cons . . .

Your display should show 1.05, 1.1, 1.15, 1.2, . . . .

You can add and subtract decimals just like whole numbers. Enter money amounts as decimals.

| Problem | Key Code | Display |
|---|---|---|
| $4 - 3.16 =$ | ON/AC 4 − 3.16 = | 0.84 |
| $3.59 + 1.41 =$ | ON/AC 3.59 + 1.41 = | 5 |
| $\$4.55 - \$2.25 =$ | ON/AC 4.55 − 2.55 = | 2.3 |

The display for the last equation shows 2.3. You know that 2.3 = 2.30. Two and three tenths dollars is $2.30.

## Activity

Choose from the amounts in the box.

| |
|---|
| $16.20 |
| $3.90 |
| $12.85 |
| $7.29 |
| $4.77 |
| $9.38 |

1. Find two amounts that total $16.67.

2. $20 minus this amount is $7.15.

3. The difference between these two amounts is $3.35.

4. The total of these three amounts is $21.44.

5. If you start at $15 and count by $0.15, this amount is in the pattern.

# Fractions

To find the decimal equivalent of a fraction, you can divide the numerator by the denominator. For example, here is how to find the decimal for 3/10.

| Enter | Display |
|---|---|
| [ON/AC] 3 [÷] 10 [=] | 0.3 |

The [F⇄D] key on the Math Explorer Calculator is a fraction/decimal key. Enter the fraction 3/10 using the [/] key. Use the [F⇄D] key to change it to a decimal.

| Enter | Display |
|---|---|
| [ON/AC] 3 [/] 10 [F⇄D] | 0.3 |
| Press [F⇄D] again. | 3/10 |

Pressing [F⇄D] the second time changes the decimal 0.3 back to a fraction.

To add or subtract fractions on the Math Explorer, enter the problem just the way you write it.

| Problem | Key Code | Display |
|---|---|---|
| 1/5 + 3/5 | 1 [/] 5 [+] 3 [/] 5 [=] | 4/5 |
| 9/10 − 6/10 | 9 [/] 10 [−] 6 [/] 10 [=] | 3/10 |
| 7/8 − 2/8 | 7 [/] 8 [−] 2 [/] 8 [=] | 5/8 |
| 3/7 + 5/7 | 3 [/] 7 [+] 5 [/] 7 [=] | 8/7 |

## Activity

Find each sum or difference. Then change the answer to a decimal.

**1.** 8/10 − 4/10    **2.** 3/8 + 1/8    **3.** 1/2 + 1/2    **4.** 3/4 − 1/4    **5.** 7/6 − 4/6

# The NIM Game

Maria and Lois were playing the game of NIM with 21 counters. To play the game, players take turns picking up 1, 2, or 3 counters. The player who must take the last counter loses the game.

Suppose it is your turn to pick up counters. You see 6 counters are left.

Can you win? How many counters would you pick up? What would be your strategy?

You can play NIM on the computer using this BASIC computer program. Think about some strategies that might help you win some games.

```
********************
21 STARS LEFT

PLAYER 1
CHOOSE 1, 2, OR 3 STARS.



****** 6 STARS LEFT
```

```
10 GOSUB 140
20 S=21
30 GOSUB 110
40 FOR K=1 TO 2:PRINT:GOSUB 70:NEXT K
50 IF S>0 THEN 40
60 END
70 PRINT "PLAYER"K:INPUT"CHOOSE 1, 2, OR 3 COUNTERS. ";C
80 IF C<1 OR C>3 THEN 70
90 S=S-C
100 IF S<1 THEN PRINT"PLAYER "K" LOSES.":GOTO 60
110 PRINT:FOR N=1 TO S:PRINT"*";
120 NEXT N:PRINT" "S" STARS LEFT"
130 RETURN
140 PRINT:PRINT:PRINT"THE NIM GAME"
150 PRINT"THERE ARE 21 COUNTERS. EACH PLAYER"
160 PRINT"TAKES TURNS AND PICKS UP 1, 2, OR 3"
170 PRINT"COUNTERS. THE PLAYER WHO MUST TAKE THE"
180 PRINT"LAST COUNTER LOSES THE GAME."
190 PRINT:PRINT:INPUT"PRESS<RETURN>";Y$:PRINT:PRINT
200 RETURN
```

# Function Tables

Study the table to discover a pattern. What is the rule between the input and output numbers? What are the two missing numbers?

This computer program will give you more practice finding rules for function tables.

| Function table | |
|---|---|
| Input | Output |
| 3 | 27 |
| 7 | 63 |
| 4 | 36 |
| 2 | ? |
| 9 | ? |

```
10 FOR I = 1 TO 5:X(I) = INT(10* RND (1) + 1)
20 IF X(I) = X(I − 1) THEN 10
30 NEXTI
40 A = INT (10* RND (1) + 1)
50 PRINT:PRINT:PRINT"INPUT | OUTPUT": PRINT" ___|___": PRINT" | "
60 GOSUB 190
70 IF X(5) < 10 THEN PRINT" "X(5)" | ? ":GOTO 90
80 PRINT" "X(5)" | ?"
90 PRINT:PRINT:PRINT: "FUNCTION TABLES": PRINT "THERE IS A PATTERN
   BETWEEN THE INPUT": PRINT "AND OUTPUT NUMBERS IN THE
   FUNCTION": PRINT "TABLE. STUDY THE TABLE AND SEE IF YOU": PRINT
   "CAN DISCOVER THE PATTERN. WHAT IS THE"
100 INPUT "MISSING NUMBER IN THE TABLE?";M: PRINT: IF (J=1) AND
    (M = X(5) + A) THEN PRINT"CORRECT.":GOTO 140
110 IF (J=2) AND (M = X(5) * A) THEN PRINT "CORRECT.":GOTO 150
120 IF J=1 THEN PRINT"NOT CORRECT.":GOTO 140
130 IF J=2 THEN PRINT "NOT CORRECT.":GOTO 150
140 PRINT "THE RULE IS: INPUT + ";A;" = OUTPUT": GOTO 160
150 PRINT "THE RULE IS: INPUT × ";A;" = OUTPUT"
160 INPUT "DO YOU WANT ANOTHER FUNCTION TABLE? ";Y$
170 IF LEFT$ (Y$,1) = "Y" THEN GOTO 10
180 END
190 J = INT (2* RND (1) + 1)
200 ON J GOTO 210,230
210 FOR N = 1 TO 4: IF X(N) < 10 THEN PRINT" "X(N)" | "X(N) + A:
    NEXT N: RETURN
220 PRINT" "X(N)" | "X(N) + A: NEXT N: RETURN
230 FOR N = 1 TO 4: IF X(N) < 10 THEN PRINT" "X(N)" | "X(N)*A:
    NEXT N:RETURN
240 PRINT" "X(N)" | "X(N)*A: NEXT N: RETURN
```

493

# Estimating Sums

Amy had $15.00. She wanted to buy a T-shirt for
$7.84 and a cassette of "Greatest Hits" for
$4.94. She was not sure that she had enough
money for both so she decided to estimate the total.

What is your estimate of the total? Try rounding
each amount to the nearest dollar to make a good
estimate. Will your estimate be more or less than
the actual amount? Why?

Computers can help you practice your estimation
skills. This computer program will give you more
estimation problems to solve.

```
10 PRINT : PRINT : PRINT "ESTIMATING SUMS"
20 A = INT (995 * RND (1))/100
30 IF A < 1.1 THEN 20
40 B = INT (995 * RND (1))/100
50 IF B < 1.1 THEN 40
60 A1 = A:B1 = B
70 PRINT "ESTIMATE THE SUM BY ROUNDING TO THE"
80 PRINT "NEAREST DOLLAR. OMIT THE $ SIGN."
90 PRINT : PRINT
100 PRINT "$";A: PRINT
110 PRINT " + ";B: PRINT "_____": PRINT : PRINT
120 INPUT "WHAT IS YOUR ESTIMATE? ";E
130 A = INT (A + .5):B = INT (B + .5)
140 PRINT :S = A + B: IF E = S THEN PRINT "YOUR ESTIMATE IS
     CORRECT.":GOTO 160
150 PRINT "THE CORRECT ESTIMATE IS $ ";S"."
160 PRINT "THE EXACT SUM IS $";A1 + B1
170 INPUT "WANT ANOTHER PROBLEM? (Y/N) ";Y$
180 IF LEFT$ (Y$,1) = "Y" THEN 10
190 END
```

# Fractions of a Number

Brad drew 12 circles on his paper. He decided to color three fourths of the circles blue. How many circles should he color blue?

You can use the set of 12 circles to help you think about the problem. There are 4 circles in each row. There are 3 rows of circles. What is $\frac{3}{4}$ of 12?

**3/4 of 12 = 9**

Computers can help you learn about fractions. This computer program will give you more practice in finding fractions of a number.

```
10 N = INT (10 * RND (1) + 1)
20 D = INT (10 * RND (1) + 1): IF D < = N THEN 10: IF D = 7 OR
   D = 9 THEN 10
30 M = INT (9 * RND (1) + 1) * D
40 PRINT : PRINT : FOR R = 1 TO M/D
50 FOR C = 1 TO D
60 PRINT "*";: NEXT C: PRINT : NEXT R
70 PRINT : PRINT : PRINT "FRACTIONS OF A NUMBER": PRINT "THERE ARE
   ";M;" TOTAL STARS. THERE ARE"
80 PRINT D;" STARS IN EACH ROW. THERE ARE": PRINT M/D;" ROWS OF
   STARS. WHAT IS ";N;"/";D;" OF ";M: INPUT "STARS?";A
90 A1 = (N/D) * M: PRINT
100 IF A = A1 THEN PRINT "CORRECT."
110 IF A < > A1 THEN PRINT "SORRY."
120 PRINT N;"/";D;" OF ";M;" =";A1: PRINT
130 FOR R = 1 TO M/D
140 FOR C = 1 TO N
150 PRINT "*";: NEXT C: PRINT : NEXT R
160 PRINT : INPUT "DO YOU WANT ANOTHER PROBLEM?";Y$
170 IF LEFT$ (Y$,1) = "Y" THEN 10
180 END
```

495

# Smallest Differences

Gary and Corey were playing the game of "Smallest Difference." Garry had the four digits 2, 3, 3, 5 to put in the boxes in the subtraction problem. Corey had the digits 7, 0, 4, 3. Each player wanted the least answer possible.

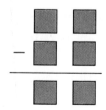

What problem would you make with each set of digits? Who would win? Play the game. Think about some strategies that may help you win.

```
10 PRINT : PRINT : PRINT "SMALLEST DIFFERENCE GAME"
20 PRINT "EACH PLAYER RECEIVES 4 DIGITS. USE THE": PRINT "4 DIGITS
   TO CREATE A SUBTRACTION"
30 PRINT "PROBLEM WITH THE SMALLEST POSSIBLE": PRINT
   "DIFFERENCE. THE PLAYER WITH THE": PRINT "SMALLEST DIFFERENCE
   WINS THE GAME."
40 FOR P = 1 TO 2
50 PRINT : PRINT "PLAYER ";P;", YOUR NUMBERS ARE:": PRINT
60 FOR C = 1 TO 4:B = INT (10 * RND (1))
70 PRINT B" ";: NEXT C: PRINT : PRINT
80 INPUT "TYPE THE TOP NUMBER.";T(P)
90 INPUT "TYPE THE BOTTOM NUMBER.";B(P)
100 IF B(P) > T(P) THEN GOTO 80
110 D(P) = T(P) − B(P): NEXT P
120 PRINT : PRINT "PLAYER 1", "PLAYER 2": PRINT
130 PRINT TAB( 5 − LEN ( STR$ (T(1))))T(1); TAB( 21 − LEN ( STR$
    (T(2))))T(2): PRINT : PRINT TAB( 2)"−"; TAB( 5 − LEN ( STR$ (B(1))))B(1);
    TAB( 18)"−"; TAB( 21 − LEN ( STR$ (B(2))))B(2)
140 PRINT TAB( 2)"_____"; TAB( 18)"_____": PRINT ; PRINT TAB
    ( 5 − LEN ( STR$ (D(1))))D(1); TAB( 21 − LEN ( STR$ (D(2))))D(2)
150 PRINT : IF D(1) = D(2) THEN PRINT "TIE SCORE OF" ;D(1):
    GOTO 180
160 IF D(1) < D(2) THEN PRINT "PLAYER 1 WINS.": GOTO 180
170 PRINT "PLAYER 2 WINS."
180 PRINT : INPUT "WANT TO PLAY ANOTHER GAME? (Y/N)";Y$
190 IF LEFT$ (Y$,1) = "Y" THEN 40
200 END
```

# Guess and Check

Ryne thought of two numbers. He told Sonja the **sum** of the two numbers was 15 and the **product** of the two numbers was 54. Sonja used the strategy of Guess and Check to find the two numbers.

She thought: $5 + 10 = 15$, but $5 \times 10 = 50$, so 5 and 10 did not check. The sum was correct, but the product was too small. Sonja thought she must be close to the two numbers, so she must guess again and then check her guess.

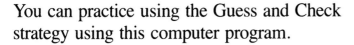

What do you think Ryne's two numbers might be?

You can practice using the Guess and Check strategy using this computer program.

```
10 PRINT : PRINT : PRINT "GUESS AND CHECK"
20 A = INT (10 * RND (1) + 1):B = INT (10 * RND (1) + 1)
30 PRINT "THE SUM OF TWO NUMBERS IS ";A + B;"."
40 PRINT "THE PRODUCT OF THE TWO NUMBERS IS ";A * B;"."
50 PRINT "USE THE GUESS AND CHECK STRATEGY TO": PRINT "FIND THE
   TWO NUMBERS."
60 PRINT : INPUT "FIRST NUMBER = ";F: INPUT "SECOND NUMBER = ";S
70 PRINT : PRINT "SUM = ";F + S, "PRODUCT =";F * S
80 IF A = F AND B = S OR A = S AND B = F THEN GOTO 110
90 PRINT : PRINT "NOT CORRECT.": INPUT "DO YOU WANT TO TRY
   AGAIN?";Y$: IF LEFT$ (Y$,1) = "Y" THEN PRINT : PRINT : GOTO 30
100 GOTO 120
110 PRINT : PRINT "YOU HAVE FOUND THE TWO NUMBERS."
120 INPUT "DO YOU WANT ANOTHER PROBLEM?";Y$
130 IF LEFT$ (Y$,1) = "Y" THEN 10
```

# Probability Spinner

Jaime and Ben wanted to find how many times heads would come up if they tossed a coin 100 times. They didn't have a coin to use. Ben said, "We can use a spinner with 2 equal parts. If a 1 comes up that is the same as heads. If a 2 comes up that is the same as tails."

Try the experiment that Jaime and Ben did. How many heads and tails will you get?

You can use this computer program to try the probability experiment. You can choose a spinner with 2, 3, 4, or any number of equal parts. You can choose as many spins as you like. The computer will quickly give the outcomes for your experiment.

```
10 PRINT "PROBABILITY SPINNER": PRINT "THIS IS AN EXPERIMENT IN
   PROBABILITY.": PRINT "CHOOSE A SPINNER WITH ANY NUMBER OF":
   PRINT "EQUAL PARTS. CHOOSE ANY NUMBER OF": PRINT "SPINS. THE
   COMPUTER WILL QUICKLY GIVE": PRINT "THE OUTCOMES."
20 PRINT : INPUT "HOW MANY SECTORS OF EQUAL SIZE? ";X
30 DIM D(X)
40 FOR N = 1 TO X:D(N) = 0: NEXT N
50 INPUT "HOW MANY SPINS DO YOU WANT? ";Y
60 FOR I = 1 TO Y:R = INT (X * RND (1)) + 1
70 LET D(R) = D(R) + 1: NEXT 1
80 PRINT "NUMBER","FREQUENCY"
90 FOR N = 1 TO X: PRINT N,D(N): NEXT N
100 END
```

# Estimating Capacity

The graph shows a water tank full of water. Each * represents one gallon of water. Count the stars to find the total number of gallons in the tank. How many gallons are there?

```
*********
*********
*********
*********
```

Now estimate the number of gallons this empty tank will hold.

You can use the computer program to try other estimation problems.

```
10 R = 10 + INT (10 * RND (1) + 1)
20 J = INT (15 * RND (1) + 1)
30 PRINT : PRINT : FOR I = 1 TO J
40 PRINT TAB ( 10)" | "; TAB( R + 1)" | ": NEXT I
50 FOR I = 11 TO R: PRINT TAB( I)" –";: NEXT I: PRINT : PRINT
60 PRINT "ESTIMATING CAPACITY": PRINT "ESTIMATE HOW MANY GALLONS
    OF WATER TO":
70 PRINT "FILL THE TANK. REMEMBER EACH * IS 1"
80 INPUT "GALLON. HOW MANY GALLONS? ";G: PRINT : PRINT
90 PRINT "WE FILL THE TANK AND COUNT THE GALLONS.": PRINT
100 FOR I = 1 TO J
110 FOR S = 10 TO R + 1
120 IF S = 10 THEN PRINT TAB( S)" | ";
130 IF S > 10 AND S < =R THEN PRINT TAB( S)"*";:G1 = G1 +1
140 IF S = R + 1 THEN PRINT TAB( S)" | "
150 NEXT S: NEXT I
160 FOR I = 11 TO R: PRINT TAB( I)" –";: PRINT : PRINT
170 PRINT : PRINT "THE CORRECT AMOUNT IS ";G1;" GALLONS."
180 PRINT "YOUR ESTIMATE IS OFF BY "; ABS (G – G1);" GALLONS."
190 G1 = 0: INPUT "DO YOU WANT ANOTHER PROBLEM? (Y/N) ";Y$
200 IF LEFT$ (Y$,1) = "Y" THEN GOTO 10: PRINT : PRINT
210 END
```

**Set A   For use after page 5.**

Add or subtract.

| **1.** | **2.** | **3.** | **4.** | **5.** | **6.** |
|---|---|---|---|---|---|
| 4 | 3 | 8 | 5 | 5 | 2 |
| +6 | +4 | −2 | +3 | −4 | +4 |

**Set B   For use after page 7.**

Add or subtract.

| **1.** | **2.** | **3.** | **4.** | **5.** | **6.** |
|---|---|---|---|---|---|
| 2 | 8 | 3 | 0 | 4 | 5 |
| +3 | +2 | −2 | +9 | −2 | −1 |

**Set C   For use after page 9.**

Add.

| **1.** | **2.** | **3.** | **4.** | **5.** | **6.** |
|---|---|---|---|---|---|
| 3 | 3 | 7 | 8 | 6 | 4 |
| +3 | +4 | +8 | +9 | +6 | +4 |

**Set D   For use after page 11.**

Add.

| **1.** | **2.** | **3.** | **4.** | **5.** | **6.** |
|---|---|---|---|---|---|
| 4 | 8 | 6 | 3 | 7 | 9 |
| +8 | +6 | +7 | +9 | +4 | +6 |

**Set E   For use after page 15.**

Subtract.

| **1.** | **2.** | **3.** | **4.** | **5.** | **6.** |
|---|---|---|---|---|---|
| 14 | 11 | 17 | 13 | 12 | 11 |
| − 9 | − 5 | − 8 | − 5 | − 4 | − 6 |

**Set F   For use after page 17.**

Subtract.

| **1.** | **2.** | **3.** | **4.** | **5.** | **6.** |
|---|---|---|---|---|---|
| 13 | 12 | 16 | 11 | 13 | 14 |
| − 6 | − 7 | − 8 | − 9 | − 8 | − 8 |

**Set A   For use after page 21.**

Add.

| **1.** | 2 | **2.** | 4 | **3.** | 8 | **4.** | 3 | **5.** | 6 | **6.** | 2 |
|---|---|---|---|---|---|---|---|---|---|---|---|
| | 4 | | 2 | | 2 | | 1 | | 1 | | 5 |
| | +5 | | +3 | | +5 | | +6 | | +2 | | +3 |

**Set B   For use after page 35.**

Give the number that comes next.

**1.** 108      **2.** 60      **3.** 111      **4.** 69      **5.** 18      **6.** 633

**Set C   For use after page 37.**

Count by twos. Give the next four numbers.

**1.** 10, 12, 14, 16, |||||, |||||, |||||, |||||      **2.** 25, 27, 29, 31, |||||, |||||, |||||, |||||

**Set D   For use after page 41.**

Write > or < for each |||||.

**1.** 90 ||||| 60      **2.** 200 ||||| 400      **3.** 40 ||||| 60      **4.** 75 ||||| 68

**Set E   For use after page 45.**

Round to the nearest ten.

**1.** 24 → |||||      **2.** 37 → |||||      **3.** 83 → |||||      **4.** 62 → |||||      **5.** 15 → |||||

**Set F   For use after page 47.**

Round to the nearest hundred.

**1.** 538 → |||||      **2.** 281 → |||||      **3.** 108 → |||||      **4.** 152 → |||||

**Set G   For use after page 69.**

Write the dates another way.

**1.** 4/2/75      **2.** 5/24/83      **3.** 8/19/65      **4.** 6/29/46

**Set A   For use after page 89.**

Find the sums. Write answers only.

**1.**   20    **2.**   70    **3.**   60    **4.**   80    **5.**   90    **6.**   20
      $+80$        $+70$        $+30$        $+50$        $+10$        $+50$

**Set B   For use after page 91.**

Estimate by rounding to the nearest ten.

**1.**   32    **2.**   86    **3.**   25    **4.**   66    **5.**   82    **6.**   55
      $+49$        $+43$        $+28$        $+47$        $+34$        $+71$

Estimate by rounding to the nearest hundred or dollar.

**7.**   480      **8.**   638      **9.**   $6.11      **10.**   $2.95
      $+520$          $+402$          $+4.95$             $+6.12$

**Set C   For use after page 97.**

Find the sums.

**1.**   12    **2.**   37    **3.**   12    **4.**   80    **5.**   58    **6.**   44
      $+29$        $+43$        $+27$        $+16$        $+24$        $+29$

**Set D   For use after page 101.**

Find the sums.

**1.**   648    **2.**   291    **3.**   260    **4.**   418    **5.**   386    **6.**   605
      $+104$         $+464$         $+917$         $+750$         $+452$         $+327$

**Set E   For use after page 103.**

Find the sums.

**1.**   652    **2.**   805    **3.**   472    **4.**   226    **5.**   753    **6.**   764
      $+189$         $+929$         $+687$         $+\;\;98$        $+685$         $+187$

**Set A** **For use after page 105.**

Find the sums. Use mental math.

**1.** $26 + 32$ __     **2.** $43 + 11$ __     **3.** $51 + 24$ __     **4.** $32 + 41$ __

**Set B** **For use after page 107.**

Find the sums.

| **1.** | **2.** | **3.** | **4.** | **5.** |
|---|---|---|---|---|
| 26 | 48 | 32 | 371 | $4.18 |
| 15 | 7 | 41 | 256 | 3.75 |
| + 34 | + 14 | + 10 | + 30 | + 1.86 |

**Set C** **For use after page 109.**

Find the sums.

| **1.** | **2.** | **3.** | **4.** | **5.** |
|---|---|---|---|---|
| 3,059 | 2,964 | 1,429 | 6,781 | $48.12 |
| + 2,717 | + 5,682 | + 6,507 | + 2,657 | + 39.68 |

**Set D** **For use after page 123.**

Find the differences. Write answers only.

| **1.** | **2.** | **3.** | **4.** | **5.** |
|---|---|---|---|---|
| 1,000 | 170 | 300 | 1,300 | 600 |
| − 300 | − 60 | − 100 | − 800 | − 100 |

**Set E** **For use after page 125.**

Estimate by rounding to the nearest ten.

| **1.** | **2.** | **3.** | **4.** | **5.** |
|---|---|---|---|---|
| 36 | 84 | 22 | 69 | 42 |
| − 19 | − 26 | − 14 | − 28 | − 33 |

Estimate by rounding to the nearest hundred
or nearest dollar.

| **6.** | **7.** | **8.** | **9.** |
|---|---|---|---|
| 418 | 630 | $8.71 | $6.89 |
| − 297 | − 149 | − 2.56 | − 1.35 |

## MORE PRACTICE BANK

**Set A   For use after page 131.**

Find the differences.

| **1.** 64 | **2.** 81 | **3.** 66 | **4.** 44 | **5.** 30 | **6.** 64 |
|---|---|---|---|---|---|
| − 27 | − 14 | − 59 | − 23 | − 19 | − 38 |

**Set B   For use after page 135.**

Find the differences.

| **1.** 559 | **2.** 348 | **3.** 442 | **4.** 312 | **5.** 559 | **6.** 677 |
|---|---|---|---|---|---|
| − 167 | − 295 | − 180 | − 91 | − 168 | − 484 |

**Set C   For use after page 137.**

Find the differences.

| **1.** 423 | **2.** 154 | **3.** 865 | **4.** 312 | **5.** 568 | **6.** 235 |
|---|---|---|---|---|---|
| − 187 | − 89 | − 467 | − 140 | − 179 | − 197 |

**Set D   For use after page 141.**

Find the differences.

| **1.** 201 | **2.** 308 | **3.** 702 | **4.** 805 | **5.** 100 | **6.** 203 |
|---|---|---|---|---|---|
| − 84 | − 169 | − 246 | − 361 | − 67 | − 186 |

**Set E   For use after page 143.**

Estimate by rounding to the nearest dollar.

| **1.** $7.95 | **2.** $5.08 | **3.** $6.98 | **4.** $8.19 | **5.** $7.25 |
|---|---|---|---|---|
| − 1.29 | − 1.95 | − 1.99 | − 3.25 | − 3.19 |

**Set F   For use after page 145.**

Find the differences.

| **1.** 9,624 | **2.** 3,855 | **3.** 4,077 | **4.** 1,026 | **5.** 6,338 |
|---|---|---|---|---|
| − 4,312 | − 1,862 | − 2,165 | − 647 | − 4,290 |

## MORE PRACTICE BANK

**Set A   For use after page 217.**

Multiply.

| **1.** 2 | **2.** 4 | **3.** 7 | **4.** 2 | **5.** 2 | **6.** 2 | **7.** 8 |
| ×6 | ×2 | ×2 | ×3 | ×5 | ×2 | ×2 |

**Set B   For use after page 219.**

Multiply.

| **1.** 5 | **2.** 5 | **3.** 2 | **4.** 6 | **5.** 8 | **6.** 5 | **7.** 3 |
| ×2 | ×5 | ×6 | ×5 | ×5 | ×4 | ×5 |

**Set C   For use after page 221.**

Multiply.

| **1.** 5 | **2.** 9 | **3.** 3 | **4.** 9 | **5.** 8 | **6.** 5 | **7.** 9 |
| ×9 | ×6 | ×9 | ×4 | ×9 | ×6 | ×7 |

**Set D   For use after page 227.**

Multiply.

| **1.** 1 | **2.** 0 | **3.** 1 | **4.** 0 | **5.** 1 | **6.** 0 | **7.** 1 |
| ×7 | ×1 | ×0 | ×4 | ×3 | ×9 | ×9 |

**Set E   For use after page 229.**

Multiply.

| **1.** 9 | **2.** 5 | **3.** 9 | **4.** 5 | **5.** 9 | **6.** 8 |
| ×3 | ×9 | ×2 | ×2 | ×9 | ×9 |

| **7.** 4 | **8.** 9 | **9.** 5 | **10.** 9 | **11.** 7 | **12.** 8 |
| ×5 | ×6 | ×6 | ×4 | ×5 | ×5 |

| **13.** 9 | **14.** 8 | **15.** 0 | **16.** 7 | **17.** 7 | **18.** 5 |
| ×1 | ×2 | ×5 | ×2 | ×9 | ×5 |

## MORE PRACTICE BANK

**Set A   For use after page 239.**

Multiply.

| **1.** 3 | **2.** 3 | **3.** 2 | **4.** 6 | **5.** 2 | **6.** 3 | **7.** 4 |
|---|---|---|---|---|---|---|
| ×5 | ×3 | ×5 | ×3 | ×3 | ×7 | ×3 |

**Set B   For use after page 241.**

Use mental math to solve.

| **1.** 13 | **2.** 16 | **3.** 19 | **4.** 11 | **5.** 17 |
|---|---|---|---|---|
| +13 | +16 | +19 | +11 | +17 |

**Set C   For use after page 243.**

Multiply.

| **1.** 7 | **2.** 4 | **3.** 4 | **4.** 5 | **5.** 2 | **6.** 2 | **7.** 8 |
|---|---|---|---|---|---|---|
| ×4 | ×4 | ×9 | ×4 | ×7 | ×4 | ×4 |

**Set D   For use after page 245.**

Multiply.

| **1.** 5 | **2.** 9 | **3.** 4 | **4.** 2 | **5.** 7 | **6.** 6 | **7.** 8 |
|---|---|---|---|---|---|---|
| ×5 | ×9 | ×4 | ×2 | ×4 | ×6 | ×8 |

**Set E   For use after page 249.**

Multiply.

| **1.** 9 | **2.** 8 | **3.** 9 | **4.** 6 | **5.** 7 | **6.** 4 | **7.** 8 |
|---|---|---|---|---|---|---|
| ×9 | ×5 | ×6 | ×7 | ×8 | ×8 | ×6 |

**Set F   For use after page 253.**

Multiply.

| **1.** 6 | **2.** 9 | **3.** 8 | **4.** 7 | **5.** 5 | **6.** 7 | **7.** 9 |
|---|---|---|---|---|---|---|
| ×8 | ×9 | ×8 | ×8 | ×9 | ×6 | ×4 |

**Set A   For use after page 255.**

Find the products.

**1.** $4 \times 2 \times 3$       **2.** $2 \times 6 \times 4$       **3.** $7 \times 3 \times 2$       **4.** $8 \times 3 \times 2$

**5.** $8 \times 4 \times 2$       **6.** $5 \times 0 \times 4$       **7.** $1 \times 9 \times 3$       **8.** $6 \times 3 \times 2$

**Set B   For use after page 257.**

Multiply.

**1.** $\begin{array}{r} 7 \\ \times 8 \\ \hline \end{array}$    **2.** $\begin{array}{r} 2 \\ \times 9 \\ \hline \end{array}$    **3.** $\begin{array}{r} 1 \\ \times 6 \\ \hline \end{array}$    **4.** $\begin{array}{r} 4 \\ \times 8 \\ \hline \end{array}$    **5.** $\begin{array}{r} 4 \\ \times 5 \\ \hline \end{array}$    **6.** $\begin{array}{r} 9 \\ \times 5 \\ \hline \end{array}$    **7.** $\begin{array}{r} 6 \\ \times 7 \\ \hline \end{array}$

**Set C   For use after page 301.**

Divide.

**1.** $8 \div 2 = $ _____     **2.** $10 \div 2 = $ _____     **3.** $12 \div 2 = $ _____     **4.** $14 \div 2 = $ _____

**Set D   For use after page 303.**

Divide.

**1.** $12 \div 3 = $ _____     **2.** $10 \div 2 = $ _____     **3.** $6 \div 3 = $ _____     **4.** $14 \div 2 = $ _____

**Set E   For use after page 305.**

Divide.

**1.** $16 \div 4 = $ _____     **2.** $9 \div 3 = $ _____     **3.** $32 \div 4 = $ _____     **4.** $12 \div 4 = $ _____

**Set F   For use after page 309.**

Divide.

**1.** $10 \div 5 = $ _____     **2.** $40 \div 5 = $ _____     **3.** $20 \div 4 = $ _____     **4.** $25 \div 5 = $ _____

**Set G   For use after page 315.**

Divide.

**1.** $5 \div 5 = $ _____     **2.** $4 \div 0 = $ _____     **3.** $6 \div 6 = $ _____     **4.** $7 \div 1 = $ _____

**Set A   For use after page 317.**

Find the quotients.

**1.** $3\overline{)21}$    **2.** $5\overline{)25}$    **3.** $4\overline{)36}$    **4.** $5\overline{)15}$    **5.** $4\overline{)12}$

**Set B   For use after page 327.**

Divide.

**1.** $6\overline{)12}$    **2.** $6\overline{)42}$    **3.** $6\overline{)54}$    **4.** $6\overline{)36}$    **5.** $6\overline{)6}$

**Set C   For use after page 329.**

Divide.

**1.** $7\overline{)28}$    **2.** $7\overline{)21}$    **3.** $7\overline{)49}$    **4.** $7\overline{)42}$    **5.** $7\overline{)63}$

**Set D   For use after page 335.**

Divide.

**1.** $8\overline{)24}$    **2.** $8\overline{)56}$    **3.** $8\overline{)72}$    **4.** $8\overline{)8}$    **5.** $8\overline{)40}$

**Set E   For use after page 337.**

Divide.

**1.** $9\overline{)18}$    **2.** $9\overline{)81}$    **3.** $9\overline{)27}$    **4.** $9\overline{)54}$    **5.** $9\overline{)72}$

**Set F   For use after page 339.**

Divide.

**1.** $5\overline{)45}$    **2.** $6\overline{)36}$    **3.** $8\overline{)48}$    **4.** $9\overline{)27}$    **5.** $7\overline{)56}$

**Set G   For use after page 349.**

Multiply and then add 1. Write answers only.

**1.** $2 \times 3$_____    **2.** $7 \times 5$_____    **3.** $4 \times 9$_____    **4.** $8 \times 0$_____

# MORE PRACTICE BANK

**Set A   For use after page 351.**

Find the products.

**1.** $3 \times 10$ _____   **2.** $6 \times 10$ _____   **3.** $4 \times 10$ _____   **4.** $7 \times 10$ _____

**5.** $8 \times 10$ _____   **6.** $2 \times 10$ _____   **7.** $5 \times 10$ _____   **8.** $9 \times 10$ _____

**Set B   For use after page 353.**

Estimate by rounding to the nearest ten.

**1.** $3 \times 27$ _____       **2.** $8 \times 33$ _____       **3.** $4 \times 85$ _____

**4.** $2 \times 66$ _____       **5.** $4 \times 21$ _____       **6.** $3 \times 68$ _____

**Set C   For use after page 357.**

Find the products.

| **1.** | **2.** | **3.** | **4.** | **5.** | **6.** |
|---|---|---|---|---|---|
| 14 | 32 | 25 | 18 | 14 | 12 |
| $\times\ 5$ | $\times\ 3$ | $\times\ 3$ | $\times\ 5$ | $\times\ 2$ | $\times\ 7$ |

**Set D   For use after page 359.**

Find the products.

| **1.** | **2.** | **3.** | **4.** | **5.** | **6.** |
|---|---|---|---|---|---|
| 43 | 75 | 93 | 64 | 25 | 31 |
| $\times\ 5$ | $\times\ 3$ | $\times\ 6$ | $\times\ 5$ | $\times\ 8$ | $\times\ 4$ |

**Set E   For use after page 365.**

Estimate by rounding to the nearest dollar.

**1.** $4 \times \$4.34$       **2.** $3 \times \$6.92$       **3.** $7 \times \$3.50$       **4.** $5 \times \$3.26$

_____       _____       _____       _____

**Set F   For use after page 405.**

Write an equivalent fraction.

**1.** $\dfrac{3}{6} = \dfrac{1}{\blacksquare}$       **2.** $\dfrac{1}{2} = \dfrac{2}{\blacksquare}$       **3.** $\dfrac{1}{4} = \dfrac{2}{\blacksquare}$       **4.** $\dfrac{2}{10} = \dfrac{\blacksquare}{5}$       **5.** $\dfrac{\blacksquare}{3} = \dfrac{2}{6}$

## MORE PRACTICE BANK

**Set A   For use after page 407.**

Write the sign, $>$, $<$, or $=$ for each ▥.

**1.** $\frac{1}{2}$ ▥ $\frac{1}{3}$     **2.** $\frac{1}{5}$ ▥ $\frac{1}{3}$     **3.** $\frac{1}{10}$ ▥ $\frac{1}{5}$     **4.** $\frac{1}{4}$ ▥ $\frac{1}{5}$     **5.** $\frac{2}{4}$ ▥ $\frac{5}{10}$

**Set B   For use after page 419.**

Find each sum or difference. Use fraction models
if you need help.

**1.** $\frac{1}{4} + \frac{2}{4}$     **2.** $\frac{2}{5} + \frac{2}{5}$     **3.** $\frac{3}{6} + \frac{2}{6}$

**4.** $\frac{3}{8} - \frac{2}{8}$     **5.** $\frac{7}{9} - \frac{3}{9}$     **6.** $\frac{3}{4} - \frac{1}{4}$

**Set C   For use after page 421.**

Write the way you say each decimal.

**1.** 0.5     **2.** 0.8     **3.** 0.3

**4.** 0.4     **5.** 0.2     **6.** 0.7

**Set D   For use after page 423.**

Write the way you say each decimal.

**1.** 7.6     **2.** 6.4     **3.** 25.3     **4.** 85.7     **5.** 37.8     **6.** 64.2

**Set E   For use after page 65.**

Write each time with a.m. or p.m.

**1.**

After school

**2.**

Sunrise

**3.**

Bedtime story

**Set A   For use after page 439.**

Estimate each quotient using compatible numbers.

**1.** $6)\overline{38}$ **2.** $7)\overline{54}$ **3.** $9)\overline{46}$ **4.** $4)\overline{19}$ **5.** $3)\overline{29}$

**Set B   For use after page 443.**

Divide to find the quotients and remainders.

**1.** $3)\overline{16}$ **2.** $5)\overline{36}$ **3.** $4)\overline{22}$ **4.** $2)\overline{19}$ **5.** $4)\overline{30}$

**6.** $5)\overline{19}$ **7.** $2)\overline{15}$ **8.** $3)\overline{26}$ **9.** $4)\overline{26}$ **10.** $3)\overline{19}$

**Set C   For use after page 445.**

Find the quotients. Check each answer.

**1.** $2)\overline{11}$ **2.** $5)\overline{14}$ **3.** $3)\overline{26}$ **4.** $2)\overline{15}$ **5.** $4)\overline{25}$

**Set D   For use after page 451.**

Estimate the quotients.

**1.** $\$6.30 \div 2$ **2.** $\$7.96 \div 4$ **3.** $\$5.70 \div 3$ **4.** $\$4.05 \div 4$

**Set E   For use after page 77.**

Decide if the change is correct. If the change is
not correct, tell the correct change.

**1.** Lou bought a pen that cost 60¢.
He gave the clerk 75¢. The clerk
gave him 2 nickels in change.

**2.** Luis bought a cup of juice that
cost 20¢. He gave the clerk 25¢.
The clerk gave him 5 pennies in
change.

## MORE PRACTICE BANK

**Set A   For use after page 19.**

Write a fact family for each pair of addends.

**1.** Addends: 8 and 5    **2.** Addends: 9 and 7    **3.** Addends: 5 and 9

**Set B   For use after page 23.**

Solve.

**1.** Tom took 9 pictures of his friends. He took 8 more of his family. How many pictures did Tom take in all?

**2.** Betty has 6 pictures of her brother. She has 7 pictures of her sister. How many does she have of both of them?

**Set C   For use after page 33.**

Write the number.

**1.** twenty-three    **2.** thirty-six    **3.** nine hundred forty-one

**4.** four hundred two    **5.** ninety-four    **6.** six hundred fifty

**Set D   For use after page 39.**

Solve.

**1.** In a jacks contest, Tina finished ahead of Jan. Sal placed between Lou and Jan. Lou finished last. Give the order in which the girls finished the contest.

**2.** Joe took his turn at the high jump between Fred and Roy. Sam was after Roy. Fred went first. In what order did the boys jump?

**Set E   For use after page 42.**

Solve.

**1.** Chris is 12th in line. How many people are ahead of him?

**2.** Terri is 16th in line. How many people are ahead of her?

**Set A   For use after page 49.**

Solve.

**1.** Music:
6 old songs
4 new songs
How many songs
are there?

**2.** Reading:
Red book—8 pages
Blue book—7 pages
How many pages
are in both?

**Set B   For use after page 51.**

Write the number. Use a comma to separate thousands.

**1.** four thousand, one hundred fifty-eight

**2.** eight thousand, three hundred seventy-nine

**3.** three thousand, eight hundred twenty-three

**4.** nine thousand, five hundred thirty

**Set C   For use after page 53.**

Write the number. Use a comma to separate thousands.

**1.** twenty-six thousand

**2.** nine hundred twenty-seven thousand

**3.** eighty-one thousand, two hundred twenty-six

**Set D   For use after page 63.**

Write each time.

**1.**    **2.**    **3.**    **4.**

### Set A    For use after page 67.

Solve.

1. The race started at 1:00. The winner finished in 1 hour and 38 minutes. What time did the winner finish?

2. Art class is right after lunch. It lasts 30 minutes. Lunch is over at 1 o'clock. What time is art class over?

### Set B    For use after page 71.

Solve. Tell what data is extra.

1. There are 15 windows on the first story of the building and 15 windows on the second story. It takes an hour to clean each window. How many windows are there on the two stories of the building?

2. Calvin has 9 marbles and Sarah has 25 marbles. They play with the marbles for an hour and a half. How many more marbles does Sarah have then Calvin?

### Set C    For use after page 73.

Write each amount.

1. 4 dollars, 3 dimes    2. 6 dollars, 2 dimes, 2 pennies    3. 8 dollars

### Set D    For use after page 75.

Tell if each amount is more than or less than $2.25. Estimate your answer.

**1.**

**2.**

**3.**

### Set A  For use after page 79.

Solve.

**1.** Charles has a white shirt and a blue shirt. He has a pair of blue jeans and a pair of white jeans. How many outfits can he wear?

**2.** Amy makes sandwiches with either cheese, jam, or meat. She uses white bread or rye bread. How many different sandwiches can she make?

### Set B  For use after page 93.

Trade 10 ones for 1 ten.

**1.**

| Tens | Ones | | Tens | Ones |
|------|------|-------|------|------|
| 1 | 15 | **Trade** | ||||| | ||||| |

**2.**

| Tens | Ones | | Tens | Ones |
|------|------|-------|------|------|
| 3 | 12 | **Trade** | ||||| | ||||| |

**3.**

| Tens | Ones | | Tens | Ones |
|------|------|-------|------|------|
| 2 | 10 | **Trade** | ||||| | ||||| |

**4.**

| Tens | Ones | | Tens | Ones |
|------|------|-------|------|------|
| 4 | 19 | **Trade** | ||||| | ||||| |

### Set C  For use after page 99.

Solve.

**1.** Ann had $8. Jan had $5. Fay had $7. Sue had $3. Two of the girls put their money together. Then they had $11. Who were the girls?

**2.** Bert had 6 cards. Joe had 4 cards. Rong had 7 cards. Habib had 8 cards. Cal had 9 cards. Three boys put their cards together. Then they had 17 cards. Who were the boys?

### Set D  For use after page 111.

Answer these questions.

**1.** You have a baseball card collection at home. A friend asks you how many cards you have. Do you estimate or do you give an exact answer? Why?

**2.** You have been asked to go to a baseball game. The ticket will cost $3.25. Do you estimate how much money you must bring to the game or figure it out exactly? Why?

## MORE PRACTICE BANK

### Set A   For use after page 113.

Use the data from page 112 to answer the questions.

**1.** You want to buy the Colored Scarves Trick and the Mind Reading Trick. How much money will you have to pay?

**2.** Can you buy the Magic Smoke, Vanishing Quarter, and Guess the Card tricks for $5.00? How much money do you need?

### Set B   For use after page 127.

Trade 1 ten for 10 ones. Tell the number of tens and ones after the trade.

**1.**

| Tens | Ones |
| --- | --- |
| 4 | 7 |

Trade

| Tens | Ones |
| --- | --- |
| ‖‖‖ | ‖‖‖ |
| 4 | 7 |

**2.**

| Tens | Ones |
| --- | --- |
| 5 | 5 |

Trade

| Tens | Ones |
| --- | --- |
| ‖‖‖ | ‖‖‖ |
| 5 | 5 |

### Set C   For use after page 133.

Copy and complete the tables to solve these problems.

**1.** Todd has the same number of pennies as nickels. He has 30 cents worth of nickels. How many pennies does he have?

| Nickel value | 5¢ | 10¢ | 15¢ |
| --- | --- | --- | --- |
| Penny value | 1¢ | 2¢ | 3¢ |

**2.** Lita sews 4 large buttons and 3 small buttons on each coat. When she has used 28 large buttons, how many small ones has she used?

| Large buttons | 4 | 8 | 12 |
| --- | --- | --- | --- |
| Small buttons | 3 | 6 | 9 |

### Set D   For use after page 147.

Choose a calculation method and solve.

**1.** In October the Sport Shop sold 355 bicycles. In November they sold 132 more bicycles than in October. How many bicycles were sold in November?

**2.** In May the store sold 414 bicycles. In June the store sold 302. How many fewer bicycles were sold in June?

**516**

### Set A   For use after page 159.

Read the graphs to answer the questions.

**1.** What is the trick that fewest children can do?

**2.** How many jumping jacks can Violet do?

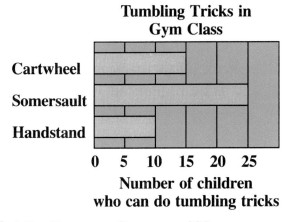

**Tumbling Tricks in Gym Class**

Cartwheel

Somersault

Handstand

0   5   10   15   20   25

**Number of children who can do tumbling tricks**

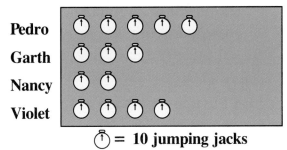

**Number of Jumping Jacks Each Child Can Do in One Minute**

Pedro

Garth

Nancy

Violet

⏱ = 10 jumping jacks

### Set B   For use after page 163.

Solve.

**1.** Show the data in a tally chart. These numbers of children were absent one day: 1st grade, 9; 2nd grade, 7; 3rd grade 4; 4th grade, 6, 5th grade, 3.

**2.** Make a bar graph of the data in Exercise 1 about absences for grades 1 through 5. Remember to choose a title for the graph.

### Set C   For use after page 165.

Solve.

**1.** Billy is doing a report about telephones. Use the data he has organized in a tally chart to make a bar graph.

| Kind of Telephone Owned by Family | |
|---|---|
| Dial | ЖЖ ЖЖ |
| Pushbutton | ЖЖ ЖЖ ЖЖ II |

**2.** Make a picture graph of Billy's data in Exercise 1. Remember to tell what each picture shows. Will the two graphs have the same or different titles?

### Set A  For use after page 169.

Solve.

**1.** Trudy started piano lessons in March when she practiced 15 minutes each day. In April she practiced 20 minutes each day, and in May, 25 minutes each day. How many minutes do you think she practiced each day in August?

**2.** Ricardo began training for a marathon in January. The first month he ran 6 miles a week. In February he ran 10 miles a week, and in March he ran 14 miles a week. How many miles a week was he running in June?

### Set B  For use after page 171.

Solve.

**1.** Make a table to compare these answers from parents and children about the best time for dinner. Parents: 5:00, 4 votes; 6:00, 9 votes; 7:00, 5 votes. Children: 5:00, 10 votes; 6:00, 6 votes; 7:00, 2 votes.

**2.** Look at your table. What can you conclude about dinnertime choices?

### Set C  For use after page 173.

Solve.

**1.** After playing a game for a while, Marco, Natasha, and Amy decided that it was unfair. How could they tell?

| Points Scored | |
|---|---|
| **Marco** | ⊦⊦⊦ ⊦⊦⊦ ⊦⊦⊦ ‖ |
| **Natasha** | ⊦⊦⊦ |
| **Amy** | ‖‖ |

**2.** Is this spinner fair? Tell why or why not.

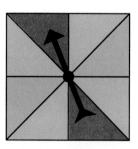

# More Practice Bank

**Set A**  **For use after page 187.**

Use a ruler to draw lines of the following lengths.

**1.** 2 inches    **2.** 5 inches    **3.** 8 inches    **4.** 12 inches

**Set B**  **For use after page 189.**

Measure a paperclip to the nearest inch. Use it to
make these estimates in inches.

**1.** the length of your foot    **2.** the width of your foot

**Set C**  **For use after page 191.**

Choose the better estimate.

**1.** How wide is a classroom door?

   **A** 1 yard    **B** 1 foot

**2.** How far can a car go in one hour?

   **A** 50 yards    **B** 50 miles

**Set D**  **For use after page 193.**

Find the perimeter for each figure.

**1.**
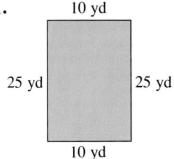

**2.**
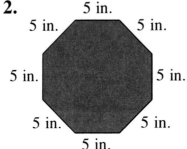

**3.**
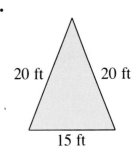

**Set E**  **For use after page 195.**

Solve.

**1.** Doug said, "The number of goldfish I have is between 25 and 35. It is more than 32. I have an odd number of goldfish." How many goldfish does Doug have?

**2.** Ito said, "The number of people in my family is more than 3 and less than 4 plus 4. It is an odd number and it is not 7." How many people are in Ito's family?

**519**

# MORE PRACTICE BANK

**Set A   For use after page 199.**

Choose the better estimate for the weight.

1. Chalkboard eraser

   **A** 3 pounds   **B** 3 ounces

2. Computer

   **A** 20 pounds   **B** 20 ounces

**Set B   For use after page 200.**

Choose the better estimate.

1. coffee pot

   **A** 10 cups   **B** 10 quarts

2. soup kettle

   **A** 1 pint   **B** 1 gallon

**Set C   For use after page 201.**

Which would you wear if this was the temperature outside?

1. 88°F   **A** tee shirt
          **B** mittens

2. 26°F   **A** boots
          **B** sandals

3. 55°F   **A** swimsuit
          **B** jacket

**Set D   For use after page 213.**

Each box has 2 shoes. How many shoes are there?

1.

2.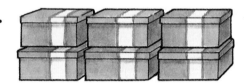

**Set E   For use after page 215.**

Each can has 3 tennis balls. How many tennis balls are there?

**Set A  For use after page 247.**

Solve.

**1.** Mandy bought 4 boxes of golf balls. Each box had 3 balls. On her first day golfing, Mandy lost 6 balls. How many balls did she have left for her second day of golfing?

**2.** Justin scored 24 points in a basketball game. He scored twice as many points as Deng. Added together, how many points did the two boys score?

**Set B  For use after page 271.**

Write **cube, sphere, cone, cylinder,** or **rectangular prism** for each object.

**1.**   **2.**   **3.**   **4.**   **5.**

**Set C  For use after page 273.**

Name each figure: **square, rectangle, circle,** or **triangle.** Then give the number of sides and corners.

**1.**   **2.**   **3.**   **4.**

**Set D  For use after page 275.**

Write **polygon** or **not a polygon** for each figure.

**1.** 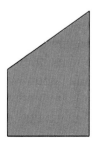  **2.**   **3.**   **4.**   **5.**

## MORE PRACTICE BANK

### Set A   For use after page 277.

Decide if the angle is a right angle, greater than a right angle, or less than a right angle.

**1.**   **2.**   **3.**   **4.**   **5.**

### Set B   For use after page 278.

Use the price list on page 278 to help you decide if the answer is reasonable. If the answer is not reasonable, tell why.

**1.** Tom went out to buy more string, glue, and tape. How much did he spend?

**2.** Safia wanted to take home a package of pipecleaners and a box of clay that were left over. How much should she pay the teacher?

### Set C   For use after page 283.

Tell if these shapes have a line of symmetry when folded. Trace and fold if you cannot tell.

**1.**   **2.**   **3.**

### Set D   For use after page 285.

Which figure is congruent to the first figure?

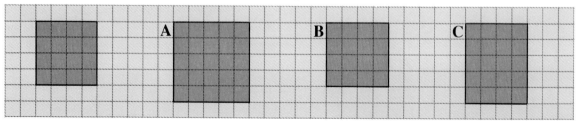

522

**Set A   For use after page 289.**

Make a graph like this. Mark and label a point on the graph for each number pair. Draw lines to connect the points. Name the figure you drew.

**A.** 1, 1
**B.** 5, 1
**C.** 3, 4

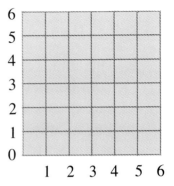

**Set B   For use after page 299.**

Solve.

**1.** Kate and 3 friends earned $20.00. They decided to share the money equally. How much money did each person get?

**2.** Helen and Wang baked 30 cookies. How many could each take home if they shared the cookies equally?

**Set C   For use after page 307.**

Use the graph on page 306 to answer these questions.

**1.** How many fires were there in January?

**2.** Which month had nearly the same number of fires?

**Set D   For use after page 310.**

Solve.

**1.** There are 12 berries. The berries are shared equally by 4 children. How many berries does each child get?

**2.** Tony and 4 friends each picked 3 baskets of berries. How many baskets of berries did they pick?

**Set E   For use after page 313.**

Write the complete fact family.

**1.** $21 \div 3 = 7$          **2.** $40 \div 5 = 8$          **3.** $24 \div 4 = 6$

**Set A  For use after page 331.**

Solve. Then tell which 2 problems are related and why.

**1.** The author read aloud a chapter from her new book. She read 15 pages, took a break, and then read another 17 pages. She ended on page 211. On what page did she begin?

**2.** Roku spent part of her allowance at a fruit stand. She bought strawberries for $0.49 and a peach for $0.24. She had $0.27 left. How much was her allowance?

**3.** Roku shared her strawberries with Seth. Seth ate 4 more berries than Roku. Together they ate 16 berries. How many berries did Seth eat?

**Set B  For use after page 363.**

Choose a calculation method and solve.

**1.** Socks cost $5 for 3 pairs. On Friday, 21 pairs were sold. What was the total cost?

**2.** City Youth Club had a membership drive during June. They signed up 10 new members every day for 30 days. How many new members did the club have at the end of June?

**Set C  For use after page 367.**

Use a calculator to solve these problems.

**1.** Jenna packed eggs in cartons for shipping. She packed 4 cartons with 12 eggs in each carton. She also packed 3 cartons with 24 eggs in each carton. She kept 12 eggs to make eggnog. How many eggs did she have at first?

**2.** One winter Bartlett Orchard sold pears in gift boxes. They sold 33 boxes with 6 pears in each box, 57 boxes with 9 pears in each box, and 48 boxes with 12 pears in each box. How many pears did they sell?

## MORE PRACTICE BANK

**Set A   For use after page 379.**

Choose the unit, **centimeter, meter, or kilometer,** that makes the answer reasonable.

**1.** Joe's height is 150 _____.

**2.** The distance from Dallas to Atlanta is 1,160 _____.

**Set B   For use after page 381.**

Find the area of each shape in square centimeters.

**1.**
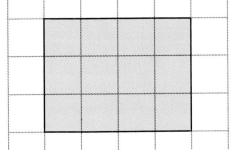

**2.**
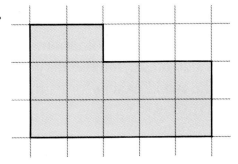

**Set C   For use after page 383.**

Find the volume of each figure in cubic centimeters.

**1.**
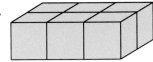

**2.**
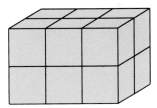

**Set D   For use after page 384.**

Give the reading for each Celsius thermometer.

**1.**
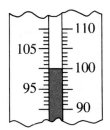

**2.**
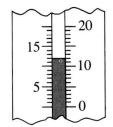

**3.**
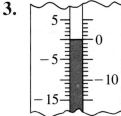

**Set A   For use after page 389.**

Choose the better estimate for the weight.

**1.** Baseball bat

    **A** 1 g    **B** 1 kg

**2.** Brick

    **A** 1 g    **B** 1 kg

**3.** Apple

    **A** 250 g   **B** 250 kg

**Set B   For use after page 390.**

For which will you estimate or have an exact measure?

**1.** Judy and her mother are building a bookcase. Should they measure the wood exactly or estimate? Why?

**2.** Bill is helping his mother cook a roast. Should they guess the correct oven temperature or set it exactly? Why?

**Set C   For use after page 401.**

Draw a graph-paper figure for each.

**1.** 5 equal parts

**2.** 7 equal parts

**3.** Thirds

**Set D   For use after page 403.**

What fraction is shaded?

**1.**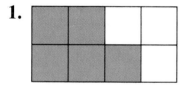

**2.**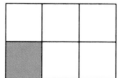

**3.**

**Set E   For use after page 409.**

Solve.

**1.** Alex has 8 stamps. He used 5 of the stamps to mail some letters. What fraction of the stamps did he use?

**2.** Cindy has 10 stamps. She used 7 of the stamps on a package. What fraction of the stamps does she have left?

## MORE PRACTICE BANK

**Set A  For use after page 411.**

Solve.

1. Andy's Cub Scout pack has 16 members. He said $\frac{1}{4}$ of them went to camp. How many Cub Scouts went to camp?

2. Mori has 6 toy horses. She painted $\frac{1}{2}$ of them brown. How many of Mori's horses are brown?

**Set B  For use after page 415.**

Estimate the fractional part.

1. 2. 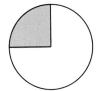 3.

**Set C  For use after page 417.**

Use a ruler to draw lines of these lengths.

1. $1\frac{1}{2}$ inches

2. $2\frac{1}{4}$ inches

3. $4\frac{1}{4}$ inches

**Set D  For use after page 429.**

Solve.

1. Pat's teacher said, "I'm not 30 yet, but I'm more than 25. If you can count by 3s, you'll say my age." How old is Pat's teacher?

2. The vases are red, blue, and green. The flowers are pink, yellow, and white. How many ways can you choose a vase with a flower?

**Set E  For use after page 446.**

Answer these problems carefully.

1. Iona can only fit 5 oranges into a bag. How many bags does she need to hold 18 oranges?

2. Mark can make a prize from 3 pieces of ribbon. How many can he make from 29 pieces?

# TABLE OF MEASURES

| Metric System | | Customary System | |
|---|---|---|---|

## Length

| | | | |
|---|---|---|---|
| 1 centimeter (cm) | 10 millimeters (mm) | | |
| 1 decimeter (dm) | ⎰ 100 millimeters (mm)<br>⎱ 10 centimeters (cm) | 1 foot (ft) | 12 inches (in.) |
| | | 1 yard (yd) | ⎰ 36 inches (in.)<br>⎱ 3 feet (ft) |
| 1 meter (m) | ⎰ 1,000 millimeters (mm)<br>⎨ 100 centimeters (cm)<br>⎱ 10 decimeters (dm) | 1 mile (mi) | ⎰ 5,280 feet (ft)<br>⎱ 1,760 yards (yd) |
| 1 kilometer (km) | 1,000 meters (m) | | |

## Area

| | | | |
|---|---|---|---|
| 1 square meter ($m^2$) | ⎰ 100 square decimeters ($dm^2$)<br>⎱ 10,000 square centimeters ($cm^2$) | 1 square foot ($ft^2$) | 144 square inches ($in.^2$) |

## Volume

| | | | |
|---|---|---|---|
| 1 cubic decimeter ($dm^3$) | ⎰ 1,000 cubic centimeters ($cm^3$)<br>⎱ 1 liter (L) | 1 cubic foot ($ft^3$) | 1,728 cubic inches ($in.^3$) |

## Capacity

| | | | |
|---|---|---|---|
| | | 1 cup (c) | 8 fluid ounces (fl oz) |
| 1 teaspoon | 5 milliliters (mL) | 1 pint (pt) | ⎰ 16 fluid ounces (fl oz)<br>⎱ 2 cups (c) |
| 1 tablespoon | 12.5 milliliters (mL) | 1 quart (qt) | ⎰ 32 fluid ounces (fl oz)<br>⎨ 4 cups (c)<br>⎱ 2 pints (pt) |
| 1 liter (L) | ⎰ 1,000 milliliters (mL)<br>⎨ 1,000 cubic centimeters ($cm^3$)<br>⎨ 1 cubic decimeter ($dm^3$)<br>⎱ 4 metric cups | 1 gallon (gal) | ⎰ 128 fluid ounces (fl oz)<br>⎨ 16 cups (c)<br>⎨ 8 pints (pt)<br>⎱ 4 quarts (qt) |

## Weight

| | | | |
|---|---|---|---|
| 1 gram (g) | 1,000 milligrams (mg) | 1 pound (lb) | 16 ounces (oz) |
| 1 kilogram (kg) | 1,000 grams (g) | | |

## Time

| | | | |
|---|---|---|---|
| 1 minute (min) | 60 seconds (s) | | |
| 1 hour (h) | 60 minutes (min) | 1 year (yr) | ⎰ 365 days<br>⎨ 52 weeks<br>⎱ 12 months |
| 1 day (d) | 24 hours (h) | | |
| 1 week (w) | 7 days (d) | 1 decade | 10 years |
| 1 month (mo) | about 4 weeks | 1 century | 100 years |

**a.m.** A way to indicate the times from 12:00 midnight to 12:00 noon.

**addend** One of the numbers to be added.

Example: $\begin{array}{r} 3 \\ + 5 \\ \hline 8 \end{array}$ addends

**addition** An operation that gives the total number when you put together two or more numbers.

**angle** Two rays from a single point.

**area** The measure of a region, expressed in square units.

**associative (grouping) principle** When adding (or multiplying) three or more numbers, the grouping of the addends (or factors) can be changed and the sum (or product) is the same.

Examples: $2 + (8 + 6) = (2 + 8) + 6$
$3 \times (4 \times 2) = (3 \times 4) \times 2$

**bar graph** A graph that uses bars to show quantities.

**benchmark** A known measure usually used to estimate other measures.

**calendar** A chart that shows months, days, and dates.

**capacity** The amount of space that can be filled in a container; the measure of content.

**centimeter (cm)** A unit of length in the metric system. 100 centimeters equal 1 meter.

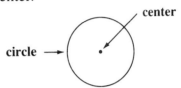

1 centimeter

**change** The amount of money you receive back when you pay with more money than something costs.

**circle** A plane figure in which all the points are the same distance from a point called the center.

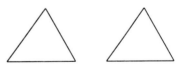

**clustering** The estimation technique of finding addends close in value to the same number, then multiplying that number by the number of addends.

**commutative (order) principle** When adding (or multiplying) two or more numbers, the order of the addends (or factors) can be changed and the sum (or product) is the same.

Examples: $4 + 5 = 5 + 4$
$2 \times 3 = 3 \times 2$

**compatible numbers** Numbers in basic division facts that can be used to estimate quotients for other numbers close in value.

**congruent figures** Figures that have the same size and shape.

congruent triangles

**constant** A number on a calculator that stays the same.

**coordinates** Number pairs used in graphing.

**corner** The point or place where converging lines, edges, or sides meet.

**529**

# GLOSSARY

**cube** A space figure that has squares for all of its faces.

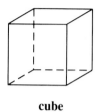

cube

**cup (c)** A unit for measuring liquids. 1 quart equals 4 cups.

**cylinder** A space figure that has congruent circles for 2 faces.

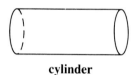

cylinder

**decimal** A number that shows tenths by using a decimal point.

3.2 ← decimal
↑
decimal point

**decimeter (dm)** A metric unit of length equal to $\frac{1}{10}$ meter.

**degree Celsius (°C)** A unit for measuring temperature in the metric system.

**degree Fahrenheit (°F)** A unit for measuring temperature in the customary system of measurement.

**difference** The number obtained by subtracting one number from another.

**digits** The symbols used to write numerals: 0, 1, 2, 3, 4, 5, 6, 7, 8, and 9.

**display** The window on a calculator that shows the numbers as they are entered and the results of the calculations.

**dividend** A number to be divided.

$$\overset{4}{7)\overline{28}} \leftarrow \text{dividend}$$

**division** An operation that tells how many sets or how many in each set.

**divisor** The number by which a dividend is divided.

$$\text{divisor} \rightarrow \overset{4}{7)\overline{28}}$$

**END** An instruction in a computer program that tells the computer to stop.

**equation** A number sentence involving the use of the equality symbol.

Examples: 9 + 2 = 11
8 − 4 = 4

**equivalent fractions** Fractions that name the same amount.

Example: $\frac{1}{2}$ and $\frac{2}{4}$

**estimate** To find an answer that is close to the exact answer.

**even number** A whole number that has 0, 2, 4, 6, or 8 in the ones place.

**fact family** A group of related facts using the same set of digits.

**factors** Numbers that are multiplied together to form a product.

Example: 6 × 7 = 42
↑     ↑
factors

**flat face** A level, unbroken surface.

**flowchart** A chart that shows a step-by-step way of doing something.

**foot (ft)** A unit for measuring length. 1 foot equals 12 inches.

**530**

**fraction** A number that expresses parts of a whole or a set.

Example: $\frac{3}{4}$

**front-end estimation** The estimation technique of adding the leading digits of addends, then using the other places to adjust the estimate.

**gallon (gal)** A unit of liquid measure. 1 gallon equals 4 quarts.

**gram (g)** The basic unit for measuring weight in the metric system. A paper clip weighs about 1 gram.

**graph** A picture that shows information in an organized way.

**greater than** The relationship of one number being larger than another number.

Example: $6 > 5$, read "6 is greater than 5"

**inch (in.)** A unit for measuring length. 12 inches equal 1 foot.

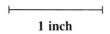

1 inch

**intersecting lines** Two lines that have exactly one point in common.

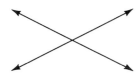

intersecting lines

**key codes** An arrangement of letters and numbers to show the order you use to press the keys of a calculator to find an answer.

**kilogram (kg)** A unit of weight in the metric system. 1 kilogram is 1,000 grams.

**kilometer (km)** A unit of length in the metric system. 1 kilometer is 1,000 meters.

**length** The measure of distance from one end to the other end of an object.

**less than** The relationship of being smaller than another number.

Example: $5 < 6$, read "5 is less than 6"

**line** A straight path that is endless in both directions.

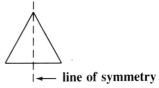

**line of symmetry** A line on which a figure can be folded so that the two parts fit exactly.

line of symmetry

**LIST** A copy of the set of instructions that tells a computer what to do.

**liter (L)** A metric unit used to measure liquids. 1 liter equals 1,000 cubic centimeters.

**logic error** The message on a calculator display showing that an operation is not logical.

**measure** A number indicating the relation between a given object and a suitable unit.

**memory keys** The keys marked with an M that control the memory features of a calculator.

**meter (m)** A unit of length in the metric system. 1 meter is 100 centimeters.

**mile (mi)** A unit for measuring length. 1 mile equals 5,280 feet.

# GLOSSARY

**milliliter (mL)** A metric unit of capacity equal to $\frac{1}{1000}$ liter.

**minus (−)** Used to indicate the subtraction operation, as in 7 − 3 = 4, read, "7 minus 3 equal 4."

**mixed number** A number that has a whole number part and a fractional part, such as $2\frac{3}{4}$.

**multiplication** An operation that combines two numbers, called factors, to give one number, called the product.

**negative number** A number that is less than zero.

**number line** A line that shows numbers in order.

Example:

7  8  9  10

**number pair** Two numbers that are used to give the location of a point on a graph.

Example:  (3, 2)

**number sentence** A way to express a relationship between numbers.

Examples:  3 + 5 = 8
6 ÷ 2 = 3

**numeral** A symbol for a number.

**odd number** A whole number that has 1, 3, 5, 7, or 9 in the ones place.

**order** A sequential arrangement of mathematical elements.

**ordinal number** A number that is used to tell order.

Example:  first, fifth

**ounce (oz)** A unit for measuring weight. 16 ounces equal 1 pound.

**overflow error** The message on a calculator display showing that a number is too large for the display window.

**p.m.** A way to indicate the times from 12:00 noon to 12:00 midnight.

**parallel lines** Lines in the same plane that do not intersect.

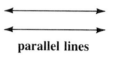

**parallel lines**

**perimeter** The distance around a figure.

**period** A group of three digits set off by commas in a large number.

Example:  432,567,098

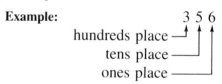

**This is the thousands period.**

**picture graph** A graph that uses pictures to show quantities.

**pint (pt)** A unit for measuring liquid. 2 pints equal 1 quart.

**place value** The value given to the place a digit occupies in a number.

Example:

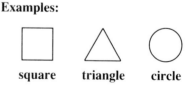

hundreds place
tens place
ones place

**plane figures** Figures that lie on a flat surface.

Examples:

square    triangle    circle

**plus (+)** Used to indicate the addition operation, as in 4 + 3 = 7, read "4 plus 3 equal 7."

**point** A single exact location, often represented by a dot.

**polygon** A closed figure consisting of straight line segments joined end to end.

These are polygons.

**pound (lb)** A unit for measuring weight. 1 pound equals 16 ounces.

**PRINT** An instruction in a computer program that tells the computer to print something.

**probability** The likelihood that an event will occur.

**product** The result of the multiplication operation.

Example: $6 \times 7 = 42$
product

**program** The set of instructions that tells a computer what to do.

**quart (qt)** A unit for measuring liquids. 1 quart equals 4 cups.

**quotient** The number (other than the remainder) that is the result of the division operation.

Examples: $45 \div 9 = 5$
quotient

$$\begin{array}{r} 6 \leftarrow \text{quotient} \\ 7\overline{)45} \\ -42 \\ \hline 3 \end{array}$$

**rectangle** A plane figure with 4 sides and 4 right angles.

rectangle

**rectangular prism** A space figure with six faces. It has the shape of a box.

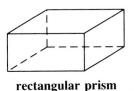

rectangular prism

**reference point** A rounded number used in estimation for comparison.

**remainder** The number less than the divisor that remains after the division process is completed.

Example:
$$\begin{array}{r} 6 \\ 7\overline{)47} \\ -42 \\ \hline 5 \end{array} \leftarrow \text{remainder}$$

**right angle** An angle that has the same shape as the corner of a square.

**Roman numerals** Numerals used by the Romans.

Examples:   $I = 1$
$V = 5$
$VI = 6$

**rounding** Replacing a number with a number that tells about how many.

Example:   23 rounded to the nearest 10 is 20.

**RUN** A message that appears on the video screen when a computer program is used.

**segment** A straight path from one point to another.

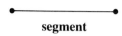

segment

**skip-counting** Counting by a number other than 1.

> Example: 0, 5, 10, 15
> Skip-counting by fives

**space figure** A figure that is not flat but that has volume.

Examples:

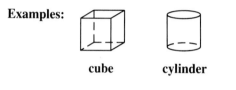

cube          cylinder

**sphere** A space figure that has the shape of a round ball.

sphere

**square** A plane figure that has four equal sides and four equal corners.

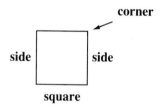

square

**subtraction** An operation that tells the difference between two numbers, or how many are left when some are taken away.

**sum** The number obtained by adding numbers.

> Example:  3
>         + 2
>         ――――
>           5 ← sum

**times (×)** Used to indicate the multiplication operation, as in 3 × 4 = 12, read "3 times 4 equal 12."

**trading** To make a group of ten from one of the next highest place value, or one from ten of the next lowest place value.

> Examples: one hundred can be traded for ten tens; ten ones can be traded for one ten.

**triangle** A plane figure with three segments as sides.

triangle

**unit** An amount or quantity used as a standard of measurement.

**volume** The number of units of space that a space figure holds.

**yard (yd)** A unit for measuring length. 1 yard equals 3 feet.

## ACKNOWLEDGEMENTS

### Illustration Acknowledgements

**Angela Adams** p. 14-376, 14-377, 14-379, 16-442, 16-443, 16-446, 16-450

**Steve Cieslawski** p. 9-252

**Linda Cook** p. 469A, 474A, 478A, 482A

**Susan Clee** p. 2-44A, 2-45A, 2-49A, 3-70A, 3-73AB, 3-78A, 3-79A, 8-218A, 8-219A, 8-227A, 8-228A, 8-229AB, 9-253A 9-257A, 9-258A, CPB-484A, MP-520ABCD

**Marie Dejohn** p. 1-26B, 1-27AB, 1-56A, 4-116A, 5-152A, 5-154A, 14-368, 14-389, 16-441ABC, 16-449ABC

**Andrea Fong** p. 477A, 483AB

**Betty Gee** p. 481A

**Ignacio Gomez** p. 11-308A, SR-464A, SR-465A, SR-466A, SR-467A, CPB-495B, CPB-497A, CPB-498A

**Jeff Hukill** p. 1-11A, 1-15A, 2-55A, 3-68AB, 3-69B, 3-71C, 3-74A, 3-75

**Pamela Johnson** p. 4-96A, 5-134A, 5-141A, 5-147A, 7-186, 7-187, 7-189, 7-192, 7-200, 7-201

**Susan Lexa** p. 8-214, 8-215, 8-233, 9-240, 9-247, 9-250, 9-251, 9-260, 9-261, 11-300, 11-301, 11-314, 11-315, 13-361, 13-364, 13-368

**Lyle Miller** p. 1-6A, 1-7A, 1-12A, 1-16A, 1-17A, 1-28C, 2-34A, 2-35B, 2-36A, 2-42A, 2-46A, 2-47A, 3-67A, 3-72A, 11-322A, 13-372A, 14-387A, 14-378A, 14-396A, 15-418A, 15-418A, 16-440A, CB-485A, CB-487SA, SR-460A, SR-461A, SR-462, SR-463A

**Deborah Morse** p. 475A

**Nancy Munger** p. 1-8A, 1-9AB, 1-20A, 1-21A, 2-40AB, 2-41A, 3-63I, 3-70A, 3-76, 3-77AB, 4-89A, 4-97A, 4-98A, 4-101A, 4-106A, 4-107, 4-108A, 4-109A, 4-122B, 5-125A, 5-131A, 5-132A, 5-140A, 5-145A, 5-146A, 8-212A, 8-219A, 8-230C, 8-231A, 9-238A, 9-239A, 9-241A, 9-'243A, 9-249A, 9-254A, 9-255A, 9-259A, 11-298B, 11-305A, 11-309A, 11-310A, 11-311AB, 11-313A, 11-316A, 12-329A, 12-336A, 12-337A, 12-339A, 15-402AB, 15-403, 15-410, 15-411, 15-414BCDE, 15-430A

**Pronk & Associates**

    **Grahm Bardell** p. 10-284

    **Jack McMaster** p. 4-99AB, 5-142A, 7-190AB

    **Margo Stahl** p. 4-91A, 4-94B, 4-95A, 4-100A, 4-103A, 4-112A, 4-113A, 5-124A, 5-126A, 5-132A, 5-143A, 7-191A, 7-192A, 10-270B, 10-271, 10-272, 10-273, 10-277ABC, 10-278, 10-279, 10-282DEF, 10-283, 10-284, 10-286, 10-287, 10-288, 10-289, 10-291, 13-355, 15-408, 15-409, 15-416, 15-417, 15-422A, 15-424A

**Rick Sams** p. 476A

**Ed Sauk** p. 1-4, 1-5, 1-18B, 2-32A, 2-33A, 2-53A, 3-64A, 8-217A, 11-302A, 11-303A, 11-306A, 11-307A, 11-316A

**Joel Synder** p. 7-198, 7-199

**Rosaland Solomon** p. 468A, 472A, 473A, 479A, 480AB

**Nancy Spier** p. 491A

**Andrea Tachiera** p. 6-161A, 6-166A, 6-168B, 6-172C, 6-180A, 6-181A, 6-183A, 7-206, 8-232A, 9-264A, 9-266AB, 11-321A, 11-322A, 12-328B, 13-348A, 13-349A, 13-351A, 13-362-A, 15-440A, 15-446A

**Pat Traub** p. 9-242A, 6-168A, 13-357, 13-358, 13-359

**Joe Veno** p. 12-326, 12-327, 13-363, 13-368, 15-415, 15-423

### Photo Acknowledgements

**Chapter 1:** 2–3 E.R. Degginger/Bruce Cloeman Inc.

**Chapter 2:** 30–31 Paul von Stroheim/West Light; Wayland Lee*/Addison-Wesley Publishing Company

**Chapter 4:** 86–87 Coco McCoy/Rainbow

**Chapter 5:** 120–121 The Betteman Archive/Bettmann Newsphotos; 130 Wayland Lee*/Addison-Wesley Publishing Company; 137 Wayland Lee*/Addison-Wesley Publishing Company

**Chapter 6:** 156–157 Bob Burch/Bruce Coleman Inc.; 176 Wayland Lee*/Addison-Wesley Publishing Company

**Chapter 7:** 184–185 Wesley Frank/Woodfin Camp & Associates; 194B Breck P. Kent/Animals, Animals; 202 Dr. Ronald Cohn/The Gorilla Foundation; 203 Dr. Ronald Cohn/The Gorilla Foundation

**Chapter 8:** 210–211 Bill Ross/West Light; 213 A Satterthwaite/The Stock Market; 221 Geoff Juckes/The Stock Market; 224 Hugh K. Loester/Tom Stack & Associates

**Chapter 9:** 236–237 Steve Solum/Bruce Coleman Inc.; 248 Geoffrey Clifford/Wheeler Pictures; 256 Wayland Lee*/Addison-Wesley Publishing Company; 262–263 Norman Tomalin/Bruce Coleman Inc.

**Chapter 10:** 268–269 Wayland Lee*/Addison-Wesley Publishing Company

**Chapter 11:** 296–297 Mickey Gibson/Animals, Animals; 298 Leonard Lee Rue III/Bruce Coleman Inc.; 299 Brian Parker/Tom Stock & Associates

**Chapter 12:** 324–325 Sepp Seitz/West Light; 327 Wayland Lee*/Addison-Wesley Publishing Company; 340–341 Wayland Lee*/Addison-Wesley Publishing Company

**Chapter 13:** 357 Wayland Lee*/Addison-Wesley Publishing Company; 366 Wayland Lee*/Addison-Wesley Publishing Company; 367 Wayland Lee*/Addison-Wesley Publishing Company

**Chapter 14:** 374–375 B. Daemmrich/Stock, Boston

**Chapter 15:** OSF/Animals, Animals

**Chapter 16:** 436–437 Craig Aurness/West Light

**Data Bank:** 470 Giraudon/Art Resource-Beeldrechrt/VAGA; 470B National Gallery of Art, Washington-Chester Dale Collection

All other photographs taken by Janice Sheldon*

Special thanks to California School for the Deaf; Chabot Elementary School; Hawthorne Elementary School; Whitton School; Bill's Trading Post; East Bay Regional Park District; Nature Company; Sweet Dreams; The Flower Cart; Tiffany's Pet Shop; Velo Sport.

*Photographs provided expressly for publisher